Green Electronics: Environment Friendly Products

Green Electronics: Environment Friendly Products

Edited by **Jeremy Giamatti**

WILLFORD **P**RESS

New York

Published by Willford Press,
118-35 Queens Blvd., Suite 400,
Forest Hills, NY 11375, USA
www.willfordpress.com

Green Electronics: Environment Friendly Products
Edited by Jeremy Giamatti

International Standard Book Number: 978-1-68285-102-9 (Hardback)

Printed in the United States of America.

Contents

Preface

Green electronics are designed and manufactured with an aim for low consumption of power and improving quality of environment by using environment-friendly components and sources for power production. Some of the significant topics for research in this area are ultralow power technologies, green electronic devices, low power applications, energy harvesting, etc. There has been rapid progress in this field and its applications are finding their way across multiple industries. It will help the readers in keeping pace with the rapid changes in this field.

The information shared in this book is based on empirical researches made by veterans in this field of study. The elaborative information provided in this book will help the readers further their scope of knowledge leading to advancements in this field.

Finally, I would like to thank my fellow researchers who gave constructive feedback and my family members who supported me at every step of my research.

Editor

Fully Integrated Solar Energy Harvester and Sensor Interface Circuits for Energy-Efficient Wireless Sensing Applications

Naser Khosro Pour *, François Krummenacher and Maher Kayal

Ecole Polytechnique Fédérale de Lausanne, CH-1015 Lausanne, Switzerland;
E-Mails: françois.krummenacher@epfl.ch (F.K.); maher.kayal@epfl.ch (M.K.)

*   Author to whom correspondence should be addressed; E-Mail: naser.khosropour@epfl.ch

Abstract: This paper presents an energy-efficient solar energy harvesting and sensing microsystem that harvests solar energy from a micro-power photovoltaic module for autonomous operation of a gas sensor. A fully integrated solar energy harvester stores the harvested energy in a rechargeable NiMH microbattery. Hydrogen concentration and temperature are measured and converted to a digital value with 12-bit resolution using a fully integrated sensor interface circuit, and a wireless transceiver is used to transmit the measurement results to a base station. As the harvested solar energy varies considerably in different lighting conditions, in order to guarantee autonomous operation of the sensor, the proposed area- and energy-efficient circuit scales the power consumption and performance of the sensor. The power management circuit dynamically decreases the operating frequency of digital circuits and bias currents of analog circuits in the sensor interface circuit and increases the idle time of the transceiver under reduced light intensity. The proposed microsystem has been implemented in a 0.18 μm complementary metal-oxide-semiconductor (CMOS) process and occupies a core area of only 0.25 mm^2. This circuit features a low power consumption of 2.1 μW when operating at its highest performance. It operates with low power supply voltage in the 0.8V to 1.6 V range.

Keywords: analog integrated circuits; solar energy harvesting; ultra-low power circuits; power management circuits; sensor interface circuits; wireless sensor networks

1. Introduction

There is an increasing demand for energy-efficient wireless sensor networks in different sensing and monitoring applications. These emerging sensors that mostly rely on energy harvesting, incorporate energy source, energy storage and electronic circuits for power management, sensing and communication into a miniaturized system. In order to have autonomous operation, different energy harvesting sources, including piezoelectric, electro-magnetic, thermoelectric, radio frequency and solar sources can be used. Solar energy harvesting can be a good energy source for autonomous microsystems in the order of a few cubic centimeters' size. However, as energy harvested from solar cells is intermittent, an energy storage device, such as a rechargeable microbattery or a supercapacitor, should be used for reliable operation of electronic circuits that are used for power management, sensor interfacing and wireless data transmission. The system level design of energy harvesting system starts by selecting an appropriate energy harvesting source and energy storage device, and the energy harvester circuit is designed according to the target energy source and energy storage. The proposed area- and energy-efficient solar energy harvester circuit harvests energy from a photovoltaic (PV) module consisting of nanowire solar cells in series [1] and stores it in a Varta V6HR NiMH microbattery [2]. A miniaturized PV module should have high efficiency to provide enough power, while other characteristics, such as open circuit voltage, short circuit current and maximum power point of the PV module, should be also considered to design a highly efficient solar energy harvester. The microbattery not only provides power for the proposed sensor interface and power management circuits, but also provides the required supply current during wireless data transmission in the target TZ1053 chipset [3].

The DC voltage that a PV module provides depends on the environmental conditions, including illumination level and temperature. Meanwhile, the battery voltage may change from its nominal voltage during battery charging and discharging. At room temperature, the voltage of the target NiMH battery reaches to an end-of-discharge voltage (V_{EOD}) of 0.9 V when the battery is discharged and reaches an end-of-charge voltage (V_{EOC}) of 1.5 V when the battery is fully charged. As a result, the DC voltage that the PV module provides may differ from the voltage of the target battery, depending on the illumination level and remaining charge of the battery. Different DC-DC converters have been proposed to harvest energy from a miniaturized PV module to charge a battery, including switched-capacitor (SC) [4] and inductive DC-DC converters [5]. These DC-DC converters either use external capacitors or inductors or operate at high frequencies to have high efficiency. However, by using higher frequencies, the power consumption of energy harvester circuits increases and a high efficiency is not achievable under a reduced illumination level when input power is only a few micro-watts. In addition, since under reduced light intensity, harvested energy from the PV module may be lower than the power consumption of the microsystem, the energy management circuit should dynamically reduce power consumption of the sensor to avoid complete discharge of the battery during system operation. As a result, the microsystem can continue its autonomous operation at a lower speed. The proposed energy harvester circuit reliably measures energy stored in the target NiMH microbattery and scales the power consumption of the microsystem up or down according to the energy stored in the battery. This circuit consumes less than 350 nW and achieves more than 90% efficiency during battery charging, using the direct charging method already proposed in [6].

As the use of hydrogen fuels becomes more common, an increasing demand for miniaturized hydrogen sensors is expected. Miniaturized palladium (Pd) nanowire hydrogen sensors [7] that can be used at room temperature have good sensitivity to H_2 concentration, thanks to their large surface-to-volume ratio. These sensors are good candidates for ultra-low power (ULP) hydrogen sensing, as they are very low power, while maintaining a small form factor. As these sensors are very sensitive to temperature, temperature effects should be compensated to increase measurement accuracy. A grid of 14 Pd nanowires, fabricated on a silicon wafer, has been used for gas sensing [8]. Half of these sensors are only sensitive to temperature and have been used as reference nanowires. The remaining sensors are sensitive to both temperature and H_2 concentration and have been used as sensing nanowires. As reference and sensing nanowires have a similar temperature dependence, temperature cross-sensitivity is cancelled by deploying a differential approach in the readout circuit. The readout circuit proposed in [9] has been used to measure the change in conductivity of these nanowires upon hydrogen exposure. Upon hydrogen exposure, the conductance change of sensing nanowires in comparison with reference nanowires is measured and converted to a digital value using an analog-to-digital converter. This paper makes the following contributions: (a) a new solar energy harvester circuit is proposed that scales the power consumption and performance of a wireless sensing platform, according to the harvested solar energy. This circuit reconfigures the power and speed of a fully integrated gas sensor interface circuit and data transmission duty cycle, according to the energy stored in the battery; (b) we present the design and implementation of an energy-efficient wireless sensing platform that uses the proposed solar energy harvester circuit to guarantee autonomous operation of the proposed hydrogen gas sensor. System level design parameters that affect autonomous operation of this sensor, including selection of the appropriate energy storage option and wireless data transceiver, have been discussed thoroughly. The remainder of this paper is organized as follows. Section 2 presents the system architecture, including the realized integrated circuit and required external components. Section 3 presents the system implementation of the realized integrated circuits. In Section 4, measurement and simulation results have been discussed to evaluate the performance and autonomous operation of the proposed hydrogen sensor, and finally, Section 5 concludes the paper.

2. System Architecture

In order to realize a miniaturized autonomous sensor, a miniaturized low power gas sensor and wireless data transceiver should be used and the power delivered to the sensor should be maximized. Appropriate energy harvesting source and the energy storage device should be selected for this purpose, and the energy harvester circuit should be designed accordingly. In this section, all main blocks of the target sensor, including the circuit architectures that have been proposed for solar energy harvesting and sensor interfacing, and external components that have been selected as the energy storage device, energy harvesting source and wireless transceiver will be discussed thoroughly.

2.1. Energy Storage Option

Selecting an appropriate energy storage device is the first step in designing an energy harvester circuit, and based on the selected energy storage device, the power management circuit should be designed with the target of optimizing the overall power efficiency. The systems targeted for micro-power energy harvesting applications typically have additional stringent constraints on die area and usage of external components. As energy harvested from the PV module is intermittent and the maximum power that it can provide may be much less than the needed peak power during sensing and data transmission, an energy storage device is needed. Although the proposed H_2 sensor may consume less than 20 µA on average for sensing and data transmission, the peak power consumption during wireless data transmission is much higher. The target wireless transceiver consumes 3.3 mA and 2.8 mA during data transmission and reception [3]. Miniaturized supercapacitors, NiMH microbatteries and lithium-based batteries are different solutions that have been used in wireless sensor nodes frequently. The battery should have enough capacity to provide enough power for a few hours' operation of the sensor, even without energy harvesting. In addition to capacity, other characteristics, such as size, peak discharge current, nominal operating voltage, V_{EOC}, V_{EOD}, cycle life and leakage, are very important factors to design an energy-efficient sensor that relies on energy harvesting.

Although miniaturized supercapacitors, such as GZ115F of CAP-XX [10], can provide much higher peak charge and discharge currents and have much longer cycle lives in comparison with the rechargeable batteries, they have some disadvantages that make them inappropriate for our application. The main disadvantage of supercapacitors is their high leakage currents that can be even larger than the current provided by the PV cell under reduced illumination level. In addition, they have much lower energy capacity compared with rechargeable batteries of similar sizes. Among lithium batteries, state-of-the-art thin film lithium batteries, such as MEC125 of Infinite Power Solutions [11], are the best candidates for micro-power energy harvesting applications thanks to their low leakage currents, long cycle lives and high discharge currents. The main disadvantage of these batteries is their high nominal voltage levels. As thin film lithium batteries have nominal voltages of more than 3.8 V, additional DC-DC converters are required to use these batteries for ULP applications. The target V6HR microbattery has a nominal voltage of 1.2 V and can provide a peak discharge current of 18 mA that is much higher than the required current during data transmission. This battery has a large nominal capacity of 6.2 mAh, while its diameter and height are 6.8 mm and 2.15 mm, respectively. As in the proposed gas sensor, all electronic circuits, including the external wireless transceiver, can work with a sub-1.2 V supply voltage, so additional step-down or step-up circuits are not required. Table 1, compares the main characteristics of the proposed V6HR battery with GZ115F and MEC125. The main disadvantage of this NiMH battery is its lower cycle life compared to supercapacitors and thin film lithium batteries. In addition, this battery has a relatively low leakage current. It loses at most 20% of its charge during the first month, which is much lower than the discharge current of supercapacitors.

Table 1. Comparing energy storage options.

Energy Storage Option/Specification	CAP-XX (GZ115F) [10]	Infinite Power Solutions (MEC125) [11]	Varta (V6HR) [2]
Technology	Supercapacitor	Thin film lithium battery	NiMH battery
Nominal operating voltage	2.3 V	4.1 V	1.2 V
Size	$20 \times 15 \times 1.25$ mm^3	$12.7 \times 12.7 \times 0.7$ mm^3	D: 6.8 mm, H: 2.15 mm
Energy capacity	4 µAh	200 µAh	6200 µAh
Peak discharge current	30 A	7.5 mA	18 mA
Cycle life	30,000+ hours	10,000	1000

2.2. Solar Energy Harvester

The block diagram of the target H_2 gas sensor, including the required external components and the main blocks of the proposed solar energy harvester, is depicted in Figure 1. Four miniaturized nanowire solar cells have been connected in series to be used as the energy source. The main tasks of a solar energy harvester block are battery charging, battery management and energy management. The direct charging method proposed in [6] has been used for charging the target NiMH microbattery with very high efficiency. In this approach, battery voltage (V_{bat}) is compared with open circuit voltage of the PV module (V_{pv}) using a dynamic comparator, and if it is lower, the battery is connected to the PV module to store harvested energy. If V_{pv} drops below V_{bat}, the switch between V_{pv} and V_{bat} is kept turned off to avoid battery discharge through the PV module. In the direct charging method that has been deployed for solar energy harvesting, the maximum power point voltage (V_{mpp}) of the PV module should be close to the V_{EOC} of the battery to achieve high efficiency. Although P_{mpp} and I_{SC} of the PV module changes considerably in different lightning conditions, V_{oc} and V_{mpp} of the PV module does not change significantly. In Figure 2, deliverable power of the PV module in different illumination levels has been simulated using the electrical model of the target PV module [12]. In AM1.5, solar intensity is almost 1 mW/mm^2. Although, in 10% of AM1.5, deliverable power is reduced by roughly the factor of 10; V_{oc} drops from 2.35 V to 1.95 V.

The second main task of the energy harvester circuit is battery management. As the charging current is limited by the PV module, the battery is charged in either standard or trickle charging modes, depending on the illumination conditions [2]. In trickle charging mode, the battery is charged by a small current, and it can be continuously charged after reaching V_{EOC}. However, in standard charging mode, the battery charging should be stopped after reaching V_{EOC} to avoid permanent damage to the battery. In addition, to avoid full discharge of the battery that reduces the cycle life of the battery, the battery should be disconnected from the sensor when it drops to V_{EOD}. In order to avoid overdischarge or overcharge of the battery in standard charging mode, V_{EOC} and V_{EOD} voltage levels are detected by this circuit.

Figure 1. Block diagram of the proposed hydrogen gas sensor.

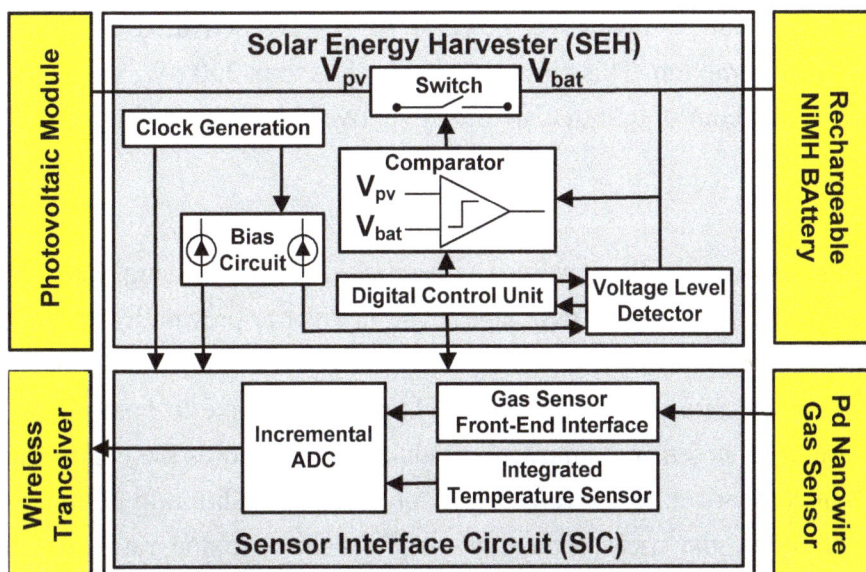

Figure 2. Deliverable power of the photovoltaic (PV) module in different lightning conditions.

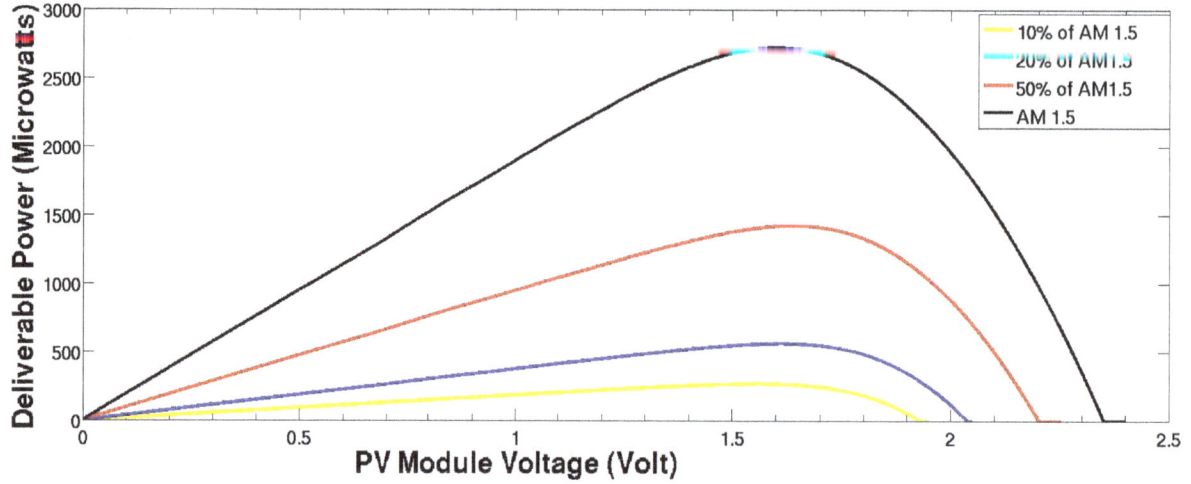

Finally, the last task of the energy harvester circuit is energy management. Since, under reduced light intensity, the power delivered by the PV module may be lower than the average power consumption of the sensor, the energy harvester circuit should reduce power consumption of the microsystem to avoid full discharge of the battery during this period. Under reduced light intensity, the microsystem continues its autonomous operation at a lower speed and with a lower duty cycle. The proposed energy harvester circuit reliably measures energy stored in the target battery and scales the power consumption of the microsystem up or down according to the measurement results. In order to reduce the power consumption and speed of digital circuits, operating frequency is scaled down and data transceiver is activated by a lower duty cycle. Bias currents are also scaled down to reduce power consumption and the speed of analog circuits. In order to reliably estimate the energy stored in the battery, battery voltage is measured when the battery is discharged by a high discharge current.

The PV module, with the total area of 28 mm^2, can provide a maximum power P_{mpp} of 2.88 mW at its V_{mpp} in AM1.5 illumination [1]. However, as harvested solar energy may be as small as a few micro-watts under reduced light intensities, the power consumption of this circuit should not be higher than a few hundred nanowatts to have high efficiency during these periods. As neither V_{pv} nor V_{bat} changes rapidly, comparator and voltage level detector blocks are activated every few seconds to minimize average power consumption. This circuit consumes less than 350 nW, which is mainly due to the clock generator and digital control unit that are always active.

2.3. Sensor Interface Unit

Pd nanowire grid fabricated on a silicon wafer has been used for gas sensing. Each grid consists of 14 Pd nanowires, half of which are covered with a passivation layer to prevent hydrogen from reaching the nanowire [8]. These coated nanowires, which are only sensitive to temperature, have been used as reference nanowires, while the remaining nanowires, which are sensitive to both temperature and H$_2$ concentration, have been used as sensing nanowires. This circuit measures the conductance change of the sensing nanowires in comparison to the reference nanowires. In addition to eliminating the effects of temperature by first order, the measurement is only sensitive to the ratio of conductance of nanowires, instead of their absolute values, and as a result, a higher accuracy is achievable. Although the temperature effect is eliminated by first order, there are still second order effects, as the temperature coefficient of nanowire resistance changes according to H_2 concentration. In order to compensate this second order effect, temperature is measured using an integrated temperature sensor [9]. H_2 sensing accuracy can be increased by incorporating the measured temperature during sensor calibration. An incremental analog to digital converter (ADC) [13] converts the measured temperature and H_2 concentration to 12-bit digital values. Power consumption and conversion time of this ADC can be reconfigured according to the energy stored in the battery to match the amount of available power, resulting in an adaptive, autonomous sensor. The power management circuit dynamically reduces the operating frequency of digital circuits and the bias currents of analog circuits in this sensor interface circuit under reduced light intensity. However, ADC conversion time increases as a result.

2.4. Wireless Transceiver

Emerging wireless sensors that are powered by micro-power energy harvesting sources have more stringent energy requirements compared to traditional wireless sensors. Apart from minimizing total power consumption of the sensor, which is mainly achieved by duty cycling operation of the sensor, they should have low peak power and ultra-low standby current. Low peak power ensures that miniaturized batteries with limited peak discharge currents can be used to power up the circuit. Ultra-low standby current guarantees that the average power consumption of the sensor can be minimized by heavily duty cycling sensing and data transmission. The main factors impacting power consumption of a wireless transceiver are supply voltage, carrier frequency and receiver sensitivity. The power consumption of transceiver can be reduced by operating at lower supply voltages. Although most of the wireless transceivers work with at least 1.8 V supply voltage, ultra-low power wireless transceivers with sub-1.2 V voltage, such as TZ1053 [3] or ZL70250 [14], are more suited to our application, thanks to their lower power consumption during data transmission and reception. These transceivers operate at sub-1 GHz frequency bands and have much lower peak-power and standby power, compared to state-of-the-art 2.4 GHz transceivers.

The second important factor is the carrier frequency. Some of the factors that affect choice of carrier frequency are operation range, power consumption, transmission data rates and antenna size. Although 2.4 GHz protocols, such as Zigbee, have been used extensively for wireless sensing applications, sub-1 GHz wireless systems offer several advantages for ultra-low power, low data rate applications. These transceivers have less power consumption for the same operating range, thanks to reduced attenuation rates and blocking effects at lower frequencies. Besides requiring higher power for the same link budget, higher traffic in the 2.4 GHz frequency band increases the interference in this frequency band.

Finally, receiver sensitivity also affects the power efficiency. A narrower bandwidth creates higher receiver sensitivity and allows efficient operation at lower transmission rates. Overall, all radio circuits running at higher frequencies, including low-noise amplifiers, power amplifiers, mixers and synthesizers, need more current to achieve the same performance as lower frequencies.

Although these sub-1 GHz transceivers have some disadvantages, such as larger size of antennas and lower data rates, overall, they are more suited for our application. A higher data rate can improve the energy efficiency (energy per bit) of the transceiver for high bandwidth applications with large data payloads. In fact, overall power consumption of a wireless transceiver is not only a factor of physical layer items, such as radio architecture, carrier frequency and antenna choice, but is also a function of the amount of time that the radio needs to run in order to transport the payload data over the air. Transmission time depends on the data rate and the protocol overhead to establish and maintain the communication link. For example, a Zigbee transceiver that has a 250 kbps data rate can send 1 MB of data almost five-times faster than a TZ1053 that has a 50 kbps data rate. However, in ultra-low power applications, typically, data payloads are small and a 50 kbps data rate is more than enough. In addition, as standard protocols, such as Zigbee or Bluetooth, offer highly sophisticated link and network layers and have a large protocol overhead, they are not very efficient for sending small data payloads.

3. System Implementation

3.1. Solar Energy Harvester

The proposed energy harvester is based on the direct charging scheme [6], in which a dynamic comparator compares V_{bat} with V_{pv} in a timely manner. The dynamic comparator and required control signals can be seen in Figure 3.

Figure 3. Circuit diagram of dynamic comparator.

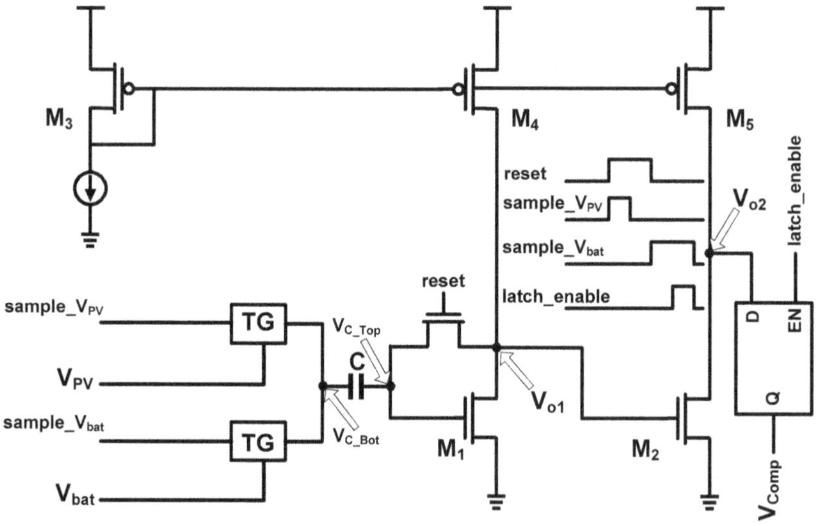

After turning off the switch between V_{pv} and V_{bat}, the first inverting gain stage gets reset by activating the reset control signal, to establish a common mode voltage of V_{cm} at the V_{C_Top} node. Meanwhile, V_{C_Bot} is connected to V_{pv} by activating the Sample_V_{pv} control signal, and the difference between V_{pv} and V_{cm} is sampled on the sampling capacitor. In the next step, by deactivating the reset and activating Sample_V_{bat} control signal, V_{C_Bot} is connected to V_{bat}, and as a result, V_{C_Top} changes to $V_{cm} + (V_{bat} - V_{pv})$. Outputs of the first and second gain stages (V_{o1} and V_{o2}) change according to the new V_{C_Top}, and V_{o2} is latched and stored as V_{Comp}, by enabling the latch_enable control signal. If V_{Comp} gets 1, it means that V_{bat} is higher than V_{pv}, and the switch should be kept turned off to avoid battery discharge, but if V_{Comp} gets 0, the PV module starts to charge the battery by turning on the switch. After the switch is turned on, V_{pv} will follow V_{bat} during battery charging.

As the operating voltage of the PV module is determined by the battery, end-to-end efficiency from the PV module to the battery is reduced when battery voltage diverges from the V_{mpp} of the PV module during battery charging. Different maximum power point tracking (MPPT) strategies have been deployed in energy harvesting applications to maximize the amount of harvested energy. As the V_{mpp} of the PV module varies with incident light conditions, an efficient MPPT scheme ensures that the maximum power is extracted from a PV module at any given time. Many complex and accurate MPPT schemes have been investigated for large solar harvesting systems [15]; however, when the PV module is small and can only provide a few micro-watts under reduced light intensity, only low-overhead schemes that incur very little power overhead can be good candidates. In [16], several low-overhead MPPT approaches, including fractional open-circuit voltage (FOC), fractional short circuit current (FSC) and hill-climbing techniques, have been discussed that can be deployed in inductive and SC DC-DC converters. Although in direct charging V_{pv} always follows V_{bat} and no MPPT scheme can be deployed, nevertheless, it achieves even higher efficiency than competing SC and inductive DC-DC converters [6]. In addition, the die area is much smaller than SC and inductive DC-DC converters, as large passive components, such as pumping capacitors, or inductors have not been used in this architecture.

When the target V6HR NiMH microbattery is fully charged, the battery voltage is close to V_{EOC}; as the battery is discharged by a low current, V_{bat} drops to its nominal value and remains almost constant, up to getting close to the fully discharged state. However, if the battery is discharged by a high current, as can be seen in the discharge curve of the battery in Figure 4 [2], V_{bat} drops immediately and voltage drop depends on the remaining charge of the battery. During wireless data transmission, the battery is discharged by a high current, close to 5 CA (C being the 1 hour charge or discharge current). As can be seen in Figure 4, if discharge capacity is less than 20%, V_{bat} is higher than 1.1 V during this period; however, if discharge capacity is higher than 20%, V_{bat} drops from 1.1 V down to 900 mV, depending on the remaining charge of the battery.

Figure 4. Typical discharge curves of the target NiMH microbattery at room temperature [2].

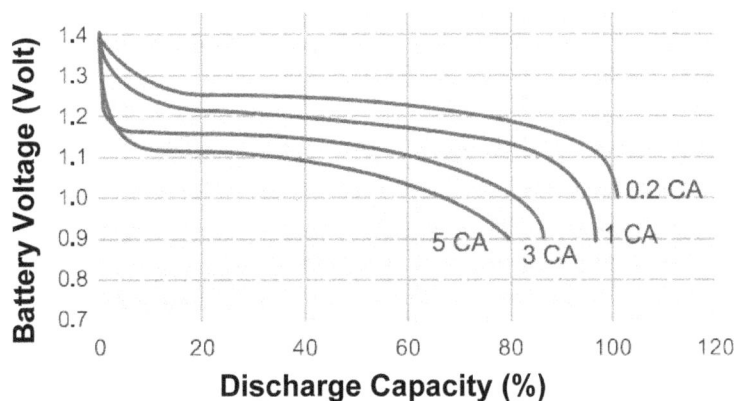

In our circuit, battery voltage is detected during this period to accurately estimate the energy stored in the battery. Depending on the detected V_{bat}, the power-performance of the integrated electronic circuits and operation duty cycle of the wireless transceiver are reconfigured to guarantee autonomous operation of the sensor. If V_{bat} is high enough, the sensor can operate at its highest performance: solar energy harvester and sensor interface circuits work at their highest speed, and measurement results are sent to the base station every 10 seconds. If V_{bat} is not high enough, the sensor interface circuit and wireless transceiver are activated with a lower duty cycle to minimize total power consumption of the sensor. If V_{bat} is close to V_{EOD}, these blocks should get deactivated temporarily to avoid full discharge of the battery.

The voltage level detector (LD) in Figure 5 is used to determine the battery voltage level [6]. After circuit startup, battery voltage is checked to make sure that it is more than the end of discharge threshold voltage (V_{EOD}) and can power up the circuit. If V_{bat} is less than V_{EOD}, it means that the battery is fully discharged and cannot provide enough power for the electronic circuits. In this situation, the PV module continuously charges the battery by keeping the switch closed. As soon as V_{bat} passes V_{EOD}, LD starts its normal operation, comparing V_{bat} with specified threshold voltages in a timely manner and updating the V_{bat_level} as a result. In order to generate a bandgap reference voltage, 25 substrate PNP transistors have been used as Q_1 and Q_2 in a common-centroid layout [6]. These transistors are biased with a 100 nA current source to generate V_{BE1} and V_{BE2} in non-overlapping Φ_1 and Φ_2 phases, and the SC circuit sums up V_{BE1} and ($V_{BE1}-V_{BE2}$) with appropriate coefficients. When V_{bat} reaches V_{EOC}, the switch between the PV module and the battery is turned off to avoid overcharge of the battery.

Figure 5. Circuit diagram of voltage level detector.

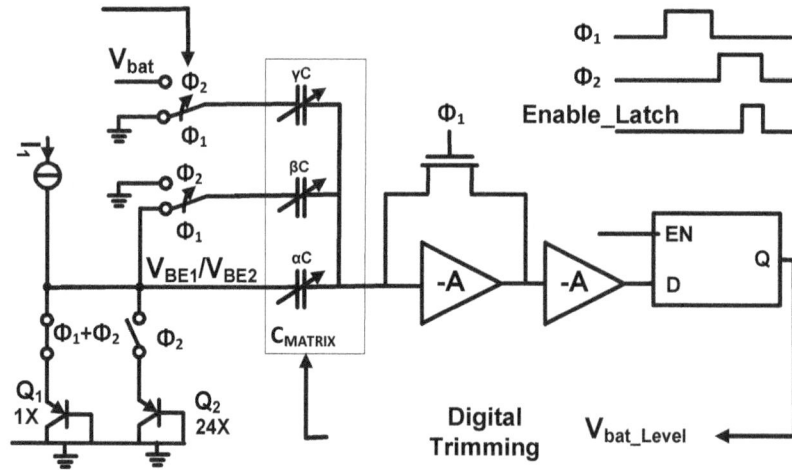

This SC circuit detects when V_{bat} passes the V_L specified in Equation (1) by setting the V_{bat_level} output. By using a variable γC capacitor, different voltage levels between V_{EOD} and V_{EOC} can be detected to estimate the remaining charge of the battery. In Equation (1), α, β and γ coefficients are the ratios of tunable αC, βC and γC capacitors. These variable capacitors have been implemented using a matrix of 10 fF metal-insulator-metal (MIM) capacitors that have been used as unity capacitors.

$$V_L = [\alpha \times (V_{BE1} - V_{BE2}) + \beta \times V_{BE1}]/(\gamma) = V_{ref}/(\gamma) \tag{1}$$

As V_{BE1} is complementary to absolute temperature (CTAT) and ($V_{BE1}-V_{BE2}$) is proportional to absolute temperature (PTAT), different CTAT, PTAT or temperature-independent reference voltages (V_{ref}) can be built by proper selection of αC and βC capacitors. After that, by modifying the γC

capacitor, target voltage levels can be detected. As can be seen in Table 2, by modifying βC and γC capacitors, different battery voltages, starting from V_{EOD} of 0.9 V up to V_{EOC} of 1.5 V, can be detected. In order to have a temperature-independent voltage reference, αC should be modified according to the selected βC capacitor. By detecting the battery voltage between 900 mV and 1.1 V, the remaining charge of the battery can be estimated according to the discharge curve of the battery in Figure 4.

Table 2. Detected voltage levels in voltage level detector.

Voltage Level (V_L)	αC Capacitor	βC Capacitor	γC Capacitor
907 mV (V_{EOD})	170 fF	1230 fF	230 fF
947 mV (V_{L0})	170 fF	1230 fF	220 fF
1002 mV (V_{L1})	180 fF	1300 fF	220 fF
1051 mV (V_{L2})	180 fF	1300 fF	210 fF
1104 mV (V_{L3})	190 fF	1370 fF	210 fF
1509 mV (V_{EOC})	160 fF	1160 fF	130 fF

After measuring V_{bat} and estimating the energy stored in the battery, the operating frequency of digital circuits and bias currents of analog circuits are reconfigured according to the remaining charge of the battery. In the current-starved ring oscillator in Figure 6a, the frequency is determined by R_L and C_L and can be specified as Equation (2). In Equation (2), K_1 is a constant value that depends on the number of inverter stages in the ring oscillator [17]. By using a digital resistive trimming network to modify R_L, F_{osc} is reconfigured by an energy harvesting circuit, according to the measured V_{bat}. When V_{bat} is low, R_L is increased to decrease F_{osc}. Four target operating frequencies have been considered, corresponding to four voltage levels that are detected by the LD block, ranging from 125 kHz, corresponding to V_{L0}, to 1 MHz corresponding to V_{L3}.

$$F_{osc} = K_1/(R_L \times C_L) \tag{2}$$

In addition to a fixed 10 nA beta-multiplier (BM) current reference that has been used to provide the required bias current for a solar energy harvester, a switched-capacitor beta-multiplier (SCBM) current source has been designed to generate frequency-proportional bias currents. In this circuit, which can be seen in Figure 6b, SC resistors have been used instead of regular resistors to generate frequency-proportional bias currents [18]. For example, I_{ADC} can be specified as in (3):

$$I_{ADC} = K_1 \times K_2 \times F_{osc} \times (C_3 + C_4) \tag{3}$$

In Equation (3), K_1 and K_2 are constant values that depend on the ratio between the widths of transistors in Figure 6b. K_1 depends on the ratio between M_7 and M_4, and K_2 depends on the ratio between M_5 and M_6. The current ripple of I_{bias} is minimized by using a large decoupling capacitor, C_2, and two complementary branches charging and discharging C_3 and C_4 capacitors in non-overlapping $\emptyset_1$ and $\emptyset_2$ clock phases with F_{osc} frequency [18]. As F_{osc} is determined by the oscillator, I_{bias} scales dynamically by changing the frequency of the oscillator. In addition, this current source can be easily deactivated by turning off the $\emptyset_1$ and $\emptyset_2$ clocks.

In ultra-low power analog circuits, such as the discrete time incremental ADC that has been used here, bias currents should be high enough to guarantee correct operation at the target operating frequency. Generated bias currents (I_{ADC} and I_{Temp}) have been used for the sensor interface circuit. If

V_{bat} is low, F_{osc} is decreased to reduce the power consumption of digital circuits in the ADC. When the system is operating at a lower frequency, lower bias currents can be used to reduce the power consumption of analog circuits in the sensor interface circuit. By using these frequency-proportional bias currents, power consumption of the sensor can be scaled down dynamically by reducing F_{osc}.

Figure 6. (**a**) Reconfigurable current-starved ring oscillator; (**b**) switched-capacitor beta-multiplier (SCBM) current source.

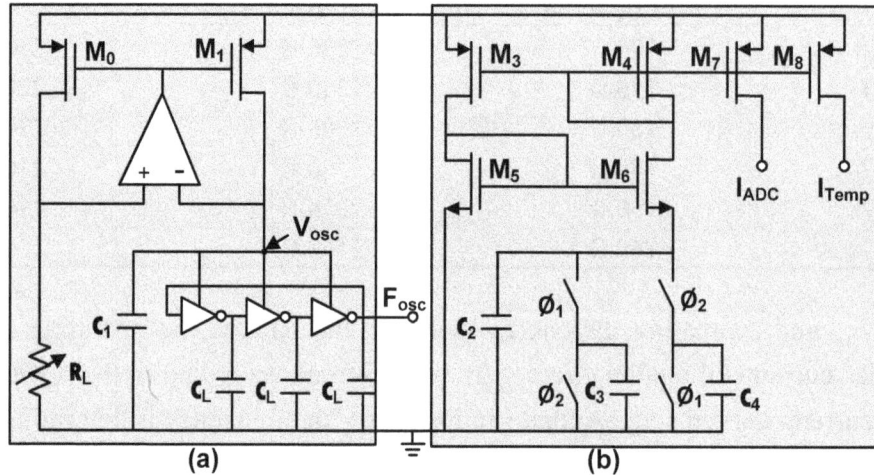

The digital control unit (DCU) activates the comparator and LD blocks in a timely manner, to check the charging status of battery. As neither V_{bat} nor V_{pv} change rapidly, these blocks are activated every few seconds to minimize their average power consumption. After activating each block, DCU generates the required control signals for related SC circuits. Standard complementary metal-oxide-semiconductor (CMOS) logic has been used to implement DCU. The power consumption of DCU is mainly determined by the low frequency 20-bits counter that activates the comparator and LD in a timely manner. Table 3 presents die area and power consumption of individual blocks, operating at 1 MHz as its highest clock frequency. As can be seen, the simulated total power consumption of the circuit is mainly determined by the clock generator, bias circuit and DCU blocks, which are always active. Although comparator and LD blocks consume considerable power during their active time, nevertheless, as they are activated every 10 seconds, their average power consumption is less than 1 nW. When V_{bat} is low, the average power consumption of the clock generator and DCU blocks is reduced by operating at a lower clock frequency. By reducing the operating clock frequency, the simulated power consumption of clock generator and DCU is reduced roughly linearly, reaching to 23 nW and 13 nW, respectively, at 125 kHz.

Table 3. Die area and active power of main blocks in solar energy harvester circuit.

Block	Die Area (μm^2)	Active Power (nW)	Average Power (nW)
Clock generator	14,400	165	165
Digital control unit	38,115	90	90
Bias circuit (10 nA)	2,832	35	35
Comparator	759	55	<1
Level detector	31,960	520	<1
SCBM	6,300	640	<1

3.2. Sensor Interface Circuit

The proposed sensor interface circuit is depicted in Figure 7. An incremental ADC [13] measures H_2 concentration and temperature at different times. Individual Pd nanowires that have between 7 kΩ and 9 kΩ resistance should be biased, with a minimum bias voltage of 50 mV for 10 seconds. These nanowires have been represented by NW_{ref} and NW_{sense} in Figure 7. The conductance of NW_{sense} nanowires may change by 20% in comparison with NW_{ref} nanowires as H_2 concentration varies from zero to 30% [7]. The voltage around the reference nanowire, $V_R = (V_1 - V_2)$, is used as the reference voltage, while the voltage of the sensing nanowire (V_2) is used as the input voltage for the following incremental ADC in consecutive non-overlapping clock phases. This ADC can achieve 12 bits resolution by using 200 fF metal-insulator-metal (MIM) capacitors, as C_3, C_4 and C_5, and a conventional two-stage amplifier with at least 75 dB gain for the integrator [9]. As a 1 MHz clock has been used for ADC and since this ADC needs 2^N cycles for N-bits conversion, a 12-bit conversion takes nearly 4 ms. After finishing gas sensing, temperature is measured to further increase the accuracy of gas sensing using an integrated temperature sensor. In the proposed integrated temperature sensor, substrate PNP transistors have been used to generate proportional to absolute temperature (V_{ptat}) and temperature-independent reference (V_{ref}) voltages, and the ADC converts (V_{ptat}/V_{ref}) to a 12-bit digital value. In Figure 7, by using a 50 fF MIM capacitor as C_1 and a 360 fF MIM capacitor as C_2, a temperature-independent reference voltage of approximately 310 mV has been generated. Similar to gas sensing, n-bit digital representation of (V_{ptat}/V_{ref}) is stored in the ADC counter and sent to a wireless transceiver after temperature sensing. The accuracy of temperature sensing is mainly limited by mismatch between Q_1 and Q_2 and nonlinearity in temperature dependence of V_{BE1} and ($V_{BE1}-V_{BE2}$). Although these errors can be minimized by using dynamic methods presented in [19] to reach ±0.1 °C accuracy, such power-consuming techniques are not needed here. The proposed low power temperature sensor can achieve ±1 °C accuracy by only calibrating C_2 and I_{Temp} at room temperature [9]. When the sensor interface circuit operates at a lower frequency, Q_1 and Q_2 transistors are biased with a lower I_{Temp} bias current to reduce the average power consumption of the circuit.

Figure 7. Circuit diagram of the proposed sensor interface.

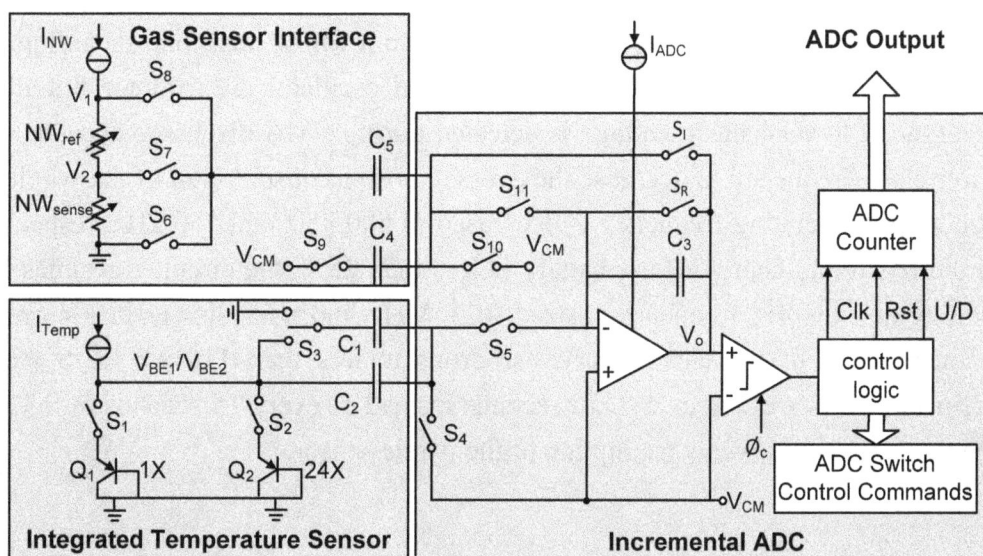

4. System Integration and Performance

The circuit has been implemented in a 0.18 μm CMOS process with 0.25 mm^2 total area, as can be seen in the chip microphotograph in Figure 8. The main blocks, including the digital and analog blocks of solar energy harvester and sensor interface circuits, have been specified separately. When operating at 1 MHz, the whole circuit consumes almost 2.1 μW; while the average power consumption of the energy harvester is less than 350 nW. The wireless transceiver sends the measurement results for both temperature and H$_2$ concentration to a base station every 15 seconds.

Figure 8. Chip microphotograph of the whole microsystem.

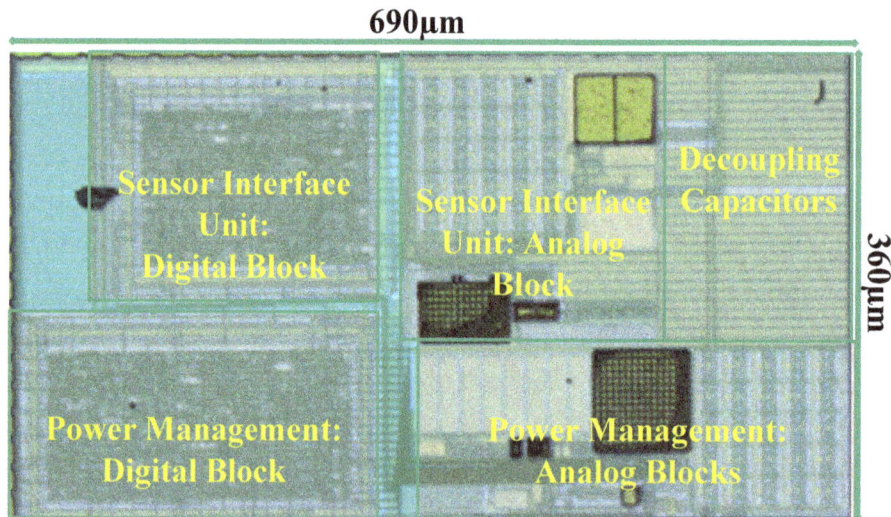

Table 4 presents measurement and simulation results for power consumption of the energy harvester and sensor interface circuits in different system operation modes. These operation modes have been defined according to the remaining charge of the battery, which is estimated accurately by detecting the battery voltage between 900 mV and 1.1 V during wireless data transmission, and the battery is discharged by a high current, close to 5 CA. Battery discharge capacity is determined according to the discharge curve of the battery shown in Figure 4. By using highly resistive poly resistors, total layout area has been decreased compared to [9]; however, as these embedded resistors are sensitive to process variations, absolute values are important for all resistors, including resistors that have been used in the beta multiplier current source, and oscillator are trimmed initially. If the battery gets discharged, a lower battery voltage is detected during 5 CA discharge and the system is switched to a lower clock frequency to decrease the average power consumption of the whole system. In S_{L2} and S_{L1} modes, the operating frequency is decreased to 500 kHz and 250 kHz, respectively. In S_{L0} mode, when the remaining charge of the battery is less than 25%, the circuit consumes less than 0.6 μW by operating at 125 kHz frequency instead of 1 MHz and using lower bias currents. The average power consumption of the energy harvester drops to less than 110 nW in S_{L0} mode. In addition, measurement results are sent every 120 seconds instead of every 15 seconds in S_{L3} mode, to further reduce the total average power consumption of the whole sensor.

Table 4. System performance and power consumption in different system operation modes.

Block		S_{L3}	S_{L2}	S_{L1}	S_{L0}
Detected V_{bat} during 5CA discharge (mV)		1,104	1,051	1,002	947
Battery discharge capacity (%)		<20%	<50%	<65%	<75%
Battery threshold voltage (V)	Measured	1.114	1.055	1.003	0.955
	Simulated	1.122	1.063	1.010	0.962
Clock frequency (kHz)	Measured	950	488	249	127
	Simulated	980	502	257	132
Time interval of sensing and data transmission (seconds)		15	30	60	120
Power consumption of clock generator (nW)		165	85	44	23
Power consumption of digital control unit (nW)		90	46	24	13
Average power consumption of energy harvester (nW)	Measured	346	210	142	103
	Simulated	293	169	106	74
Average power consumption of sensor interface circuit (nW)	Measured	1730	1120	790	640
	Simulated	1360	870	625	500
Average power consumption of the integrated circuit (nW)		2076	1330	932	707
Average current consumption of gas sensor bias circuit (µA)		4.67	2.33	1.16	0.58
Average current consumption of the wireless transceiver (µA)		9.4	7.7	6.6	5.55
Average current consumption of the complete system (µA)		16	11	8.5	6.7

In order to estimate total power consumption of the sensor, the average power consumption of sensor biasing circuit and wireless transceiver should be calculated. Pd nanowires should be biased with a 7 µA bias current for 10 seconds, before measuring H_2 concentration. TZ1053 consumes 5 µA during standby and consumes 3.3 mA during a period of 20 ms to send a sample with the minimum payload size of 55 bytes [3]. In S_{L3} mode, samples are sent every 15 seconds, and the average current consumption of Pd nanowire sensors and wireless transmission are 4.67 µA and 9.4 µA, respectively. By sending the samples every 120 seconds in S_{L0} mode, these values will be reduced to 0.58 µA and 5.6 µA, respectively. So, the total average current consumption of the whole sensor is less than 16 µA, operating at its highest performance in S_{L3} mode, and is reduced to less than 7 µA, operating in S_{L0} mode.

Four nanowire solar cells presented in [1] have been connected in series to provide an appropriate V_{mpp} voltage, close to the V_{EOC} of the V6HR NiMH battery. The power delivered to the battery by the PV module can be simulated using the equivalent circuit model of the PV module. This PV module, with total area of 28 mm^2, can provide a maximum power of 2.88 mW at its V_{mpp} under AM1.5 illumination level [1]. In order to evaluate autonomous operation of the sensor, end-to-end efficiency from the PV module to the battery has been simulated in different illumination levels. As can be seen in Table 5, even in 1% light intensity, the PV module delivers average power of 18.4 µW to the battery with 89.8% average efficiency. The power that can be delivered to the battery in 1% can be seen in Figure 9. When battery voltage is close to V_{EOD}, 15.9 µW power is delivered to the battery with 76.8% efficiency. As in this situation, the system is operating in S_{L0} mode, its average power consumption is less than 8 µW and the battery gets charged. By increasing the battery voltage, efficiency is improved and more power is delivered to the battery. As a result, even in 1% light intensity, the harvested energy is enough for autonomous operation of the whole system. Meanwhile, as the whole sensor node consumes less than 16 μA for sensing and data transmission, the target 6 mAh battery can provide enough power for approximately 375 hours operation, even without energy harvesting.

Table 5. Die area and active power of solar energy harvesting blocks.

Illumination Level	V_{mpp} (V)	P_{mpp} (μW)	P_{charge} (μW)	Efficiency
100% of AM1.5	1.586	2,718	2,240.8	82%
50% of AM1.5	1.622	1,422	1,127.3	79.2%
20% of AM1.5	1.604	561.6	448.4	79.8%
10% of AM1.5	1.549	269.5	221.4	82.1%
1% of AM1.5	1.289	20.3	18.4	89.3%

Figure 9. Power delivered to the battery under simulated 1% of AM1.5 illumination.

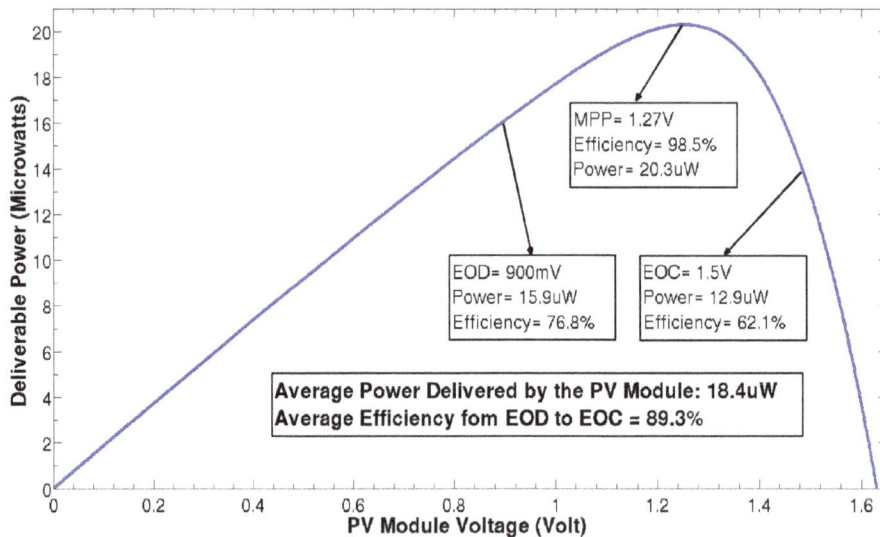

5. Conclusions

An ultra-low power energy-efficient solar energy harvesting and sensing microsystem has been proposed for wireless sensing applications. The circuit has been realized in a 0.18 μm CMOS process with only a 0.25 mm^2 die area to measure both hydrogen concentration and temperature, using fully integrated solar energy harvesting and sensor interfacing circuits. An external wireless data transceiver is used to transmit the measurement results to a base station. Four nanowire solar cells have been connected in series, and the proposed area- and power-efficient energy harvester circuit stores the energy harvested from the PV module in a NiMH microbattery. In addition, this circuit scales the power consumption and performance of the sensor by reconfiguring the operating frequency, bias currents and time interval of sensing and data transmission to guarantee autonomous operation of the sensor. This circuit consumes only 2.1 μW when operating at its highest performance, which is further reduced to less than 0.7 μW when operating at its lowest speed. Simulation results show that even under reduced illumination, the harvested energy is enough for autonomous operation of the sensor.

Acknowledgments

This work is supported by European project SiNAPS under contract number 257856. The authors would like to thank Fritz Falk and Jia Goubin from the Institute of Photonic Technology, Jena (IPHT-Jena) for providing miniaturized nanowire solar cells.

References

1. Jia, G.; Steglich, M.; Sill, I.; Falk, F. Core-shell heterojunction solar cells on silicon nanowire arrays. *Sol. Energy Mater. Sol. Cells* **2012**, *96*, 226–230.

2. Varra V6HR Datasheet. Available online: http://www.varta-microbattery.com (accessed on 3 December 2012).

3. Toumaz TZ1053 Datasheet. Available online: http://www.toumaz.com/page.php?page=telran (accessed on 3 December 2012).

4. Chen, G.; Ghaed, H.; Haque, R.; Wieckowski, M.; Yejoong, K.; Gyouho, K.; Fick, D.; Daeyeon, K.; Mingoo, S.; Wise, K.; Blaauw, D.; Sylvester, D. A Cubic-Millimeter Energy-Autonomous Wireless Intraocular Pressure Monitor. In *Proceedings of IEEE International Solid-State Circuits Conference*, San Francisco, CA, USA, 20–24 February 2011; pp. 310–312.

5. Qiu, Y.; Liempd, C.V.; Veld, B.O.H.; Blanken, P.G.; Hoof, C.V. 5 µW-to-10 mW Input Power Range Inductive Boost Converter for Indoor Photovoltaic Energy Harvesting with Integrated Maximum Power Point Tracking Algorithm. In *Proceedings of IEEE International Solid-State Circuits Conference*, San Francisco, CA, USA, 20–24 February 2011; pp. 118–120.

6. Khosro Pour, N.; Krummenacher, F.; Kayal, M. Fully integrated ultra-low power management system for micro-power solar energy harvesting applications. *Electron. Lett.* **2012**, *48*, 338–118.

7. Offermans, P.; Tong, H.D.; Van Rijn, C.J.M.; Merken, P.; Brongersma, S.H.; Crego-Calama, M. Ultralow-power hydrogen sensing with single palladium nanowires. *Appl. Phys. Lett.* **2009**, *94*, 223110–223113.

8. Van der Bent, J.F.; van Rijn, C.J.M. Ultra low power temperature compensation method for palladium nanowire grid. *Procedia Eng.* **2010**, *5*, 184–187.

9. Pour, N.K.; Krummenacher, F.; Kayal, M. A Miniaturized Autonomous Microsystem For Hydrogen Gas Sensing Applications. In *Proceedings of IEEE 10th International New Circuits and Systems Conference*, Montreal, Canada, 17–20 June 2012; pp.201–204.

10. Cap-XX Co. Website. Available online: http://www.cap-xx.com (accessed on 3 December 2012).

11. InfinitePowerSolutions Co. Website. Available online: http://www.infinitepowersolutions.com (accessed on 3 December 2012).

12. Lu, C.; Raghunathan, V.; Roy, K. Micro-Scale Energy Harvesting: A System Design Perspective. In *Proceedings of 15th Asia and South Pacific Design Automation Conference (ASP-DAC)*, Taipei, Taiwan, 18–21 January 2010; pp. 89–94.

13. Markus, J.; Silva, J.; Temes, G.C. Theory and applications of incremental ΔΣ converters. *IEEE Trans. Circuits Syst.* **2004**, *51*, 678–690.

14. Zarlink ZL70250 Datasheet. Available online: http://www.zarlink.com/zarlink (accessed on 3 December 2012).

15. Esram, E.; Chapman, P.L. Comparison of photovoltaic array maximum power point tracking techniques. *IEEE Trans. Energy Conver.* **2007**, *22*, 439–449.

16. Lu, C.; Raghunathan, V.; Roy, K. Maximum power point considerations in micro-scale solar energy harvesting systems. In *Proceedings of IEEE International Symposium on Circuits and Systems (ISCAS)*, Paris, France, 30 May–2 June 2010; pp. 273–276.

17. Pastre, M.; Krummenacher, F.; Kazanc, O.; Khosro Pour, N.; Pace, C.; Rigert, S.; Kayal, M. A solar battery charger with maximum power point tracking. In *Proceedings of 18th IEEE International Conference on Electronics, Circuits and Systems (ICECS)*, Beirut, Lebanon, 11–14 December 2011; pp. 394–397.

18. Pastre, M.; Krummenacher, F.; Robortella, R.; Simon-Vermot, R.; Kayal, M. A fully integrated solar battery charger. In *Proceedings of Joint IEEE North-East Workshop on Circuits and Systems and TAISA Conference*, Toulouse, France, 28 June–1 July 2009; pp. 1–4.

19. Pertijs, M.A.P.; Makinwa, K.A.A.; Huijsing, J.H. A CMOS smart temperature sensor with a 3σ inaccuracy of ±0.1 °C from −55 °C to 125 °C. *IEEE J. Solid State Circuits* **2005**, *40*, 2805–2815.

Exploration of Sub-V_T and Near-V_T 2T Gain-Cell Memories for Ultra-Low Power Applications under Technology Scaling

Pascal Meinerzhagen [1,*]**, Adam Teman** [2]**, Robert Giterman** [2]**, Andreas Burg** [1] **and Alexander Fish** [3]

[1] Institute of Electrical Engineering, Ecole Polytechnique Fédérale de Lausanne, Station 11, Lausanne, VD 1015, Switzerland; E-Mail: andreas.burg@epfl.ch

[2] VLSI Systems Center, Ben-Gurion University of the Negev, POB 653, Be'er Sheva 84105, Israel; E-Mails: teman@ee.bgu.ac.il (A.T.); robertgi@ee.bgu.ac.il (R.G.)

[3] Faculty of Engineering, Bar-Ilan University, Ramat-Gan 52900, Israel; E-Mail: alexander.fish@biu.ac.il

* Author to whom correspondence should be addressed; E-Mail: pascal.meinerzhagen@epfl.ch

Abstract: Ultra-low power applications often require several kb of embedded memory and are typically operated at the lowest possible operating voltage (V_{DD}) to minimize both dynamic and static power consumption. Embedded memories can easily dominate the overall silicon area of these systems, and their leakage currents often dominate the total power consumption. Gain-cell based embedded DRAM arrays provide a high-density, low-leakage alternative to SRAM for such systems; however, they are typically designed for operation at nominal or only slightly scaled supply voltages. This paper presents a gain-cell array which, for the first time, targets aggressively scaled supply voltages, down into the subthreshold (sub-V_T) domain. Minimum V_{DD} design of gain-cell arrays is evaluated in light of technology scaling, considering both a mature 0.18 μm CMOS node, as well as a scaled 40 nm node. We first analyze the trade-offs that characterize the bitcell design in both nodes, arriving at a best-practice design methodology for both mature and scaled technologies. Following this analysis, we propose full gain-cell arrays for each of the nodes, operated at a minimum V_{DD}. We find that an 0.18 μm gain-cell array can be robustly operated at a sub-V_T supply voltage of 400 mV, providing read/write availability over 99% of the time, despite refresh cycles. This is demonstrated on a 2 kb array, operated at 1 MHz, exhibiting full functionality under parametric variations. As opposed to sub-V_T operation at the mature node, we find that the scaled 40 nm node requires a near-threshold 600 mV supply to achieve at least 97% read/write availability due to higher leakage currents that limit the bitcell's retention time. Monte Carlo simulations show that a 600 mV 2 kb 40 nm gain-cell array is fully functional at frequencies higher than 50 MHz.

Keywords: embedded memory; gain cell; energy efficiency; subthreshold operation; near-threshold operation; retention time; access speed; technology scaling

---

1. Introduction

Many ultra-low power (ULP) systems, such as biomedical sensor nodes and implants, are expected to run on a single cubic-millimeter battery charge for days or even for years, and therefore are required to operate with extremely low power budgets. Aggressive supply voltage scaling, leading to near-threshold (near-V_T) or even to subthreshold (sub-V_T) circuit operation, is widely used in this context to lower both active energy dissipation and leakage power consumption, albeit at the price of severely degraded on/off current ratios (I_{on}/I_{off}) and increased sensitivity to process variations [1]. The majority of these biomedical systems require a considerable amount of embedded memory for data and instruction storage, often amounting to a dominant share of the overall silicon area and power. Typical storage capacity requirements range from several kb for low-complexity systems [2] to several tens of kb for more sophisticated systems [3]. Over the last decade, robust, low-leakage, low-power sub-V_T memories have been heavily researched [4–6]. In order to guarantee reliable operation in the sub-V_T domain, many new SRAM bitcells consisting of 8 [7,8], 9 [5,9], 10 [4], and up to 14 [2] transistors have been proposed. These bitcells utilize the additional devices to solve the predominant problems of write contention and bit-flips during read, and, in addition, some of the designs reduce leakage by using transistor stacks. All these state-of-the-art sub-V_T memories are based on static bitcells, while the advantages and drawbacks of dynamic bitcells for operation in the sub-V_T regime have not yet been studied.

Conventional 1-transistor-1-capacitor (1T-1C) embedded DRAM (eDRAM) is incompatible with standard digital CMOS technologies due to the need for high-density stacked or trench capacitors. Therefore, it cannot easily be integrated into a ULP system-on-chip (SoC) at low cost. Moreover, low-voltage operation is inhibited by the offset voltage of the required sense amplifier, unless special offset cancellation techniques are used [10].

Gain-cells are a promising alternative to SRAM and to conventional 1T-1C eDRAM, as they are both smaller than any SRAM bitcell, as well as fully logic-compatible. Much of the previous work on gain-cell eDRAMs focuses on high-speed operation, in order to use gain-cells as a dense alternative to SRAM in on-chip processor caches [11,12], while only a few publications deal with the design of low-power near-V_T gain-cell arrays [13–15]. A more detailed review of previous work in the field of gain-cell memories, including target application domains and circuit techniques, can be found in [16]. The possibility of operating gain-cell arrays in the sub-V_T regime for high-density, low-leakage, and voltage-compatible data storage in ULP sub-V_T systems has not been exploited yet. One of the main objections to sub-V_T gain-cells is the degraded I_{on}/I_{off} current ratio, leading to rather short data retention times compared with the achievable data access times. However, the present study shows that these current ratios are still high enough in the sub-V_T regime to achieve short access and refresh cycles and high memory availability, at least down to $0.18\,\mu m$ CMOS nodes. While gain-cells are considerably smaller than robust sub-V_T SRAM bitcells, they also exhibit lower leakage currents, especially in mature CMOS nodes where sub-V_T conduction is the dominant leakage mechanism. Recent studies for above-V_T, high-speed caches show that gain-cell arrays can even have lower retention power (leakage power plus refresh power) than SRAM (leakage power only) [17]. However, a direct power comparison between gain-cell eDRAM and SRAM is difficult and not within the scope of this paper; for example, an ultra-low power sub-V_T SRAM implementation [2] employs power gating of all peripheral circuits and of the read-buffer in the bitcell, while most power reports for gain-cell eDRAMs include the overhead of peripherals. Compared with SRAM, gain-cells are naturally suitable for two-port memory implementation, which provides an advantage in terms of memory bandwidth, and enables simultaneous and independent optimization of write and read reliability. Finally, while local parametric variations directly compromise the reliability of the SRAM bitcell (write contention, and data loss during read), such parametric variations only impact the access and retention times of gain-cells, which is not a severe issue when targeting the typically low speed requirements of ULP applications, such as sub-V_T sensor nodes or biomedical implants.

To start with, we consider sub-V_T gain-cell eDRAM design in a mature $0.18\,\mu m$ CMOS node, which is typically used to: (1) easily fulfill the high reliability requirements of ULP systems; (2) reach the highest energy-efficiency of such ULP systems, typically requiring low frequencies and duty cycles [18]; and (3) achieve low manufacturing costs. In a second step, we investigate the feasibility of sub-V_T gain-cell eDRAMs under the aspect of technology scaling. In particular, in addition to the mature $0.18\,\mu m$ CMOS node, we analyze low voltage gain-cell operation in a $40\,nm$ CMOS technology node. We show that deep-nanoscale gain-cell arrays are still feasible, despite the reduced retention times inherent to these nodes. Due to high refresh rates, we identify that the minimum supply voltage (V_{DDmin}) that ensures an array availability of 97% is in the near-V_T domain.

1.1. Contributions:

The contributions of this work can be summarized as follows:

- We investigate the minimum achievable supply voltage for ultra-low power gain-cell operation.
- We analyze gain-cell arrays from a technology scaling perspective, examining the design trade-offs that arise due to the inherent characteristics of various technology nodes.
- For the first time, we present a fully functional gain-cell array at a deeply scaled technology node, as low as $40\,nm$.
- For the first time, we present a gain-cell array operated in the sub-V_T domain.

1.2. Outline:

The remainder of this article is organized as follows. Section 2 explains the best-practice 2T gain-cell design in light of technology scaling, emphasizing the optimum choices of the write access transistor, read access transistor, storage node capacitance, and word line underdrive voltage for different nodes. Sections 3 and 4 present detailed implementation results of a 2 kb gain-cell memory in a 0.18 μm and in a 40 nm CMOS node, respectively. Section 5 summarizes the findings of this article.

2. Two-Transistor (2T) Sub-V_T Gain-Cell Design

Previously reported gain-cell cell topologies include either two or three transistors and an optional MOSCAP or diode [16]. While the basic two-transistor (2T) bitcell has the smallest area cost, it limits the number of cells that can connect to the same read bitline (RBL) due to leakage currents from unselected cells masking the sense current [19]. However, as many ULP systems require only small memory arrays with relatively few cells per RBL, in the following section, we consider the implementation of a 2T bitcell as a viable low-voltage option and propose a best-practice 2T bitcell design for the considered technology nodes (0.18 μm and 40 nm).

2.1. 2T Gain-Cell Implementation Alternatives

Figure 1 shows the four basic options for implementing a 2T gain-cell, allowing both the write transistor (MW) and the combined storage and read transistor (MR) to be implemented with either an NMOS or a PMOS device. These standard topologies require the following control schemes to achieve robust write and read operations. A boosted write wordline (WWL) voltage is required during write access due to V_T drop across MW; above V_{DD} for the NMOS option (V_{BOOST}) and below V_{SS} for the PMOS option (V_{NWL}). For a read operation with a PMOS MR, the parasitic RBL capacitance is pre-discharged, and the read wordline (RWL) is subsequently raised. If the selected bitcell's storage node (SN) holds a "0", MR is conducting and charges RBL past a detectable sensing threshold. If SN holds a "1", MR is cut off, such that RBL remains discharged below the sensing threshold. Using an NMOS transistor to implement MR provides the exact opposite operation, *i.e.*, RBL is pre-charged and RWL is lowered to initiate a read.

In the considered 0.18 μm CMOS technology, both MW and MR can be implemented with either standard-V_T core or high-V_T I/O devices. In more advanced technology nodes, typically starting with the 130 nm or 90 nm node for most semiconductor foundries, several V_T options become available for core devices, most commonly low-V_T (LVT), standard-V_T (SVT), and high-V_T (HVT) devices. One of the primary considerations for gain-cell implementation is achieving high retention time, *i.e.*, the time it takes for the level stored on SN to deteriorate through leakage currents. In mature, above-100 nm CMOS nodes, subthreshold conduction is the dominant leakage mechanism, compromising data retention in any 2T gain-cell through the channel of MW, as shown in Figure 2(a). Therefore, the primary selection criterion for the device type of MW is to minimize subthreshold conduction. Note that subthreshold conduction of MW weakens both a logic "1" and a logic "0" level, whenever the write bitline (WBL) voltage is opposite to the SN voltage.

Figure 1. 2T gain-cell implementation options including the schematic waveforms.

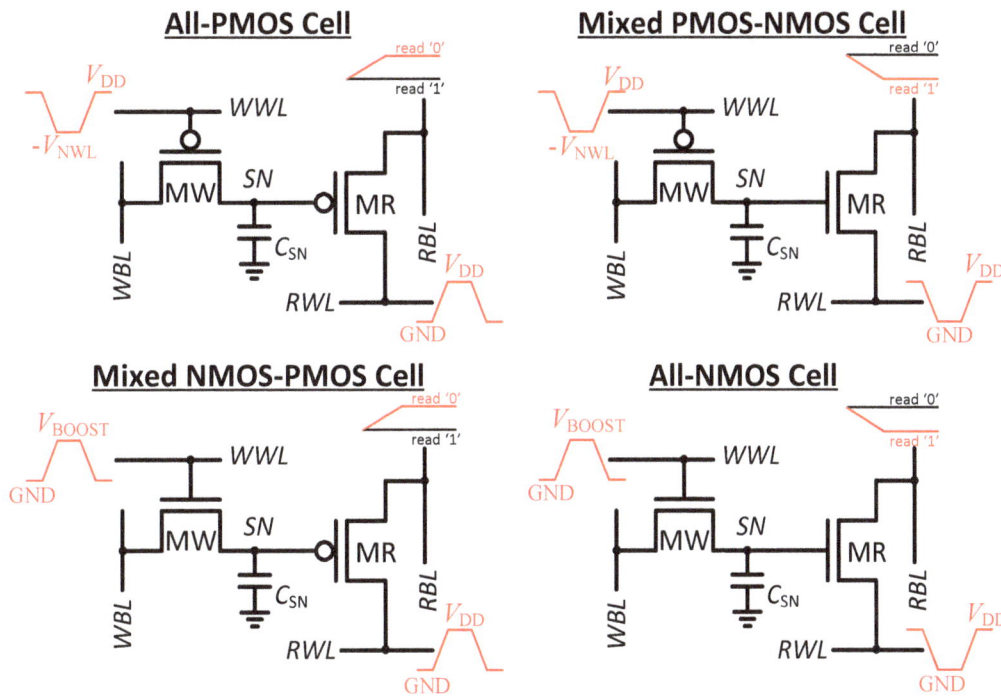

Figure 2. Leakage components that are considered for the choice of the best-practice write and read transistor implementations, for (**a**) Mature CMOS nodes; and (**b**) Scaled CMOS nodes.

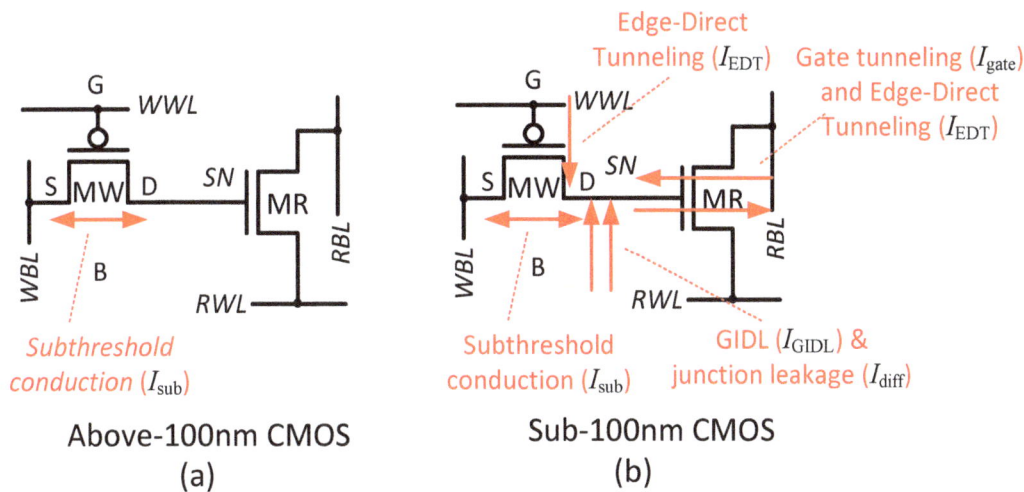

In more advanced, sub-100 nm CMOS nodes, there are other significant leakage mechanisms that can compromise data integrity. (Note that in the sub-V_T region, these mechanisms are still negligible, as compared with subthreshold conduction. However, as shown in Section 2.3.2, at near-V_T supplies, some of the mechanisms must be considered). Only leakage components that bring charge onto the SN or take charge away from SN need to be considered in terms of retention time, while other leakage components are merely undesirable in terms of static power consumption. Figure 2(b) schematically shows the main leakage components that can compromise the stored level in sub-100 nm nodes, including reverse-biased pn-junction leakage (I_diff), gate-induced drain leakage (I_GIDL), gate tunneling leakage (I_gate), edge-direct tunneling current (I_EDT), and subthreshold conduction (I_sub). When employing a

PMOS MW, the bulk-to-drain leakages (I_{diff} and I_{GIDL}) weaken a logic "0" and strengthen a logic "1", but have the opposite impact (strengthen a logic "0" and weaken a logic "1") when MW is implemented with an NMOS device. During standby, MW is always off and has no channel; therefore, forward gate tunneling (I_{gate}) from the gate into the channel region and into the two diffusion areas that would occur in a turned-on MOS device is of no concern here. Only the edge-direct tunneling current, from the diffusion connected to the SN in the absence of a strongly inverted channel, compromises data integrity. When using an NMOS MW, edge-direct tunneling discharges a logic "1", while it charges a logic "0" for a PMOS MW.

The only leakage through MR that affects the stored data level is gate tunneling. During standby, there is no channel formation in MR, no matter what the stored data level is. For example, if using an NMOS MR, both RWL and RBL are charged to V_{DD} during standby, such that even a logic "1" level results in zero gate overdrive. In this case, both diffusion areas of MR are at the same potential as the SN, eliminating tunneling currents between the diffusions and the gate ($I_{\mathrm{EDT}} = 0$). However, tunneling might occur from the gate directly into the grounded bulk (I_{gate}), weakening a logic "1". If the same cell stores a logic "0", tunneling between the gate and bulk is avoided ($I_{\mathrm{gate}} = 0$), while reverse tunneling from the diffusions (I_{EDT}) into the gate can charge the logic "0" level. The exact opposite biasing conditions and corresponding tunneling mechanisms are found when implementing MR with a PMOS.

2.2. Best-Practice Write Transistor Implementation

2.2.1. Mature 0.18 μm CMOS Node

For the ULP sub-V_{T} applications, long retention times that minimize the number of power-consuming refresh cycles are of much higher importance than fast write access. Therefore, low subthreshold conduction becomes the primary factor in the choice of a best practice write transistor in the 0.18 μm node. The subthreshold conduction of NMOS and PMOS, core and I/O devices offered in this process are shown in Figure 3(a). Clearly, the I/O PMOS device has the lowest subthreshold conduction I_{sub} ($V_{\mathrm{GS}} = 0\,\mathrm{V}$, $V_{\mathrm{DS}} = -V_{\mathrm{DD}}$) among all device options and across all standard process corners, leading to the longest retention time. At a 400 mV sub-V_{T} V_{DD}, the on-current I_{on} ($V_{\mathrm{GS}} = -V_{\mathrm{DD}}$, $V_{\mathrm{DS}} = -V_{\mathrm{DD}}$) of this preferred I/O PMOS device is still four orders of magnitude larger than I_{sub}, as shown in Figure 3(b), which results in sufficiently fast write and refresh operations compared with the achievable retention time. This holds for temperatures up to 37 °C, which is considered a maximum, worst-case temperature for ULP systems that are often targeted at biomedical applications, typically attached to the human body, and hardly suffer from self-heating due to low computational complexity. Nevertheless, for temperatures as high as 125 °C, a sufficiently high $I_{\mathrm{on}}/I_{\mathrm{sub}}$ ratio of four orders of magnitude is still achieved at a slightly higher supply voltage of 500 mV.

Figure 4(a) shows the worst-case time dependent data deterioration after writing into a 2T gain-cell with a PMOS I/O write transistor under global and local variations. The blue (bottom) curves show the deterioration of a logic "0" level with WBL tied to V_{DD}, and the red (top) curves show the deterioration of a logic "1" level with WBL tied to ground. The plot was simulated with a sub-V_{T} 400 mV V_{DD} assuming a storage node capacitance of 2.5 fF. A worst-case retention time of 40 ms can be estimated from this figure, corresponding to the minimum time at which the "0" and "1" levels intersect. It is clear that a logic "0" level decays much faster than a logic "1" level, corresponding with previous reports for the above-V_T domain [11,13]. In fact, the decay of a "1" level is self-limited due to the steady increase of the reverse gate overdrive ($V_{GS,MW} = V_{DD} - V_{SN}$) and the increasing body effect ($V_{BS,MW} = V_{DD} - V_{SN}$) of MW with progressing decay. Both of these effects suppress the device's leakage. Furthermore, the charge injection (CI) and clock feedthrough (CF) that occur at the end of a write access (when MW is turned off) cause the SN voltage level to rise, strengthening a "1" and weakening a "0" level [16,20]. Therefore, careful consideration must be given to the initial state of the "0" level following a write access, as will be discussed in Section 2.4.

Figure 3. (**a**) Subthreshold conduction of different transistor types in an 0.18 µm node; and (**b**) I/O PMOS I_{on}/I_{sub} current ratio as a function of V_{DD} for the typical-typical (TT) process corner at different temperatures.

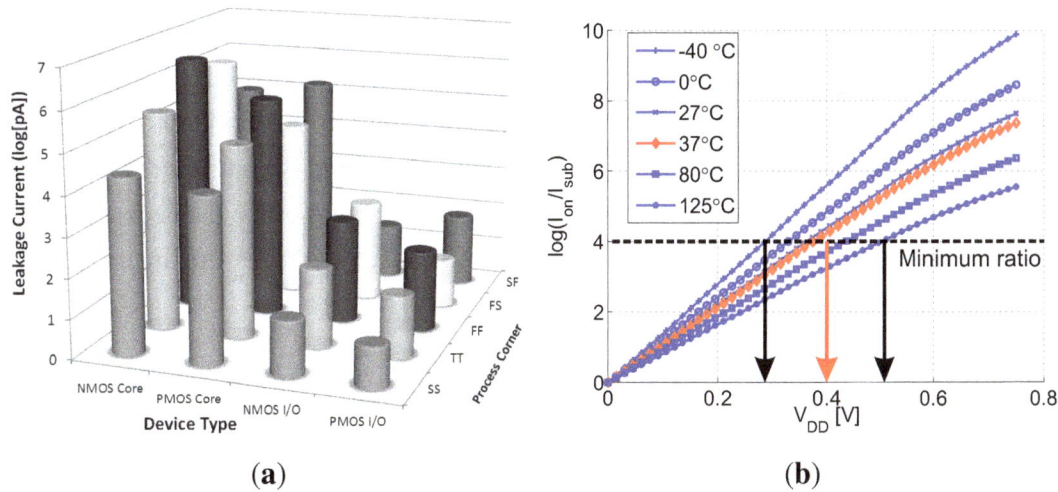

(**a**)   (**b**)

Figure 4. (**a**) Worst-case retention time estimation of 0.18 µm sub-V_T gain-cell with $V_{DD} = 400$ mV; (**b**) Best-practice gain-cell for sub-V_T operation in 0.18 µm CMOS.

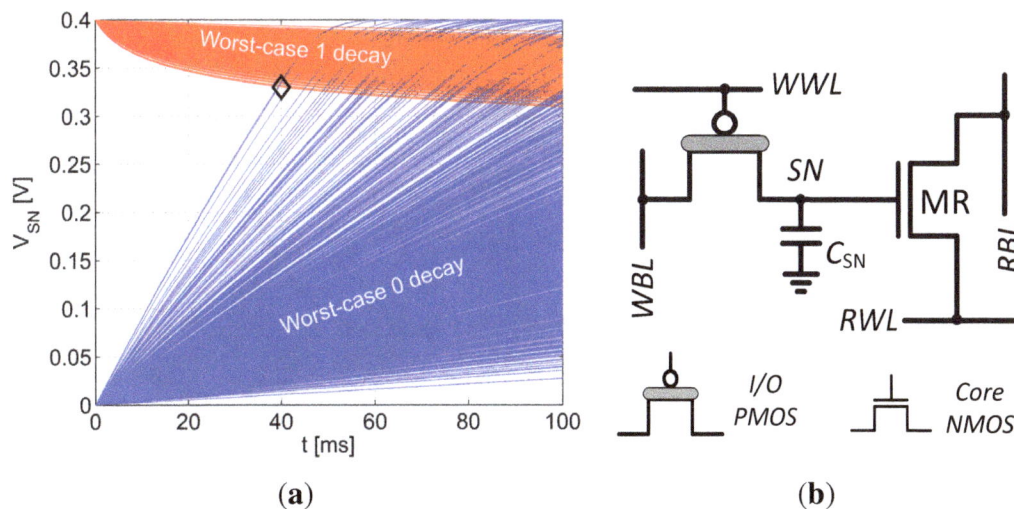

(**a**)   (**b**)

2.2.2. Scaled 40 nm CMOS Node

While choosing the best device option for MW, subthreshold conduction must again be kept as small as possible, as it affects both a "1" and a "0" level. The diffusion leakage, the GIDL current, and the edge-direct tunneling current weaken one logic level, while they strengthen the other. However, all three leakage components work against the logic level that has already been weakened through CI and CF at the end of a write pulse. For example, with a PMOS MW, the logic "0" level is weakened through a positive SN voltage step when closing MW, while I_{GIDL}, I_{diff}, and I_{EDT} further pull up SN, deteriorating the stored "0". Therefore, in order to protect the already weaker level, the optimum device selection aims at minimizing all of these leakage components. Figure 5(a) shows the leakage components of minimum sized devices provided in the 40 nm process (the LVT devices were left out of the figure for display purposes, as their leakage is significantly higher than the leakage of other devices) at a near-V_T supply voltage of 600 mV. This figure clearly shows that despite the increasing significance of other leakage currents with technology scaling, I_{sub} is still dominant at this node (some of the leakage components are not modeled for the I/O devices; however, this does not impact our analysis, as the PMOS HVT already provides the lowest total leakage). However, the advantage of using an I/O device is lost, and a more compact HVT PMOS device provides the lowest total leakage. This trend is confirmed when evaluating the leakage components of intermediate process nodes, as well, showing that the leakage benefits of using an I/O device deteriorate to the point where the area versus leakage trade-off favors the use of an HVT device at around the 65 nm node.

Figure 5.   (a) Leakage components of various devices in the considered 40 nm node at a near-V_T supply voltage of 600 mV; (b) Worst-case $I_{on}(\text{weak}'1')/I_{off}(\text{weak}'0')$ of MR, implemented with LVT, SVT, and HVT devices. Both plots were simulated under typical conditions.

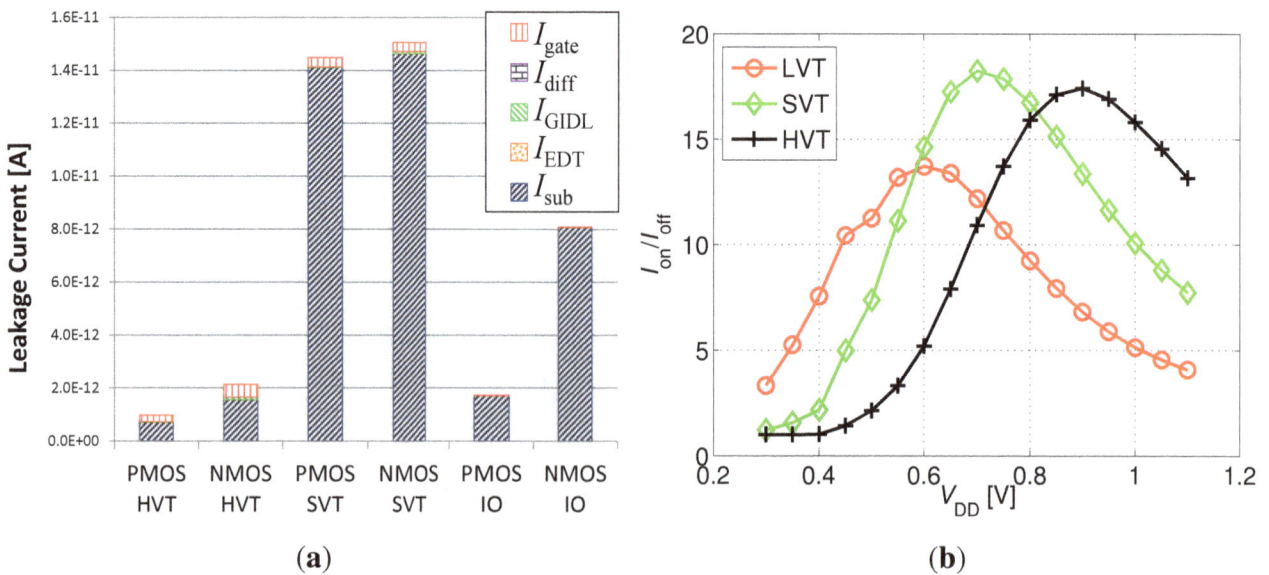

2.3. Best-Practice Read Transistor Implementation

2.3.1. Mature 0.18 μm CMOS Node

At the onset of a read operation, capacitive coupling from RWL to SN causes a voltage step on SN [20]. Our analysis from the previous section showed that MW should be implemented with a PMOS device, resulting in a strong logic "1" and a weaker logic "0". Therefore, it is preferable to implement MR with an NMOS transistor that employs a negative RWL transition for read assertion. The resulting temporary decrease in voltage on SN counteracts the previous effects of CI and CF, thus improving the "0" state during a read operation (effect is reversed upon deassertion of the RWL). As a side effect, this negative SN voltage step also lowers the "1" level and therefore slightly slows down the read operation; however, this level is already initially boosted due to deassertion of the WWL. An additional and perhaps more significant reason to choose an NMOS device for readout is that NMOS devices are approximately an order-of-magnitude stronger than their PMOS counterparts at sub-V_T voltages. Therefore, implementing MR with an NMOS device provides a fast read access, which not only results in better performance but also is essential for ensuring high array availability. As mentioned, the considered 0.18 μm process provides only core and I/O devices, and considering the three-orders-of-magnitude higher on-current for core devices at sub-V_T, the choice of an NMOS core MR is straightforward.

To summarize, the most appropriate 2T gain-cell for sub-V_T operation in an above-100 nm CMOS node comprises an I/O PMOS write transistor and a core NMOS read transistor, as illustrated in Figure 4(b). The resulting hybrid NMOS/PMOS gain-cell shares the n-well on three sides between neighboring cells [19] to keep the area cost low, as discussed in Section 3.

2.3.2. Scaled 40 nm CMOS Node

When considering the best device type for scaled nodes, the large number of options presents some interesting trade-offs for the implementation of MR. The increasing gate leakage currents (I_{gate} and I_{EDT}) at scaled nodes could potentially present an advantage for a thick oxide I/O device due to its reduced gate tunneling. However, at low voltages, the tunneling currents are small in comparison with the subthreshold conduction through MW, as shown in Figure 5(a). In addition, I_{gate} and I_{EDT} actually appear in opposite directions, as the stored "0" level rises, further reducing their impact. On the other hand, the two primary considerations for the above-100 nm nodes are even more relevant at scaled nodes. The achievable retention time in the 40 nm process turns out to be approximately three orders-of-magnitude lower than that of the 0.18 μm node. Therefore, the negative step caused by RWL coupling to SN is even more important, and fast reads are essential to provide sufficient array availability, despite the high refresh rates. To further enhance the read step, layout techniques can be implemented to increase the capacitive coupling between RWL and SN. However, when considering read access times, additional trade-offs arise. For maximum read performance, MR could be implemented with an LVT device. At the 40 nm node, an LVT NMOS provides an 8× increase in on-current at 400 mV compared with an SVT NMOS. However, as the supply voltage is increased, this benefit reduces to 3× at 600 mV. The superior on-currents of LVT devices, as compared with SVT or HVT options, come at the expense of much higher off-currents, as well as increased process variations. When choosing the read device, this trade-off must be taken into consideration, as it is mandatory to correctly differentiate between the discharged level of RBL due to a stored "1" and the depleted level due to a weak stored "0". Furthermore, the unselected cells on the same column of a selected cell storing a "1" will start to counteract the discharge of RBL during a read, as $V_{\text{GS,MR}}^{\text{unselected}} = V_{\text{DD}} - V_{\text{RBL}}$. In effect, this limits the speed and minimum discharge level of RBL, according to the drive strength of the unselected MR devices. When considering sub-V_{T} operation in the 40 nm node, the relatively low subthreshold conduction of the SVT, HVT, and I/O devices renders the LVT the only feasible option for MR to achieve a reasonable RBL discharge time. However, as V_{DD} is increased into the near-V_{T} region, an SVT device provides sufficient on-current, while the higher V_{T} and lower leakage enable better reliability under process variations, as well as improved array availability.

Figure 5(b) shows the worst case current ratio $I_{\text{on}}/I_{\text{off}}$ of the NMOS read transistor MR, implemented with different device types as a function of V_{DD}. I_{on} is given for a weak "1" level, estimated as the steady state high voltage of SN when tying WBL to V_{DD} ($V_{\text{SN}} = 0.85V_{\text{DD}}$). I_{off} is given for a weak "0" level, estimated at $V_{\text{SN}} = 0.4V_{\text{DD}}$, which would provide a sufficient margin to differentiate between the two levels (this is verified for the chosen implementation at the minimum feasible bias in Section 4). For supply voltages below 600 mV, the LVT device has the highest current ratio and is therefore preferred, as it provides the best achievable array availability. Likewise, the SVT device is preferred for V_{DD} between 600 and 800 mV, while the HVT device is the best option for even higher V_{DD}.

2.4. Storage Node Capacitance and WWL Underdrive Voltage

2.4.1. Mature 0.18 µm CMOS Node

To close the design of the 2T bitcell, two important design parameters must be taken into consideration. First, the storage node capacitance (C_{SN}), primarily made up of the diffusion capacitance of MW and the gate capacitance of MR, is typically around 1 fF for minimum device sizes. However, we find that by applying layout techniques, such as metal stacking, this value can be extended by over 5×, providing a configurable design parameter. Second, to address the V_{T} drop across MW especially affecting the write "0" operation (but also the write "1" operation in the sub-V_{T} regime), an underdrive voltage (V_{NWL}) needs to be applied to WWL, the magnitude of which affects the write access time and the SN voltage.

Figure 6(a) shows the storage node voltage (V_{SN}) after a write "0" access as a function of C_{SN} and V_{NWL}, before and after closing MW. Figure 6(b) emphasizes the impact of CI and CF by showing the voltage step ΔV that occurs while closing MW. It is clear that any V_{NWL} above -650 mV already results in a degraded logic "0" transfer prior to turning off MW. ΔV can be reduced by increasing C_{SN} and by decreasing the magnitude of V_{NWL}. Therefore, on the one hand, V_{NWL} must be low enough to ensure a proper logic "0" transfer, while, on the other hand, it should be as high as possible to minimize ΔV. The optimum value for V_{NWL} leading to the strongest "0" state after a completed write operation is found to be -650 mV, as shown in Figure 6(a). The optimum value for C_{SN} is clearly the maximum displayed value of 2.5 fF.

Figure 6. Following a write "0" operation: (**a**) V_{SN} before and after closing MW, as a function of C_{SN} and V_{NWL}; (**b**) ΔV due to charge injection from MW and due to capacitive coupling from WWL to SN.

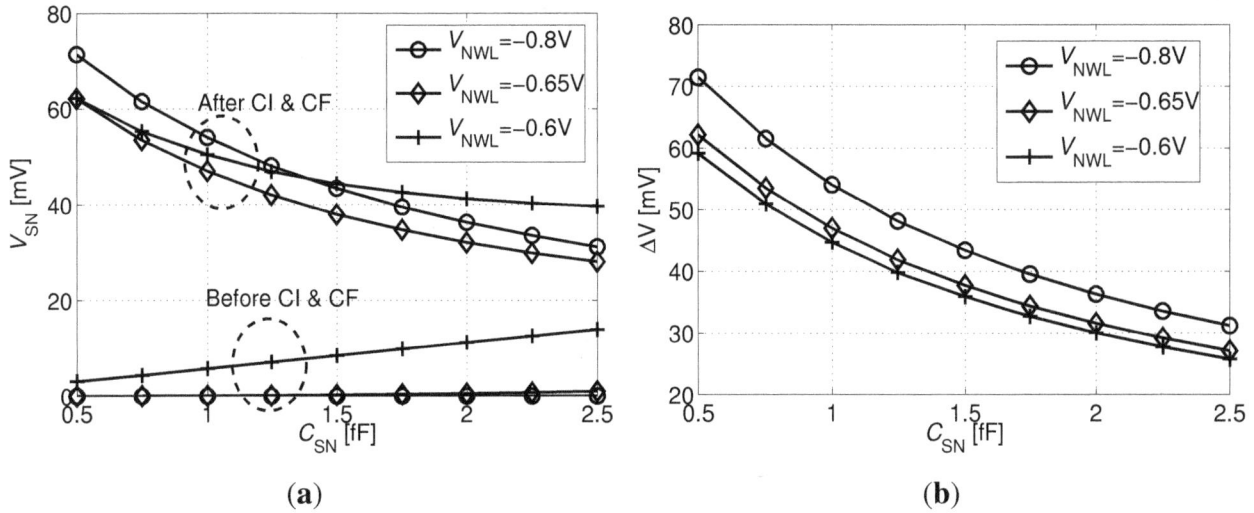

(**a**)   (**b**)

2.4.2. Scaled 40 nm CMOS Node

It is clear that the storage node capacitance should always be as big as possible, regardless of the technology node. This not only results in an improved initial "0" level, as shown above, but also provides more stored charge and thus extends the retention time. A general characteristic of scaled CMOS nodes is the increased number of routing layers, which in the case of gain-cell design, can be used to build up the storage node capacitor. Here, we assume that all available metal layers can be used at no additional cost, as the memory is going to be embedded in a system-on-chip that already uses all the metal layers. Moreover, with technology scaling, the aspect ratio of metal wires changes to narrower but higher, and wires can be placed closer to each other, which is beneficial in terms of side-wall parasitic capacitance. However, much of this benefit is offset by the lower dielectric constants of the insulating materials (*low-k*) integrated into digital processes with technology scaling. In addition, the absolute footprint of the bitcell shrinks with technology, making it more challenging to allocate many inter-digit fingers for a high capacitance. In fact, in the considered 40 nm node, the footprint of a gain-cell containing only two core devices is so small that the minimum width and spacing rules for medium and thick metals are too large to exploit for increasing the capacitance of the SN. Therefore, our layout of the 40 nm cell is limited to 5 routing layers, and the overall SN capacitance is much lower than that achieved in the 0.18 μm node. Figure 7(a) summarizes the achievable storage node capacitance according to the number of thin metal layers provided by the two considered technology nodes.

Figure 7(b) shows the 40 nm SN voltage step ΔV that occurs during the positive edge of WWL for a logic "0" transfer. As already observed for the 0.18 μm node, ΔV decreases with increasing SN capacitance and with decreasing WWL step size (*i.e.*, with decreasing absolute value of the underdrive voltage, V_{NWL}). While the charge injected from the large channel area of the selected I/O PMOS write transistor in the mature technology node results in a large voltage step severely threatening data integrity, the problem is slightly alleviated in more advanced nodes where small core transistors are preferred. The resulting voltage steps of 10 to 45 mV are rather small compared with the minimum V_{DD} where high array availability is achieved (as will be shown in Section 4). Moreover, it is worth mentioning that strong "0" levels are transferred to SN even with the least aggressive underdrive voltage of -0.4V (however, at the expense of write access time). Therefore, the ΔV values in Figure 7(b) also correspond to the final SN voltage right after the write access. The final choice of V_{NWL} for the 40 nm node needs to account for the write access time, which must remain short to guarantee high array availability in a node with high leakage and short retention time (see Section 4). Therefore, Figure 7(c) shows the final V_{SN} after CI and CF, as a function of the write pulse width. Over a large range of pulse widths as short as several ns, an underdrive voltage of -700 mV results in the strongest "0" levels, and is therefore preferred. Less underdrive, e.g., -500 mV, would result in weak "0" levels for pulse widths that are shorter than 3 ns.

Figure 7. (**a**) Storage node capacitance versus number of employed metal layers; (**b**) ΔV due to CI and CF, as a function of C_{SN} and V_{NWL}, for $V_{DD} = 700$ mV; (**c**) V_{SN} after CI and CF versus write pulse width.

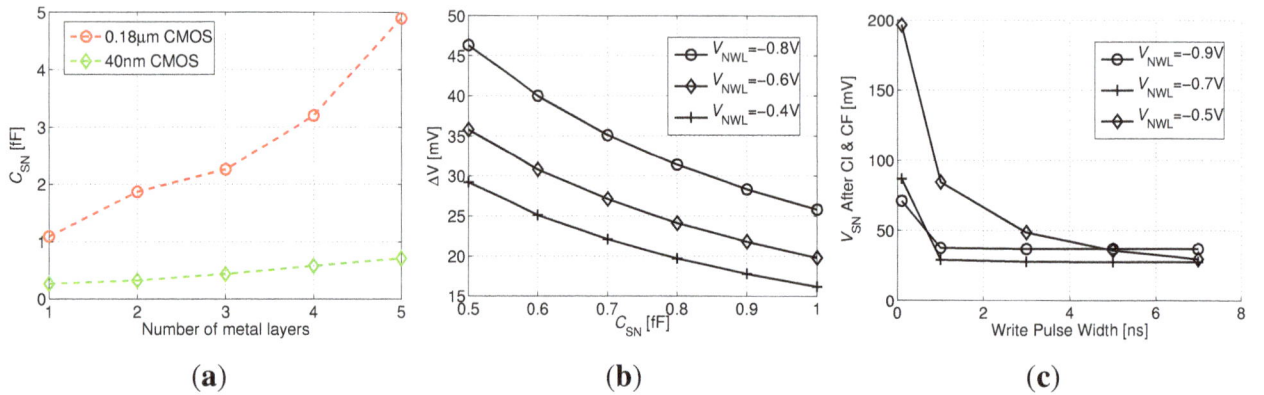

(**a**)   (**b**)   (**c**)

3. Macrocell Implementation in 0.18 μm CMOS

This section presents a 64×32 bit (2 kb) memory macro based on the previously elaborated 2T gain-cell configuration (Figure 4(b)), implemented in a bulk CMOS 0.18 μm technology. The considered V_{DD} of 400 mV is clearly in the sub-V_T regime, as V_T of MW and MR are -720 mV and 430 mV, respectively. Special emphasis is put on the analysis of the reliability of sub-V_T operation under parametric variations. While the address decoders and the sense buffers are built from combinational CMOS gates and operate reliably in the sub-V_T domain [21], the analysis focuses on the write-ability, data retention, and read-ability of the gain-cell. All simulations assume a 1 μs write and read access time (1 MHz operation); a 3-metal SN capacitance of 2.5 fF, providing a retention time of 40 ms (according to previously presented worst case estimation); a temperature of 37 °C and account for global and local parametric variations (1k-point Monte Carlo sampling).

Figure 8 plots the distribution of the bitcell's SN voltage at critical time points for the "0" and the "1" states. As expected, nominal 0 V and 400 mV levels are passed to SN just before the positive edge of the write pulse. CI and CF cause the internal levels to rise by 20–50 mV, resulting in a slightly degraded "0" level and an enhanced "1" level, while the distributions remain sharp. After a 40 ms retention period with a worst-case opposite WBL voltage, the distributions are spread out, but the "1" levels are still strong, while the extreme cases of the "0" levels have severely depleted, approaching 200 mV. However, the "0" and "1" levels are still well separated, and moreover, the "0" levels are improved following the falling RWL transition, resulting in a 10–20 mV decrease.

Figure 8. Distribution of the SN voltage of a logic "0" and a logic "1" at critical time points: (1) [circles] directly after a 1 μs write access (before turning off MW); (2) [squares] after turning off MW; (3) [diamonds] after a 40 ms retention period under worst-case WBL conditions; and (4) [triangles] during a read operation.

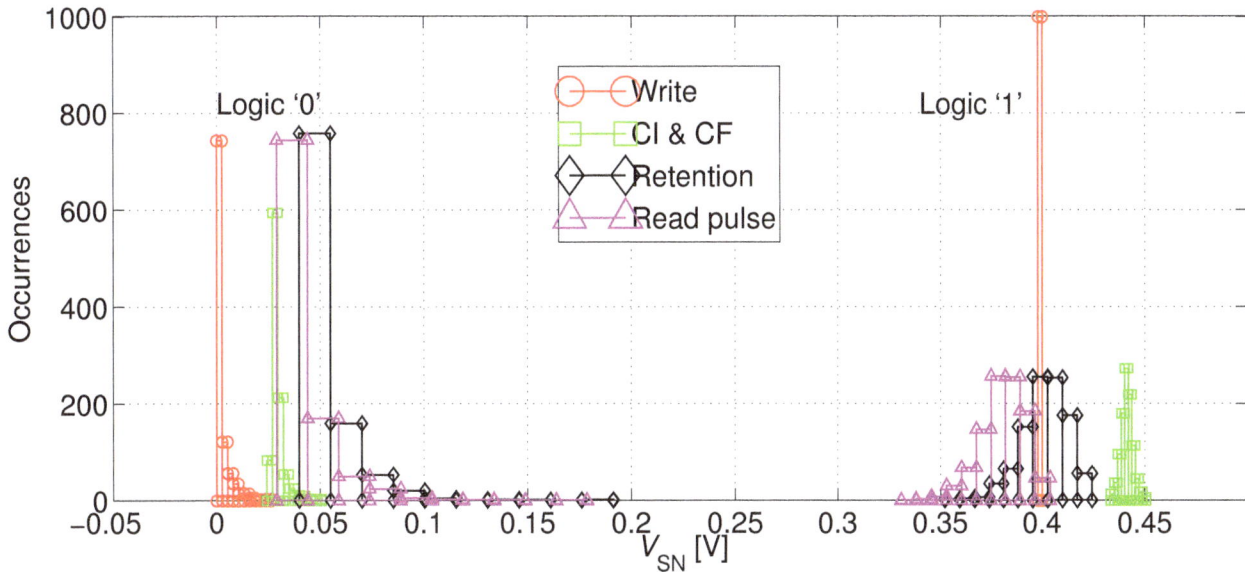

To verify the read-ability of the bitcell, Figure 9 shows the distribution of the RBL voltage (V_{RBL}) following read "0" and read "1" operations after the 40 ms retention period. In addition, the figure plots the distribution of the trip-point (V_M) of the sense buffer. While read "0" is robust in any case (RBL stays precharged), read "1" is most robust if all unselected cells on the same RBL as the selected cell store "0" (see Figure 9(a)), while it becomes more critical if all unselected cells store "1" (see Figure 9(b)), thereby inhibiting the discharge of RBL through the selected cell. This worst-case scenario for a read "1" operation is illustrated in Figure 10(a). In order to make the read operation more robust, V_M is shifted to a value higher than $V_{DD}/2$ by appropriate transistor sizing in the sense inverter. Ultimately, the V_{RBL} distributions for read "0" and read "1" are clearly separated, and the distribution of V_M is shown to comfortably fit between them, as shown in Figure 9.

The layout of the 0.18 μm 2T gain-cell, comprising a PMOS I/O MW and an NMOS core MR, is shown in Figure 10(b). The figure presents a zoomed-in view of one bitcell (surrounded by a dashed line) as part of an array. The chosen technology requires rather large design rules for the implementation of I/O devices; however, by sharing the n-well on three sides and stacking the bitlines, a reasonable area of 4.35 μm² per bitcell is achieved. In the same node, a single-ported 6T SRAM bitcell for above-V_T operation has a comparable area cost of 4.1 μm² (cell violates standard DRC rules), whereas SRAM bitcells optimized for robust operation at low voltages are clearly larger (e.g., the 14T SRAM bitcell in [2] has an area cost of 40 μm²). The depicted layout also enables metal stacking above the storage node to provide an increased SN capacitance of up to 5 fF (see Figure 7(a)).

Figure 9. Distribution of RBL voltage (V_{RBL}) after read "1" [circles] and read "0" [diamonds] operations and distribution of the trip-point V_M of the read buffer [squares], for (**a**) favorable and (**b**) unfavorable read "1" conditions.

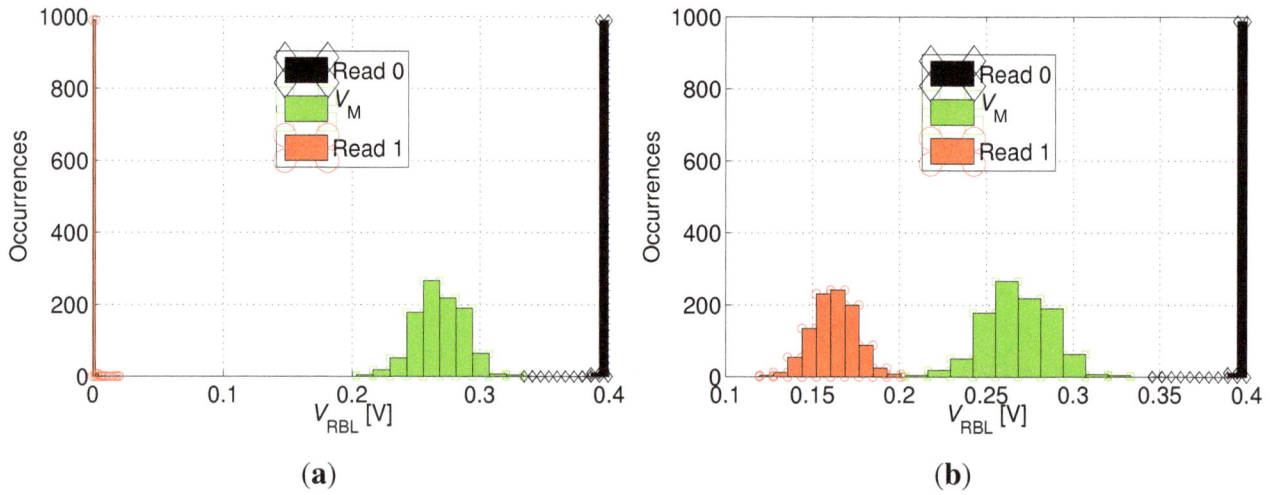

(**a**)

(**b**)

Figure 10. 180 nm gain-cell array: (**a**) Worst-case for read "1" operation: all cells in the same column store data "1"; to make the read "1" operation more robust, the sense inverter is skewed, with a trip-point $V_M > V_{DD}/2$; (**b**) Zoomed-in layout.

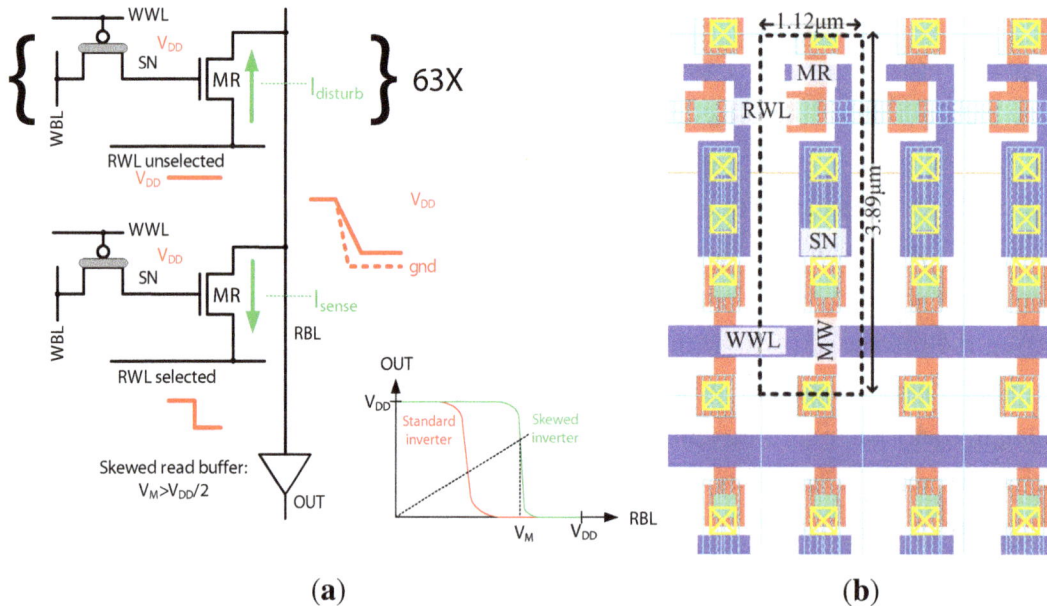

(**a**)

(**b**)

At an operating frequency of 1 MHz, a full refresh cycle of 64 rows takes approximately 128 μs. With a worst-case 40 ms retention time, the resulting availability for write and read is 99.7%. As summarized in Table 1, the average leakage power of the 2 kb array at room temperature (27 °C) is 1.95 nW, while the active refresh power of 1.68 nW is comparable, amounting to a total data retention power of 3.63 nW (or 1.7 pW/bit). This total data retention power is comparable with previous reports on low-voltage gain-cell arrays [13], given for room temperature as well.

Table 1. Figures of Merit.

Technology Node	180 nm CMOS	40 nm LP CMOS
Number of thin metal layers	5	5
Write Transistor	PMOS I/O	PMOS HVT
Read Transistor	NMOS Core	NMOS SVT
V_{DDmin}	400 mV	600 mV
Storage Node Capacitance	1.1 fF–4.9 fF	0.27 fF–0.72 fF
Bitcell Size	1.12 μm × 3.89 μm (4.35 μm^2)	0.77 μm × 0.42 μm (0.32 μm^2)
Array Size	64 × 32 (2 kb)	64 × 32 (2 kb)
Write Access Time	1 μs	3 ns
Read Access Time	1 μs	17 ns
Worst-Case Retention Time	40 ms	44 μs
Leakage Power	1.95 nW (952 fW/bit)	68.3 nW (33.4 pW/bit)
Average Active Refresh Energy	67 pJ	21.2 pJ
Average Active Refresh Power	1.68 nW (818 fW/bit)	482 nW (235.5 pW/bit)
Average Retention Power	3.63 nW (1.7 pW/bit)	551 nW (268.9 pW/bit)
Array Availability	99.7%	97.1%

4. Macrocell Implementation in 40 nm CMOS

Whereas gain-cell implementations in mature technologies have been frequently demonstrated in the recent past, 65 nm CMOS is the most scaled technology in which gain-cells have been reported to date [16]. In this section, for the first time, we present a 40 nm gain-cell implementation, and explore array sizes and the corresponding minimum operating voltages that result in sufficient array availability.

As previously described, core HVT devices are more efficient than I/O devices for write transistor implementation at scaled nodes, providing similar retention times with relaxed design rules (*i.e.*, reduced area). In addition, the multiple threshold-voltage options for core transistors provide an interesting design space for the read transistor selection, trading off on and off currents, depending on supply voltage. Two additional factors that significantly impact the design at scaled nodes are the reduced storage node capacitance, due to smaller cell area and low-k insulation materials, and severely impeded retention times, due to lower storage capacitance and increasing leakage currents. Therefore, array availability becomes a major factor in gain-cell design and supply voltage selection. For this implementation, a minimum array availability of 97% was defined.

Considering a minimum array size of 1 kb (32 × 32), sufficient array availability is unattainable with the LVT MR implementation for a supply voltage lower than 500 mV, suitable for this device according to Figure 5(b). Therefore, an SVT device was considered with near-threshold supply voltages above 500 mV. Figure 11(a) shows the array availability achieved under varying supply voltages, considering array sizes from 1 kb to 4 kb. The red dashed line indicates the target availability of 97%, showing that this benchmark can be achieved with a 2 kb array at 600 mV. At this supply voltage, with a −700 mV underdrive write voltage, the write access time is 3 ns, and the worst-case read access time is 17 ns, while the worst-case retention time is 44 μs (see Table 1). Figure 12 shows the distribution of the time required to sense the discharged voltage of RBL during a read "1" operation following a full retention period (green bars). The red bars (read "0") represent an incorrect readout, caused by a slow RBL discharge through leakage, such that the read access time must be shorter than the first occurrence of an incorrect read "0". The clear separation between the two distributions shows that by setting the read access time to 17 ns, the system will be able to robustly differentiate between the two stored states.

Figure 11. 40 nm gain-cell array: (**a**) Array availability as a function of supply voltage and array size; (**b**) Zoomed-in layout.

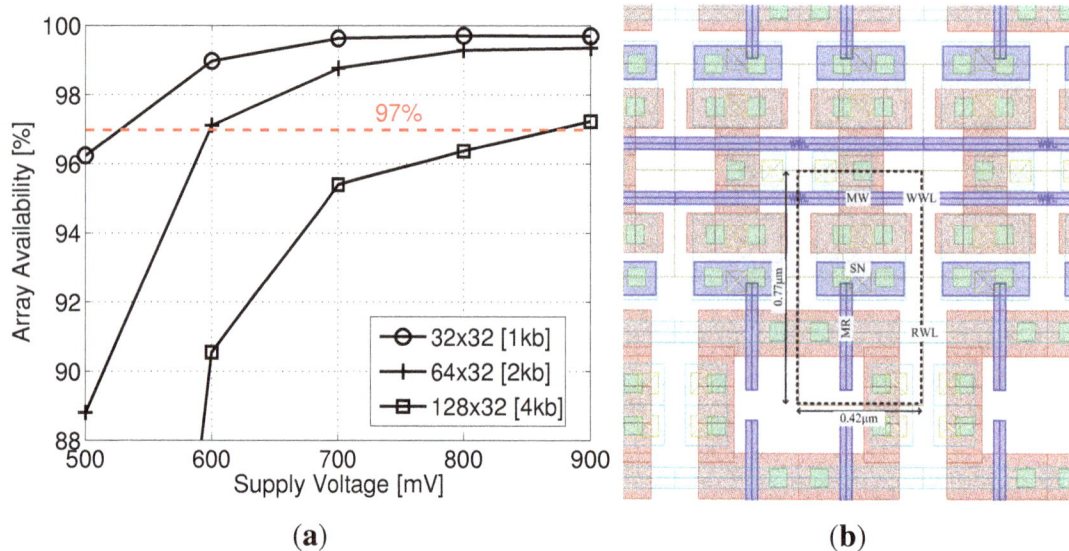

(**a**)   (**b**)

Figure 12. Read access time distribution for the 40 nm gain-cell implementation: RBL discharge time for correct data "1" sensing, and undesired RBL discharge time till sensing threshold through leakage for data "0".

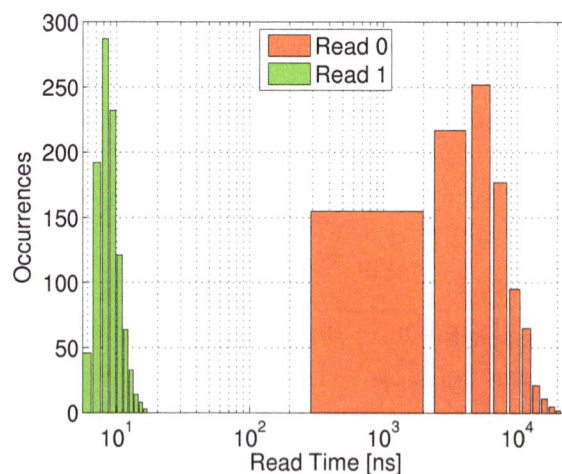

A zoomed-in layout of the 40 nm gain-cell array is shown in Figure 11(b), with a bitcell area of $0.32\,\mu m^2$ (surrounded by the dashed line). For comparison, a single-ported 6T SRAM bitcell in the same node has a slightly larger silicon area of $0.572\,\mu m^2$, while robust low-voltage SRAM cells are considerably larger (e.g., the 9T SRAM bitcell in [5] has an area cost of $1.058\,\mu m^2$). As shown in Table 1, the implemented 40 nm array exhibits a leakage power of 68.3 nW, which is clearly higher than for the

0.18 μm array. Even though the active energy for refreshing the entire array is only 21.2 pJ, the required refresh power of 482 nW is again higher than for the 0.18 μm node, due to the three orders-of-magnitude lower retention time. Consequently, the total data retention power is around 150× higher in 40 nm CMOS, compared with 0.18 μm CMOS.

5. Conclusions

This paper investigates two-transistor sub-V_T and near-V_T gain-cell memories for use in ultra-low-power systems, implemented in two very different technology generations. For mature, above-100 nm CMOS nodes, the main design goals of the bitcell are long retention time and high data integrity. In the considered 0.18 μm CMOS node, a low-leakage I/O PMOS write transistor and an extended storage node capacitance ensure a retention time of at least 40 ms. At low voltages, data integrity is severely threatened by charge injection and capacitive coupling from read and write wordlines. Therefore, the positive storage-node voltage disturb at the culmination of a write operation is counteracted by a negative disturb at the onset of a read operation, which is only possible with an NMOS read transistor. Moreover, the write wordline underdrive voltage must be carefully engineered for proper level transfer at minimum voltage disturb during de-assertion. Monte Carlo simulations of an entire 2 kb memory array, operated at 1 MHz with a 400 mV sub-V_T supply voltage, confirm robust write and read operations under global and local variations, as well as a minimum retention time of 40 ms leading to 99.7% availability for read and write. The total data retention power is estimated as 3.63 nW/2 kb, the leakage power and the active refresh power being comparable. The mixed gain-cell with a large I/O PMOS device has a large area cost of 4.35 μm^2, compared with an all-PMOS or all-NMOS solution with core devices only.

In more deeply scaled technologies, such as the considered 40 nm CMOS node, subthreshold conduction is still dominant at reduced supply voltages. Gate tunneling and GIDL currents are still small, but of increasing importance, while reverse-biased pn-junction leakage and edge-direct tunneling currents are negligible. In the 40 nm node, the write transistor is best implemented with an HVT core PMOS device, which provides the lowest aggregated leakage current from the storage node, even compared with the I/O PMOS device. A write wordline underdrive voltage of -700mV is employed to ensure strong "0" levels with a short write access time. Among various NMOS read transistor options, an SVT core device maximizes the sense current ratio between a weak "1" and a weak "0" for near-V_T supply voltages (600–800 mV) where 97% array availibility is achieved. Both the access times and the retention time are roughly three orders-of-magnitude shorter than in the 0.18 μm CMOS node, due to the increased leakage currents and smaller storage node capacitance. While the active refresh energy is low (21 pJ), the high refresh frequency results in high refresh power (482 nW), dominating the total data retention power (551 nW). As compared with the 0.18 μm CMOS implementation, the scaled down design provides better performance (17 ns read access and 3 ns write access), and a compact bitcell size of 0.32 μm^2.

To conclude, this analysis shows the feasibility of sub-V_T gain-cell operation for mature process technologies and near-V_T operation for a deeply scaled 40 nm process, providing a design methodology for achieving minimum V_{DD} at these two very different nodes.

Acknowledgments

This work was kindly supported by the Swiss National Science Foundation under the project number PP002-119057. Pascal Meinerzhagen is supported by an Intel Ph.D. fellowship. The authors would like to thank Itzik Icin and Meitav Liber for their contribution to this work.

Declaration

Based on "A sub-V_T 2T Gain-Cell Memory for Biomedical Applications", by P. Meinerzhagen, A. Teman, A. Mordakhay, A. Burg, and A. Fish which appeared in the Proceedings of the IEEE 2012 Subthreshold Microelectronics Conference. ©2012 IEEE.

References

1. Sinangil, M.; Verma, N.; Chandrakasan, A. A Reconfigurable 65 nm SRAM Achieving Voltage Scalability from 0.25–1.2 V and Performance Scalability from 20 kHz–200 MHz. In Proceedings of the IEEE European Solid-State Circuits (ESSCIRC), Edinburgh, UK, 15–19 September 2008.

2. Hanson, S.; Seok, M.; Lin, Y.S.; Foo, Z.Y.; Kim, D.; Lee, Y.; Liu, N.; Sylvester, D.; Blaauw, D. A low-voltage processor for sensing applications with picowatt standby mode. *IEEE J. Solid-State Circuit* **2009**, *44*, 1145–1155.

3. Constantin, J.; Dogan, A.; Andersson, O.; Meinerzhagen, P.; Rodrigues, J.; Atienza, D.; Burg, A. TamaRISC-CS: An Ultra-Low-Power Application-Specific Processor for Compressed Sensing. In Proceedings of IFIP/IEEE International Conference on Very Large Scale Integration (VLSI-SoC), Santa Cruz, CA, USA, 7–10 October 2012.

4. Calhoun, B.H.; Chandrakasan, A.P. A 256-kb 65-nm sub-threshold SRAM design for ultra-low-voltage operation. *IEEE J. Solid-State Circuit* **2007**, *42*, 680–688.

5. Teman, A.; Pergament, L.; Cohen, O.; Fish, A. A 250 mV 8 kb 40 nm ultra-low power 9T supply feedback SRAM (SF-SRAM). *IEEE J. Solid-State Circuit* **2011**, *46*, 2713–2726.

6. Meinerzhagen, P.; Andersson, O.; Mohammadi, B.; Sherazi, Y.; Burg, A.; Rodrigues, J. A 500fW/Bit 14fJ/Bit-Access 4 kb Standard-Cell Based Sub-Vt Memory in 65 nm CMOS. In Proceedings of the IEEE European Solid-State Circuits (ESSCIRC), Bordeaux, France, 17–21 September 2012.

7. Chiu, Y.W.; Lin, J.Y.; Tu, M.H.; Jou, S.J.; Chuang, C.T. 8T Single-Ended Sub-Threshold SRAM with Cross-Point Data-Aware Write Operation. In Proceedings of the IEEE International Symposium on Low Power Electronics and Design (ISLPED), Fukuoka, Japan, 1–3 August 2011.

8. Sinangil, M.; Verma, N.; Chandrakasan, A. A reconfigurable 8T ultra-dynamic voltage scalable (U-DVS) SRAM in 65 nm CMOS. *IEEE J. Solid-State Circuit* **2009**, *44*, 3163–3173.

9. Teman, A.; Mordakhay, A.; Fish, A. Functionality and stability analysis of a 400 mV quasi-static RAM (QSRAM) bitcell. *Microelectron. J.* **2013**, *44*, 236–247.

10. Hong, S.; Kim, S.; Wee, J.K.; Lee, S. Low-voltage DRAM sensing scheme with offset-cancellation sense amplifier. *IEEE J. Solid-State Circuit* **2002**, *37*, 1356–1360.

11. Chun, K.C.; Jain, P.; Lee, J.H.; Kim, C. A 3T gain cell embedded DRAM utilizing preferential boosting for high density and low power on-die caches. *IEEE J. Solid-State Circuit* **2011**, *46*, 1495–1505.

12. Somasekhar, D.; Ye, Y.; Aseron, P.; Lu, S.L.; Khellah, M.; Howard, J.; Ruhl, G.; Karnik, T.; Borkar, S.; De, V.K.; Keshavarzi, A. 2 GHz 2 Mb 2T Gain-Cell Memory Macro with 128 GB/s Bandwidth in a 65 nm Logic Process. In Proceedings of the IEEE International Solid-State Circuits Conference (ISSCC), San Francisco, CA, USA, 3–7 February 2008.

13. Lee, Y.; Chen, M.T.; Park, J.; Sylvester, D.; Blaauw, D. A 5.42nW/kB Retention Power Logic-Compatible Embedded DRAM with 2T Dual-Vt Gain Cell for Low Power Sensing Applicaions. In Proceedings of the IEEE Asian Solid State Circuits Conference (A-SSCC), Beijing, China, 8–10 November 2010.

14. Chun, K.C.; Jain, P.; Kim, C. Logic-Compatible Embedded DRAM Design for Memory Intensive Low Power Systems. In Proceedings of the IEEE International Symposium on Circuits and Systems, Paris, France, 30 May–2 June 2010.

15. Iqbal, R.; Meinerzhagen, P.; Burg, A. Two-Port Low-Power Gain-Cell Storage Array: Voltage Scaling and Retention Time. In Proceedings of the IEEE International Symposium on Circuits and Systems, Seoul, Korea, 20–23 May 2012.

16. Teman, A.; Meinerzhagen, P.; Burg, A.; Fish, A. Review and Classification of Gain Cell eDRAM Implementations. In Proceedings of the IEEE Convention of Electrical & Electronics Engineers in Israel, Eilat, Israel, 14–17 November 2012.

17. Chun, K.C.; Jain, P.; Kim, T.H.; Kim, C. A 667 MHz logic-compatible embedded DRAM featuring an asymmetric 2T gain cell for high speed on-die caches. *IEEE J. Solid-State Circuit* **2012**, *47*, 547–559.

18. Seok, M.; Sylvester, D.; Blaauw, D. Optimal Technology Selection for Minimizing Energy and Variability in Low Voltage Applications. In Proceedings of the ACM/IEEE International Symposium on Low Power Electronics and Design, Bangalore, India, 11–13 August 2008.

19. Meinerzhagen, P.; Andic, O.; Treichler, J.; Burg, A. Design and Failure Analysis of Logic-Compatible Multilevel Gain-Cell-Based DRAM for Fault-Tolerant VLSI Systems. In Proceedings of the IEEE Great Lakes Symposium on VLSI, Lausanne, Switzerland, 2–4 May 2011.

20. Meinerzhagen, P.; Teman, A.; Mordakhay, A.; Burg, A.; Fish, A. A Sub-V_T 2T Gain-Cell Memory for Biomedical Applications. In Proceedings of the IEEE Subthreshold Microelectronics Conference, Waltham, MA, USA, 910 October 2012.

21. Calhoun, B.; Wang, A.; Chandrakasan, A. Modeling and sizing for minimum energy operation in subthreshold circuits. *IEEE J. Solid-State Circuit* **2005**, *40*, 1778–1786.

3

A DC-DC Converter Efficiency Model for System Level Analysis in Ultra Low Power Applications

Aatmesh Shrivastava * and Benton H. Calhoun

The Charles L. Brown Department of Electrical and Computer Engineering, University of Virginia, Charlottesville, VA 22904, USA; E-Mail: bcalhoun@virginia.edu

* Author to whom correspondence should be addressed; E-Mail: as4xz@virginia.edu

Abstract: This paper presents a model of inductor based DC-DC converters that can be used to study the impact of power management techniques such as dynamic voltage and frequency scaling (DVFS). System level power models of low power systems on chip (SoCs) and power management strategies cannot be correctly established without accounting for the associated overhead related to the DC-DC converters that provide regulated power to the system. The proposed model accurately predicts the efficiency of inductor based DC-DC converters with varying topologies and control schemes across a range of output voltage and current loads. It also accounts for the energy and timing overhead associated with the change in the operating condition of the regulator. Since modern SoCs employ power management techniques that vary the voltage and current loads seen by the converter, accurate modeling of the impact on the converter efficiency becomes critical. We use this model to compute the overall cost of two power distribution strategies for a SoC with multiple voltage islands. The proposed model helps us to obtain the energy benefits of a power management technique and can also be used as a basis for comparison between power management techniques or as a tool for design space exploration early in a SoC design cycle.

Keywords: power management; modeling; DC-DC converter; DVS; DVFS; Ultra low power SoC; efficiency

1. Introduction

This paper presents a model for inductor based DC-DC converters and its application to quantify the benefits of power management techniques for ultra low power (ULP) SoCs. Various power management techniques, such as dynamic voltage and frequency scaling (DVFS), clock gating, and power gating are now commonly employed in many SoCs. However, the power benefits of these techniques cannot be quantified accurately without assessing their impact on the DC-DC converter that delivers power. For example, DVFS uses a high voltage to support higher performance and lower voltage to save power. However, changing the output voltage of a DC-DC converter incorporates significant power overhead, and the efficiency can vary widely across voltage and current loads. These overheads may offset the benefits from DVFS. There is a need to characterize the benefits of power management techniques like DVFS in conjunction with their impact on the DC-DC converter. This is particularly important for ultra-low energy near- or sub-threshold systems that operate in a very dynamic power environment, and whose power constraints are stringent. This paper presents a model [1] that enables the study of power management techniques by taking into account their impact on DC-DC converters of different topologies. The model is based on an analytical treatment of inductor based DC-DC converters, and it captures their efficiency trends with varying current load and output voltage. Since parameter selection will influence the specific behavior of the modeled converter, the model provides a rapid and effective tool for early design phase exploration of the impact of a converter by using parameters based on different prior designs or by sweeping parameters to investigate the optimal design requirements for the overall system.

Figure 1. Structure of the proposed model. (**a**) DC-DC efficiency modeling; (**b**) System energy cost model.

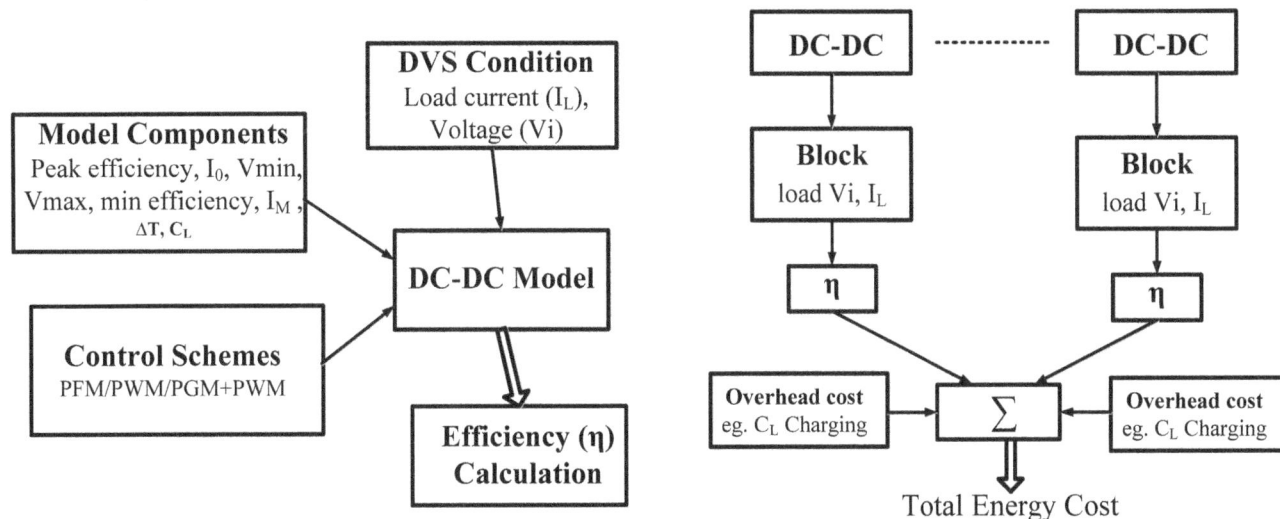

Figure 1 shows the structure of the proposed model, which is broken into two parts. First, a model of inductor based DC-DC converters (Figure 1a) takes the operating condition of a workload as an input, which includes peak efficiency, load at which the peak occurs, minimum efficiency, maximum and minimum output voltages, settling time, and the decoupling capacitance of the DC-DC converter, and generates an efficiency calculation. In order to capture the dynamic load condition of the converter in the context of the operation of a specific power management scheme, the second part of the model (Figure 1b) uses one or more of the DC-DC converter models in a larger system model that includes the input voltage (Vi), output voltage (Vo), load current (I_L), time of operation, parasitic capacitance on the block, switching frequency, and activity factor. These parameters can change dynamically in power management techniques like DVFS. Using these parameters as input, the model calculates the overhead cost and change in the efficiency of each DC-DC converter in the system and provides the total system level energy consumed while executing a power management technique for the given workload profile.

The energy savings for a power management technique are typically reported at the load circuit operating voltage and load level in literature [2]. For example, the authors in [2] report the energy savings obtained by scaling the voltage and frequency to a lower value. The paper does not calculate the total energy drawn from the original source of the supply voltage, for example a battery, which may not change linearly with load at the final circuit. Changing the operating condition of a voltage regulator causes deviation from its optimal behavior. If the output voltage of a buck converter is reduced, the efficiency of the converter degrades. Also, reducing the output voltage means that the capacitor C_L is discharged to a lower voltage by dissipating its stored energy. The actual benefits can be obtained by taking these losses and overhead into account. Figure 2 shows a block diagram of a typical inductor based DC-DC converter. It includes a bias generator and comparators that cause the static loss. The control scheme that implements the switching pattern of the power switches MP and MN can vary across topologies. The switching loss is a function of the control scheme and the load. The power switches MP and MN, parasitic resistance of inductor (L_{PAR}), and capacitor (C_{ESR}) cause the conduction loss, which is determined by the load current and output voltage. These losses are all a function of the operating condition. The proposed model accurately predicts the trends in behavior of DC-DC converters across topologies implementing both pulse width modulation (PWM) and pulse frequency modulation (PFM) control schemes.

Figure 2. Loss mechanisms inside a typical switching DC-DC converter.

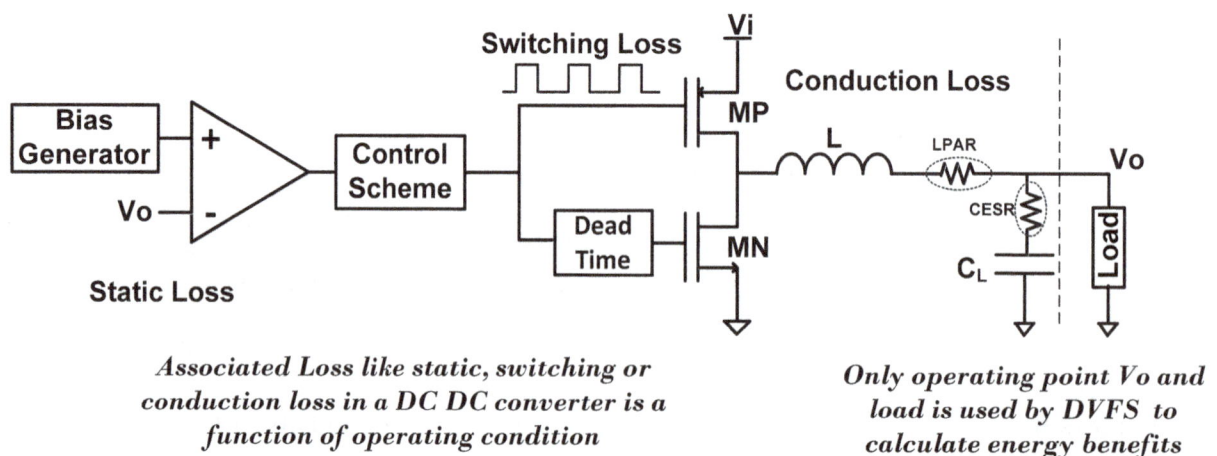

Associated Loss like static, switching or conduction loss in a DC DC converter is a function of operating condition

Only operating point Vo and load is used by DVFS to calculate energy benefits

2. DC-DC Converter Model

In order to model DC-DC converter trends and to capture the impact of DC-DC converters on power management strategies, it is essential to account for the converter efficiency, which is the power delivered to the load divided by the total power drawn by the converter. This efficiency of the DC-DC converter is a function of its output load, output voltage (Vo), the switching frequency, the switch resistance, and the parasitic resistance in the inductor and capacitor (Figure 2). In this section, we derive models for the efficiency for both PWM and PFM control schemes. Additionally, it is important to model how the converter will respond to changes in its usage. In a dynamic power environment like DVFS, Vo and load current vary dynamically, which changes the efficiency of the converter and results in energy overhead. Also, the converter can take significant time to settle from one voltage to another, resulting in timing overhead. Additional energy overhead comes in the form of charging and discharging of the decoupling capacitor. To quantify the benefit of a given power management technique like DVFS, we need to account for these overheads in addition to modeling the efficiency at a fixed load. In this section, we derive and describe the proposed model for an inductor based DC-DC converter that accounts for these overheads and that can be used in a larger system model to study specific power management techniques.

2.1. DC-DC Efficiency with Load Current

2.1.1. Model for PWM Control Scheme

Voltage and frequency are varied in DVFS to trade off power consumption with speed. This changes the load current and output voltage of the DC-DC converter, which changes its efficiency. In this section, we define a model that captures the change in efficiency that results from changing load conditions. One prior work [3] models the power loss in a DC-DC Buck converter using a PWM switching scheme with the following equation:

$$P_{Buck} = a\sqrt{(I_L^2 + \Delta i^2/3)f_s} + b\left\{\frac{I_L^2}{\Delta i f_s} + \frac{\Delta i}{3f_s} + \frac{C_L V_{DD1}{}^2}{R_{L0}\Delta i}\right\} + df_s\Delta i \tag{1}$$

where I_L is the load current; Δi is the current ripple in the converter; fs is the switching frequency; C_L is the decoupling capacitor; R_{LO} is the inductor series resistance; and a, b, and d are constants. Equation (1) represents the power loss in terms of various constants that cannot be obtained and that are non-intuitive to approximate prior to the design of converter, so it is difficult to apply this equation for design space exploration or for general modeling of DC-DC converter trends.

Instead, we propose a model that accurately captures the trends in the efficiency of the DC-DC converter in terms of the peak and minimum efficiency values of the converter, which can be either predicted, specified as targets, or pulled from prior work. To derive this simplified model, we begin by following previous work [3,4] in the observation that Equation (1) leads to an efficiency in the form

$$\eta_{I_L} = \eta_2 - (\eta_2 - \eta_1) * (\log(I_L/I_0))^2/4 \tag{2}$$

where η_2 is the peak efficiency occurring at load Io, and η_1 is the minimum efficiency at a given load. For the verification of the model proposed in Equation (2), let us consider the following cases.

For a light load condition in Equation (1),

$$I_L \sim \Delta i$$

So the power loss given by Equation (1) in the buck converter takes the form of

$$P_{Buck} = \alpha I_L + \beta/I_L$$

The efficiency of the converter will be given by power delivered (VI_L) over power drawn:

$$\eta_{I_L} = VI_L/(VI_L + \alpha I_L + \frac{\beta}{I_L}) \tag{3}$$

Expanding Equation (3) using Taylor's series we get:

$$\eta_{I_L} = V/(V + \alpha)\{1 - \frac{\beta}{(V+\alpha)I_L{}^2} + \frac{\beta^2}{(V+\alpha)^2 I_L{}^4} + \cdots\} \tag{4}$$

It is clear from Equation (3) that the efficiency decreases as load current decreases for the cases of light load conditions on the converter. The proposed equation in the model is given by:

$$\eta_{I_L} = \eta_2 - (\eta_2 - \eta_1) * (\log(I_L/I_0))^2/4$$

This expression also decreases for the light load condition, capturing the correct behavior of DC-DC converter efficiency trend. We know from Taylor's series that:

$$\ln(I_L/I_0) = -\{\left(\frac{I_0}{I_L} - 1\right) - 1/2 \left(\frac{I_0}{I_L} - 1\right)^2 + \cdots\} \tag{5}$$

Since, $I_0 \gg I_L$ and converting natural logarithm into logarithm of base 10 we get:

$$\log(I_L/I_0) = -2.303\{\left(\frac{I_0}{I_L}\right) - 1/2 \left(\frac{I_0}{I_L}\right)^2 + \cdots\}$$

So the efficiency equation in Equation (2) can be rewritten as:

$$\eta_{I_L} = \eta_2 - 2.303(\eta_2 - \eta_1)/4 * \{\left(\frac{I_0}{I_L}\right)^2 - \left(\frac{I_0}{I_L}\right)^3 + \cdots\}$$

$$\eta_{I_L} > \eta_2 - 2.303(\eta_2 - \eta_1)/4 * \left(\frac{I_0}{I_L}\right)^2 \tag{6}$$

Using, $x > \log x$

The proposed model in Equation (2) for the DC-DC converter reduces to Equation (6) under light load, which follows the behavior of DC-DC converter reported in literature [3] and compares well with Equation (4) with maximum efficiency η_2 in Equation (2) can be obtained by equating the constants in Equations (4) and (2),

$$\eta_2 = V/(V + \alpha) \tag{7}$$

Equations (2) and (6) are not bounded for the cases when I_L becomes very small, but it can predict the behavior for light load condition in a PWM control switching scheme based DC-DC converter with less than 5% error. Figure 4 shows this result.

Now consider the case when load is very high compared to the point of peak efficiency. The constant a in Equation (1) represents the resistance of the MOS transistor used for switching [3]. The model of [3] assumes it to be a constant. However, the resistance of MOS transistor increases with the increase in load current and it is given by:

$$R_D = K * (1 + \frac{I_L}{I_{DSAT}}) \tag{8}$$

where I_{DSAT} is the saturation current of the transistor. Clearly, as load current increases, the resistance increases. For light load condition, it is correct to assume that resistance does not change as I_{DSAT} is much larger compared to I_L. However, with an increase in load, the MOS resistance increases, causing elevated conduction loss. Also, at higher load the increased current in the inductor causes elevated conduction loss in the inductor's parasitic resistance. Overall, the I^2R loss increases, because of the increase in current and because of the increase in resistance caused by that increase in current. For a high load we know that:

$$I_L \gg \Delta i$$

so the power loss in the buck converter using equation (1) of [3] takes the form of:

$$P_{Buck} = \alpha_1 I_L + \beta_1 I_L^2$$

The equation for efficiency can be written as:

$$\eta_{I_L} = VI_L/(VI_L + \alpha_1 I_L + \beta_1 I_L^2) \tag{9}$$

This expression shows that efficiency decreases as the load current increases. Using Taylor series expansion, this expression can be written as,

$$\eta_{I_L} = V/(V + \alpha_1)\{1 - \frac{\beta_1 I_L}{(V+\alpha_1)} + (\frac{\beta_1 I_L}{(V+\alpha_1)})^2 + \cdots\} \tag{10}$$

In the proposed equation in this paper, Equation (2), the efficiency also decreases for the higher load. Expanding (2) using Taylor's series,

$$\eta_{I_L} = \eta_2 - (\eta_2 - \eta_1)/4 * \{k1 + k2\frac{I_L}{I_o} + k3\left(\frac{I_L}{I_o}\right)^2 + \cdots\} \tag{11}$$

Equations (10) and (11) follow closely, with constants η_2, η_1, k_1, etc. in Equation (11) can be obtained by equating with respect to the powers of I_L in Equation (10). The proposed equation matches the trend of equation reported in literature [3].

While Equations (1), (4) or (10) can be more analytical versions of the DC-DC converter efficiency formulation, they are not very useful for early design space exploration or for studying system level power management techniques because of the unknown constants. In contrast, the compact model in Equation (2) can be expressed in terms of peak efficiency and minimum efficiency, making it easy to use.

2.1.2. Model for PFM Control Scheme

In the PFM control scheme with a constant peak inductor current, the switching and conduction loss scale with frequency, and the efficiency remains flat for higher loads. The power loss is given by

$$P_{Buck} = K * P_{Out} + C \tag{12}$$

where K and C are constants. The first term indicates switching and conduction loss that scales with frequency. The second term indicates the static loss that does not scale with frequency. Therefore:

$$\eta_{I_L} = \frac{P_{Out}}{(P_{Out} + K * P_{Out} + C)} \tag{13}$$

Using $P_{Out} = VI_L$ and expanding using Taylor series,

$$\eta_{I_L} = \eta_2 - \frac{a}{I_L} \tag{14}$$

Equation (14) shows that the efficiency increases and becomes constant as load increases. This happens because power loss in PFM schemes scales with the load. At light load condition, the static loss dominates reduces the efficiency.

Figure 3. Efficiency Variation with load current in (**a**) pulse width modulation (PWM) scheme with $\eta_2 = 0.9$, $\eta_1 = 0.68$ and $I_L = 1$ mA; (**b**) pulse frequency modulation (PFM) scheme with $\eta_2 = 0.88$ and a $= 5 \times 10^{-5}$.

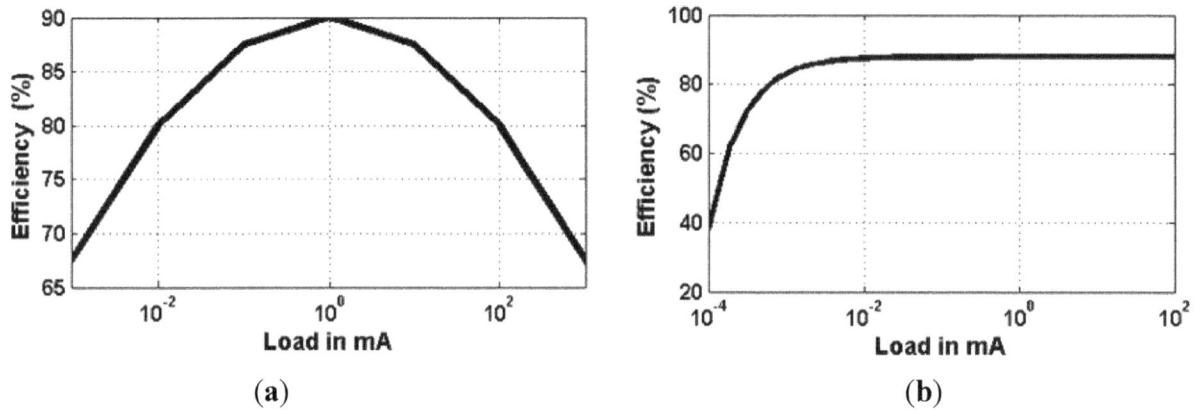

(**a**)   (**b**)

2.3. Verification of the Model

The equations we have provided accurately model the trends for PWM, PFM, and combinations of those control schemes. Figure 3 shows the variation of efficiency with respect to load when using dedicated PWM and PFM control schemes. In the PWM scheme, the efficiency degrades at high load and at light load conditions, while in the PFM scheme the efficiency remains flat as load increases and degrades at light load conditions.

Figure 4 shows the comparison of the model equations for PWM and PFM control schemes with the measured results reported in literature [5–8]. Figure 4a shows the results for the PWM scheme. For Equation (2), we have selected η_2 as the peak efficiency reported in the corresponding paper and I_O as the load for that peak efficiency. The value of η_1 is obtained experimentally based on [5–8], and we set $\eta_1 = 10$ for all the papers ([5–8]) employing the PWM control scheme. The converters [5–7] and [8] in part implement PWM. We find that the model predicts the efficiency behavior of the converters very accurately (<5% error for these papers). For [5,7,8] the error is less than 3%.

We also compared the model with reported works that employ the PFM switching scheme. Figure 4b,c shows the results. For the PFM scheme, we used η_2 as the peak efficiency reported in the corresponding paper. The constant a represents the static loss of the converter and will vary from one converter to another. It causes the degradation in efficiency at light load condition in a PFM control scheme. For this comparison, we set the value of constant a in Equation (14) to match the least efficiency reported in each paper. The model predicts the behavior of [6] correctly for the PFM scheme, while it deviates at higher load for [8]. This is because we assumed in our model that PFM scheme is used only for a light load condition, whereas [8] shows results for loads up to 400 mA.

Figure 4. Comparison of the load equation with measured work in literature. (**a**) Comparison of model with the measured PWM DC-DC converter reported in literature; (**b**) Comparison of model with [8], implementing both PFM and PWM; (**c**) Comparison of model with [6], schemes implementing both PFM and PWM.

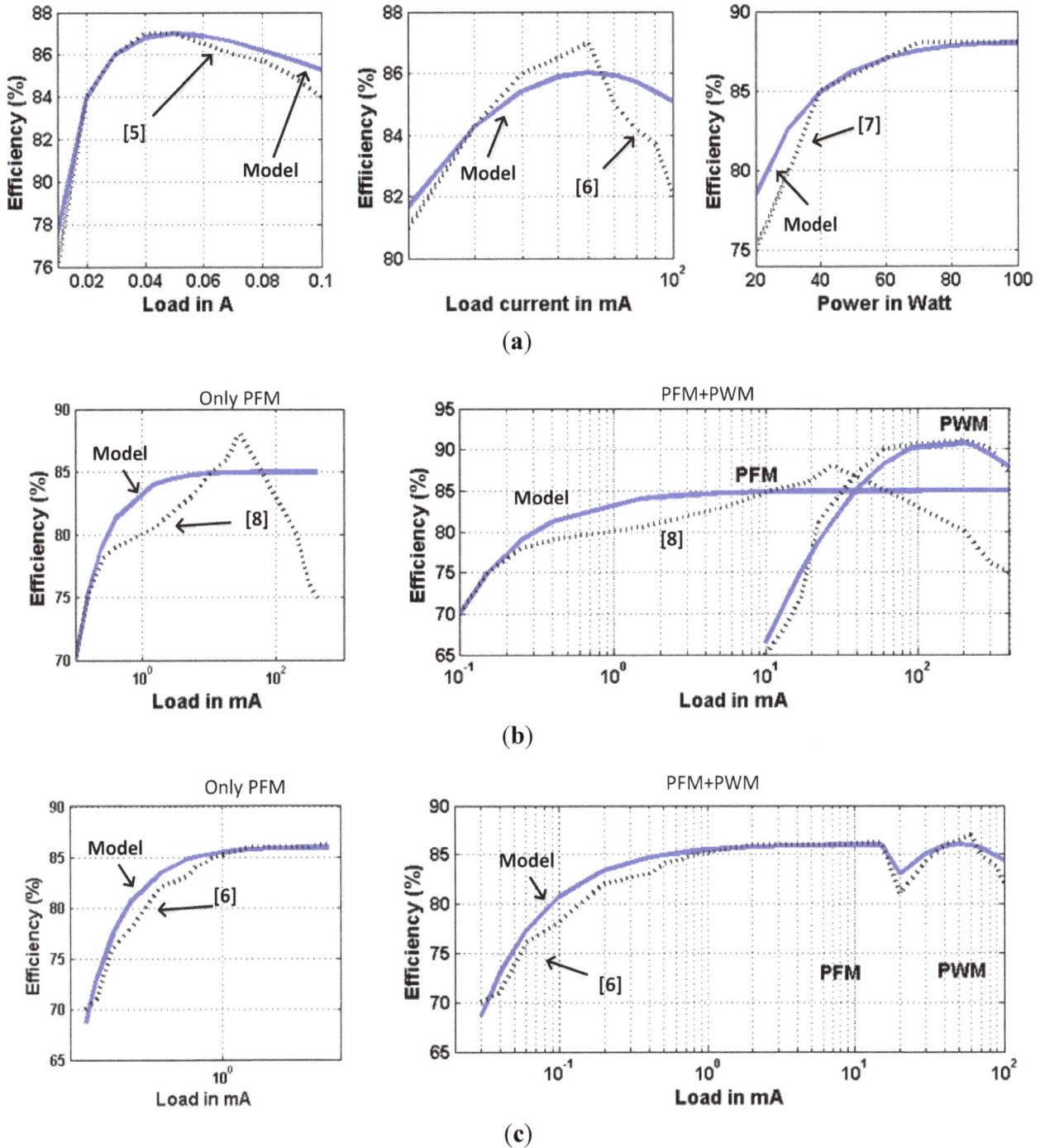

(**a**)

(**b**)

(**c**)

To further illustrate the usefulness of the model, we also compared it with the schemes where both PFM and PWM schemes are employed. This is often done to increase the range of load a converter can support. Figure 4b,c shows the results of comparison of the model with efficiencies reported in [6] and [8]. The model accurately predicts the behavior of such converters. There can be several other combinations in which the model can be used. For example, the model can be used for a segmented switch scheme if two or more curves for Equation (2) are used in conjunction, with each curve having a peak efficiency ($\eta 2$) corresponding to a different value of I_O.

The proposed model is accurate in predicting the efficiency trend with respect to the load if the control scheme is PWM or PFM. Also, an approximate peak efficiency of a DC-DC converter can be obtained very early in the design cycle as it is dependent on the technology, size of the inductor, and ripple on rails. Therefore, the model can be employed for analyzing the DC-DC converter overheads early in design while implementing power management techniques.

2.4. Efficiency with Output Voltage

The peak efficiency of an inductor based DC-DC converter decreases with a decreasing output voltage. For a given load current, the switching loss and conduction loss of the converter remain the same. The decreased output power level at lower voltages results in a decreased efficiency. The efficiency as a function of voltage can be modeled as:

$$\eta_V = \eta_1 + m(V - V_{min}) \tag{15}$$

where m is the slope of the line given by $m = (\eta_2 - \eta_1)/(V_{max} - V_{min})$. This approximate linear behavior is reported in both [3] and [4]. Plugging Equation (2) for PWM or Equation (14) for PFM and using Equation (15) lets us write the combined voltage and load efficiency as:

$$\eta = \eta_V * \eta_{I_L} \tag{16}$$

Figure 5 shows the output of the proposed model with V_O and with load current assuming a PWM control scheme. A DC-DC converter designed for a specific voltage and load will follow this trend when its load current or output voltage changes.

Figure 5. Dynamic efficiency variation with current and voltage.

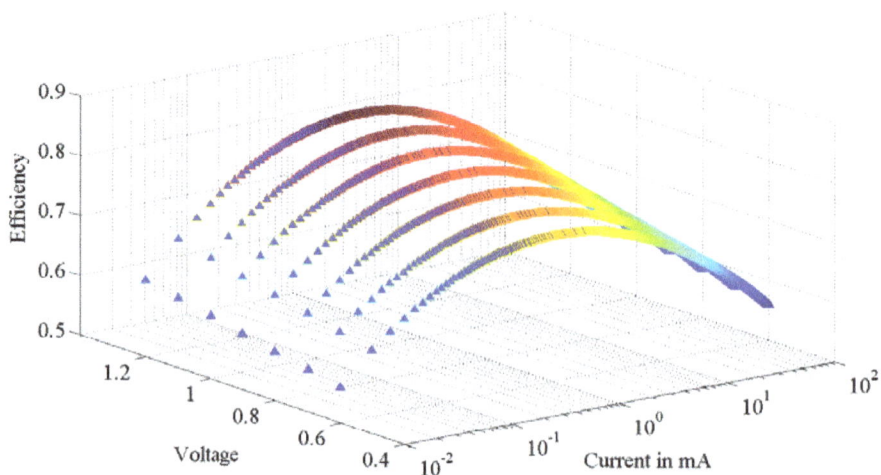

2.5. Settling Time

The settling time of a converter is the time it takes to reach the desired supply voltage. A typical converter has a large inductor and a large filter capacitor that makes the settling time very large (few μs to ms [9]). In a dynamic environment like DVFS or Ultra DVS (UDVS), the output voltage V_O is expected to change. The settling time of a converter to reach the desired voltage becomes an important overhead for these scenarios. The settling time ΔT in our proposed model is approximated as,

$$\Delta T = \frac{T}{V} * \Delta V \tag{17}$$

when output voltage is charged to V from ground, where T is the settling time of the converter. We assume that the inductor carries the same amount of current for each cycle of charging. Consequently, the rate at which the output reaches a given voltage will be linear with time. T/V is the slope of this curve and is set by assuming that at each switching cycle the inductor carries the same amount of current even for the cases when the output voltage is rising from zero.

2.6. Supply Rail Switching Energy

The change in the output voltage of a converter results in a change of the stored energy on the capacitance of the supply voltage rail of the load blocks. Some of this stored energy is dissipated if the new voltage is lower than the previous voltage, whereas energy is consumed from the source supply, V_{in}, if the new voltage is higher than the previous voltage. The additional energy overhead Ec is given by Equation (18) where V_1 and V_2 are the new and previous voltages of the converter. If V_1 is greater than V_2, work is to be done by the supply V_{in}. When V_1 is less than V_2, no work is done by V_{in} hence energy overhead will be zero.

$$E_C = V_{in} C_L \max(V_1 - V_2, 0) \tag{18}$$

In some cases, the voltage on the capacitor is not immediately discharged to lower voltage. VDD slowly discharges from V_2 to V_1 while running workloads. In such cases E_C will be lower than given by Equation (18). Equation (16) helps us to predict the losses at a given load or voltage condition, while Equations (17) and (18) give the conversion energy and timing overhead. These equations enable a framework where overheads that originate from dynamic changes to the DC-DC converter output can be calculated for techniques like DVFS to accurately measure their energy benefits.

In summary, this section has derived models for inductor based DC-DC converter efficiency for both PWM and PFM control schemes at a fixed load and provided equations for modeling the overheads that arise when the loads change. In the next section, we examine an example of how to apply these models in the evaluation of power management schemes like DVS.

3. Evaluation of DVS Techniques Using the Proposed Model

The DC-DC converter model can be used for assessing a variety of block level power management techniques. Since the model captures both the efficiency of the converter for fixed loads and the cost of making dynamic changes to the converter output voltage and load, we can use it to model system level implementations of varying complexity. In the most basic case, the model can provide additional insight into the system level cost of reducing the voltage delivered to a block or to a chip.

For example, Figure 6 shows the measured energy consumption of a microcontroller in a sub-threshold system on chip [13] across voltage. The minimum energy voltage occurs at below 0.3 V. However, when we consider the amount of energy drawn from the battery or energy storage node, including the overhead of loss in a DC-DC converter, the situation changes significantly. The top line in Figure 6 shows the energy consumed from the source (at the input to the DC-DC converter) using our model, which was fitted to low load DC-DC converter measurements from [14] to illustrate an example converter for this system. Since the output voltage and load current both vary for the block as VDD decreases, it is not accurate to assume a constant efficiency loss in the converter, and our model captures the changing efficiency across the space. This result shows that reporting block level consumption only can result in an inaccurate view of the total impact on the battery, and that the actual optimal voltage [19] for minimizing energy consumed at the battery may be higher than anticipated from block measurements alone.

Figure 6. Energy consumption for a microcontroller across voltage with and without consideration of the DC-DC converter efficiency.

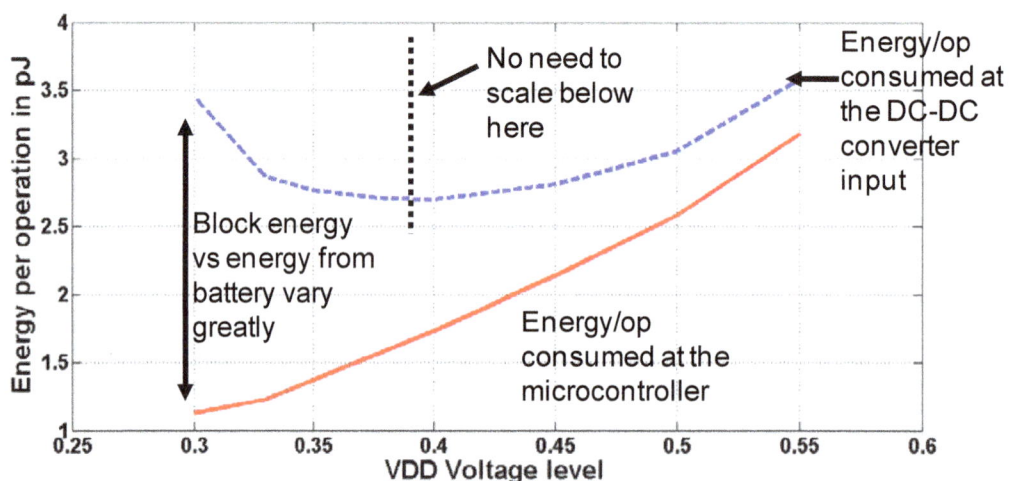

Further, the DC-DC model can help designers to choose the optimal specification for a converter to use for a given block or chip. This is especially important for embedded converters serving extremely low power systems like the one in [13], which operates from harvested energy. For example, we revisit the design in Figure 6 and consider a different converter design with lower overall peak efficiency, but whose peak efficiency occurs at a lower voltage. Figure 7a shows the efficiency versus VDD for two different converters: one is the converter used in Figure 6 and modeled on [14], and the other is a hypothetical converter. The new converter has a lower overall peak efficiency than the original design, which might lead to the misconception that it will hurt the overall energy consumption of any chip. However, its peak efficiency comes at a lower voltage, so its efficiency scales better across the lower range of voltage values. The impact of this is that, at lower voltage (and lower current) values, the second converter provides more efficient operation. Figure 7b shows the impact of using this converter alongside the curves from Figure 6. Not only is the total energy drawn from Vin much lower, the optimal voltage for minimizing energy occurs at a voltage much nearer to the optimal voltage of the block. The flexibility of our model allows a designer to experiment with different converter specifications while co-designing an embedded converter with its loading system, providing for rapid design space exploration early in the system design process.

Figure 7. (**a**) Efficiency profiles for two different converters and (**b**) the impact of a converter change on the overall energy drawn from the battery.

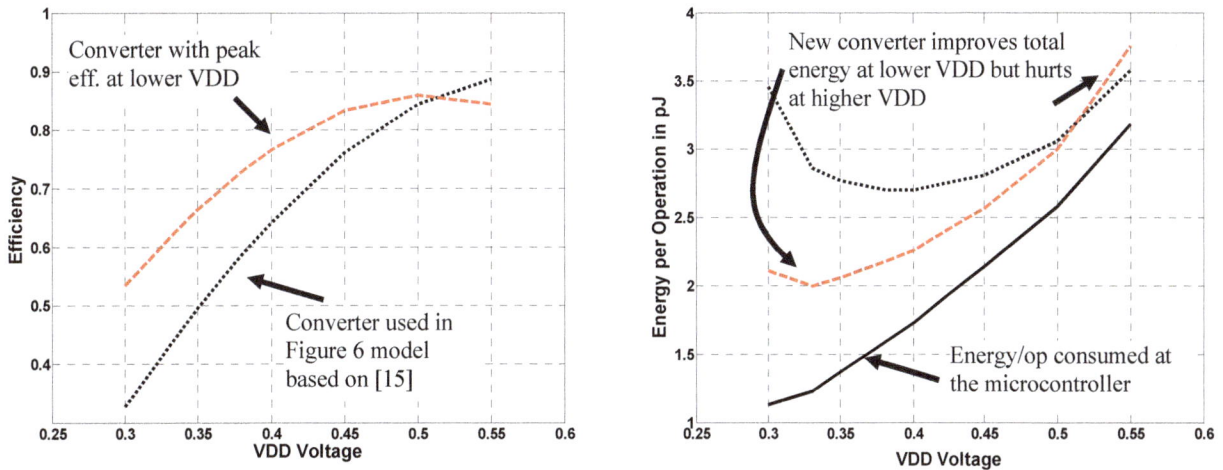

In addition to supporting the co-design of embedded DC-DC converters, our model can enable higher level comparisons of power management techniques that apply to multiple blocks. The next example illustrates the application of our model to two different power management strategies.

DVFS is commonly used to save power in a SoC by changing the supply voltage at the full chip level. Even larger energy savings can be realized by implementing block level DVFS, so that each block can use a voltage that is best matched to its own workload. Figure 8 shows the idealized implementation in which each block has a dedicated DC-DC converter. However, it may not be practically possible to implement such a system because of the area and cost of replicating DC-DC converters for each block. We study this scheme using our model to analyze its benefits by taking into account the overheads discussed earlier in the paper. Later on, we compare this with a more practical implementation of block level power management.

Figure 8. Dedicated DC DC converter per block.

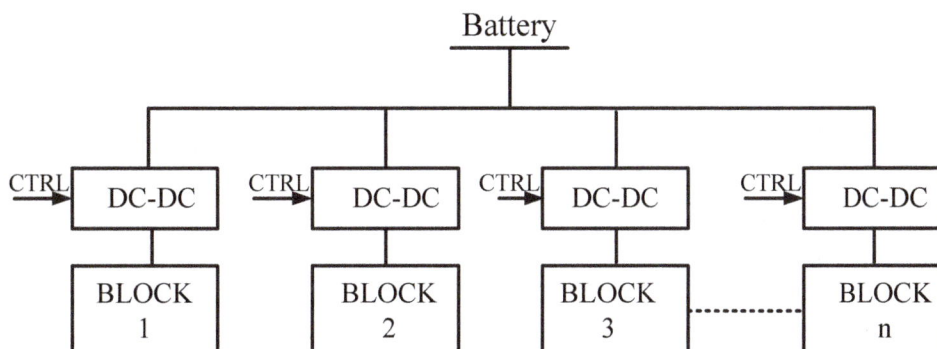

3.1. Framework for Energy Calculation in DVFS

In this section, we establish the framework for computing the system level energy consumption of a multiple block DVFS system, where individual DC-DC converters are modeled using the equations from Section 2. Figure 9 shows the operating condition of an example block that has a dedicated DC-DC converter. VDD and load will change with time. We have assumed a uniform random distribution for the power supply voltage setting in the range of 0.4 V to 1.2 V. ΔT is the settling time of the converter, and we use $T = 20$ μs and $V = 1$ V in Equation (18) [3]. The load capacitor on each block is assumed to be 200 pF.

Figure 9. Operating condition for dedicated DC DC.

VDD	Load	Time of operation
V1	i1	T1+ΔT1
V2	i2	T2+ΔT1
……	……	……

VDD	Load	Time of operation
0.9V	100μA	8μs
1.2V	900μA	8μs
……	……	……

Optimal Condition for a Block   Example Table

$$E_{op} = \frac{V_1 i_1 (T_1 + \Delta T_1)}{\eta_1} + \frac{V_2 i_2 (T_2 + \Delta T_2)}{\eta_2} + \cdots \tag{19}$$

E_{op} is the operating energy and η is calculated using Equation (3).

$$E_C = V_{in} C_L (\max(V_1 - V_2, 0) + \max(V_2 - V_3, 0) + \cdots) \tag{20}$$

$$E_{TOTAL} = E_{OP} + E_C$$

Each block is modeled as a chain of inverters with different depths. The delay of the block is calculated as its time of operation. The power supply level changes for 100 iterations. The rate at which the voltage changes to a new value is varied from 10 ns to 1 ms. The energy dissipation in each case is compared with a single VDD (always 1.2 V) block. Figure 10 shows the result of our experiment. We find that at fast rates of voltage scaling (a voltage transition every ~10 ns), the overheads of a DC-DC converter dominate, and there is an energy loss. Energy benefits can be realized for T_{OP} greater than 1μs with a maximum benefit of more than 150% achievable at slower rates of VDD transitions. This implies that, for these assumptions, the 5 VDD system would save energy relative to the single VDD system when transitioning VDD to adapt to changes in the workload that are slower than ~1 μs.

2.3. Panoptic Dynamic Voltage Scaling (PDVS)

This section applies the DVFS modeling approach to a different DVS implementation. Figure 11 shows the block diagram of a block level voltage scaling technique called panoptic dynamic voltage scaling (PDVS) [15]. In the PDVS technique, a block can switch from one voltage to another by the use of headers as shown in Figure 11. The advantage of this technique is that it enables a much faster switching. An equivalent DVFS voltage can be realized by dithering between the supplies. This scheme is more practical and has lower cost.

Figure 10. Energy Savings with rate of Voltage Scaling.

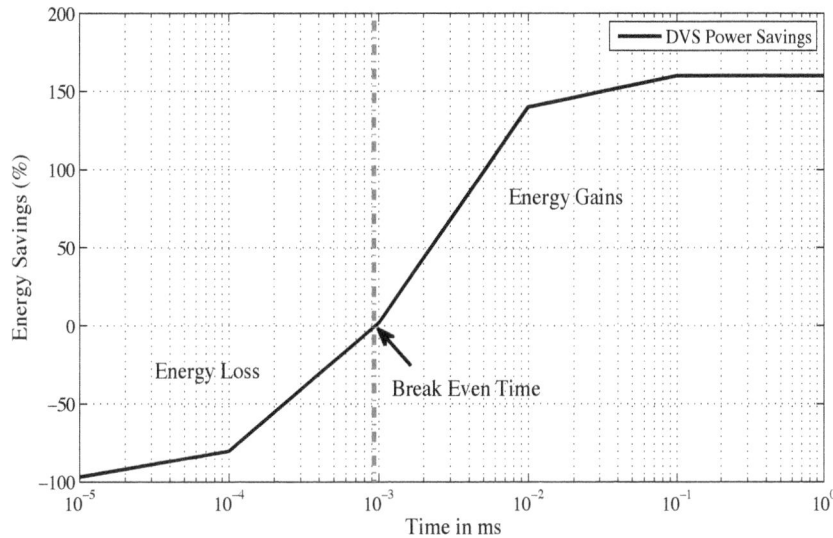

Figure 11. Panoptic dynamic voltage scaling (PDVS): Block level voltage scaling technique.

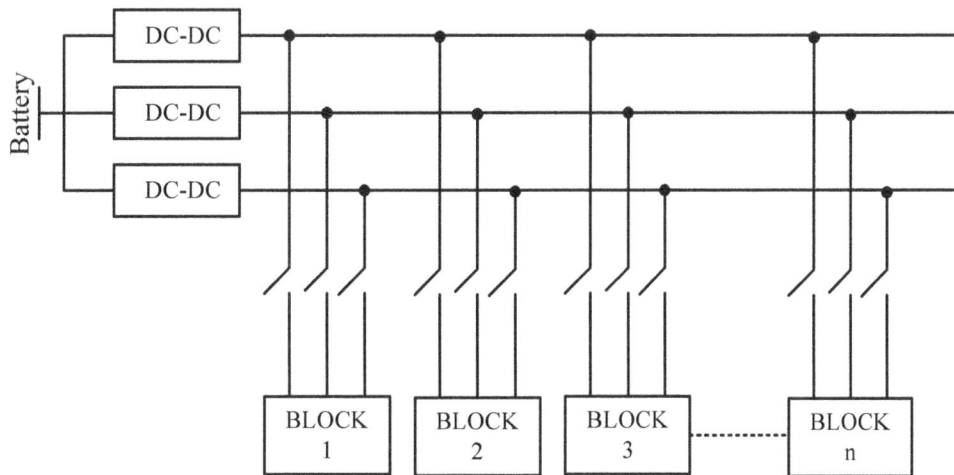

2.4. Framework for Energy Calculation in PDVS

We reproduce the operating condition of a block from Figure 9. This block operates at different voltages for different times to accomplish optimal energy operation. We break down operating condition of Figure 9 into Figure 12. If a block has to operate at V1 ($0.4 \text{ V} < V_1 < 0.8 \text{ V}$) for T1 time, PDVS accomplishes it by connecting the block to 0.4 V for T_{11} and 0.8 V for T_{12}, such that $T_{11} + T_{12} = T_1$ of Figure 9. This approach is called voltage dithering. T_{11} and T_{12} are such that the performance of the block does not change. A final operating condition is given by Figure 12b.

Figure 12. PDVS operating condition evaluation. (**a**) Operation condition; (**b**) Final load table.

VDD	Load			T_{OP}
	0.4V	0.8V	1.2V	
v1 V	i11 µA	0	0	T11
v1 V	0	i12 µA	0	T12
v2 V	0	i21 µA	0	T21
v2 V	0	0	i22 µA	T22
……	……	……..	…..	……

(**a**)

Load			dT_{OP}
0.4V	0.8V	1.2V	
$i_1(t)$	$i_2(t)$	$i_3(t)$	1e-9

(**b**)

The load on each supply will change depending on the blocks that are connected to it and results in a continuous time varying load on each supply. We include time to obtain the energy. Each supply has larger load variation which will have an impact on the overall efficiency. The total energy is given by,

$$E_{OP} = 0.4 \int i_1(t)/\eta_1(i_1)dt + 0.8 \int i_2(t)/\eta_2(i_2)dt + 1.2 \int i_3(t)/\eta_3(i_3)dt$$

$$E_C = 1.2C_L * 0.4(1 + 0 + 0 + 0 \dots) + 0.8C_L * 0.4(0 + 1 + 1 + 0 \dots)$$

(21)

$$E_{TOTAL} = E_{OP} + E_C$$

We keep the same system set-up as was used for the dedicated DC-DC converter case. It should be noted that there will be an insignificant overhead of settling time in this case. We assume a capacitive load of 20 pF on each block, since the local block virtual VDD rail switches instead of the total chip-wide VDD rail (with all of its decoupling capacitance).

4. Comparison of DVFS and PDVS using the Proposed Model

Figure 13 compares the PDVS scheme with a dedicated DC-DC converter case using the system level model that incorporates the DC-DC converter equations. The PDVS scheme has a break-even time of 30 ns compared to 1 µs in the dedicated supply case. This is because of the very small settling time in PDVS as the block is charged to the rail almost instantly. It also has lower conversion energy. The total energy benefits from the PDVS scheme is, however, lower than the dedicated DC-DC converter case, because it sees much wider load variation. The PDVS scheme though is better suited for implementing block level DVFS owing to its much smaller break-even time, allowing it to adjust to short changes in the workload. PDVS scales much better to larger numbers of blocks, since it requires only one DC-DC converter output per voltage rail rather than per block, as in the dedicated DVFS scheme. These results are of course influenced by the values of the parameters in the model, and the model makes it very easy to investigate how the results will vary when the assumptions change.

Figure 13. Energy Benefits of PDVS and Dedicated DVFS.

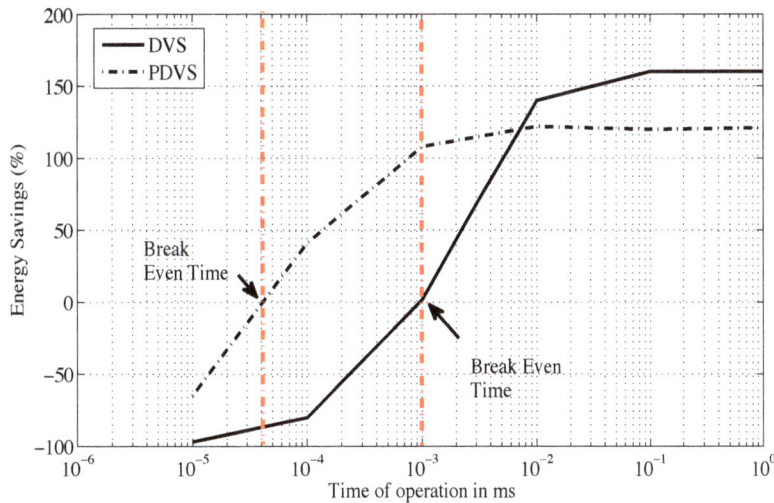

Figure 14 shows how the PDVS savings for the same scenario as before will change as a function of the capacitance of the virtual VDD rail of each PDVS block. As the capacitance of the block increases, the breakeven time moves to larger times, meaning that the workload needs to remain at a new value for a longer time to make it worthwhile to switch the voltage of that block.

Figure 14. Energy Benefits of PDVS for different values of the virtual VDD capacitance of the block.

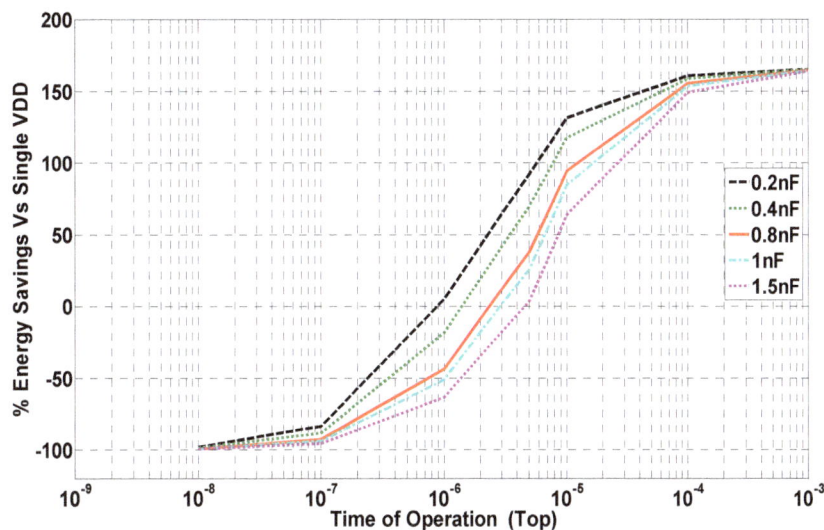

5. Conclusions

A model that can accurately capture the behavior of inductor based DC-DC converters in a dynamic environment has been presented. The converter model has been validated and compared with measured results from a variety of DC-DC converter topologies in existing literature. We use this model to study block level power management techniques for a SoC by incorporating it into a higher level model of the multiple block system. The system model predicts that there is a break-even time before the benefit of voltage scaling becomes positive, and our proposed modeling framework provides a quantitative means for comparing multiple power management techniques in a given use case scenario.

Acknowledgement

This work was supported in part by the NSF NERC ASSIST Center (EEC-1160483) and by DARPA through a subcontract with Camgian Microsystems.

References

1. Shrivastava, A.; Calhoun, B.H. Modeling DC-DC Converter Efficiency and Power Management in Ultra Low Power Systems. In Proceedings of the Sub-threshold Microelectronics Conference, Waltham, USA, 9–10 October 2012.
2. Martin, S.M.; Flautner, K.; Mudge, T.; Blaauw, D. Combined dynamic voltage scaling and adaptive body biasing for lower power microprocessors under dynamic workloads. In Proceedings of the International Conference on Computer Aided Design, San Jose, USA, 10–14 November 2002.
3. Kursun, V.; Narendra, S.G.; de, K.V.; Friedman, E.G. Efficiency Analysis of a High Frequency Buck Converter for On-Chip Integration with a Dual-VDD Microprocessor. In Proceedings of the IEEE European Solid State Circuits Conference, Firenze, Italy, 24–26 September 2002.
4. Choi, Y.; Chang, N.; Kim, T. DC–DC converter-aware power management for low-power embedded systems. *IEEE Trans. Comput. Aided Des. Integr. Circ. Syst.* **2007**, *26*, 8.
5. Yao, K.; Ye, M.; Xu, M.; Lee, F.C. Tapped-inductor buck converter for high-step-down DC-DC conversion. *IEE Trans. Power Electron.* **2005**, *20*, 775–780.
6. Bandyopadhyay, S.; Ramadass, Y.K.; Chandrakasan, A.P. 20 µA to 100 mA DC–DC converter with 2.8-4.2 V battery supply for portable applications in 45 nm CMOS. *IEEE J. Solid-State Circ.* **2011**, *12*, 2807–2820.
7. Li, W.; Xiao, J.; Zhao, Y.; He, X. PWM plus phase angle shift (PPAS) control scheme for combined multiport DC/DC converters. *IEEE Trans. Power Electron.* **2012**, *27*, 1479–1489.
8. Xiao, J.; Peterchev, A.V.; Zhang, J.; Sanders, S.R. A 4µA quiescent-current dual-mode digitally controlled buck converter IC for cellular phone applications. *IEEE J. Solid-State Circ.* **2004**, *12*, 2342–2348.
9. Kuroda, T.; Suzuki, K.; Mita, S.; Fujita, T.; Yamane, F.; Sano, F.; Chiba, A.; Watanabe, Y.; Matsuda, K.; Maeda, T.; Sakurai, T.; Furuyama, T. Variable supply-voltage scheme for low-power high-speed CMOS digital design. *IEEE J. Solid-State Circ.* **1998**, *33*, 454–462.
10. Ramadass, Y.K.; Fayed, A.A.; Chandrakasan, A.P. A Fully-Integrated Switched-Capacitor Step-down DC-DC converter with digital capacitance modulation in 45nm CMOS. *IEEE J. Solid-State Circ.* **2010**, *45*, 2557–2565.
11. Patounakis, G.; Li, Y.W.; Shepard, K.L. A fully integrated on-chip DC-DC conversion and power management system. *IEEE J. Solid-State Circ.* **2004**, *39*, 443–451.
12. Lee, C.F.; Mok, P.K.T. A monolithic current-mode CMOS DC-DC converter with on-chip current-sensing technique. *IEEE J. Solid-State Circ.* **2004**, *39*, 3–14.

13. Zhang, Y.; Zhang, F.; Shakhsheer, Y.; Silver, J.; Klinefelter, A.; Nagaraju, M.; Boley, J.; Pandey, J.; Shrivastava, A.; Carlson, E.; *et al.* A battery-less 19μW MICS/ISM-band energy harvesting body sensor node SoC for ExG applications. *IEEE J. Solid-State Circ.* **2013**, *48*, 199–213.

14. Ramadass, Y.; Chandrakasan, A.P. Minimum energy tracking loop with embedded DC-DC converter enabling ultra-low-voltage operation down to 250 mV in 65 nm CMOS. *IEEE J. Solid-State Circ.* **2008**, *43*, 256–265.

15. Shaksheer, Y.; Khanna, S.; Craig, K.; Arrabi, S.; Lach, J.; Calhoun, B.H. A 90nm Data Flow Processor Demonstrating fine Grained DVS for Energy Efficient Operation from 0.25V to 1.2V. In Proceedings of the Custom Integrated Circuits Conference, San Jose, USA, 19–21 September 2011.

16. Richelli, A.; Comensoli, S.; Kovacs-Vajna, Z.M. A DC/DC boosting technique and power management for ultralow-voltage energy harvesting applications. *IEEE Trans. Industr. Electron.* **2012**, *59*, 2701–2708.

17. Bassi, G.; Colalongo, L.; Richelli, A.; Kovacs-Vajna, Z. A 150 Mv–1.2 V fully-integrated DC-DC converter for Thermal Energy Harvesting. In Proceedings of the 2012 International Symposium on Power Electronics, Electrical Drives, Automation and Motion (SPEEDAM), Sorrento, Italy, 20–22 June 2012; pp. 331–334.

18. Richelli, A.; Colalongo, L.; Tonoli, S.; Kovacs-Vajna, Z.M. A 0.2 V DC/DC boost converter for power harvesting applications. *IEEE Trans. Power Electron.* **2009**, *24*, 1541–1546.

19. Calhoun, B.H.; Wang, A.; Chandraksan, A.P. Modeling and sizing for minimum energy operation in subthreshold circuits. *IEEE J. Solid-State Circ.* **2005**, *40*, 1778–1786.

4

Performance Limits of Nanoelectromechanical Switches (NEMS)-Based Adiabatic Logic Circuits

Samer Houri *, Christophe Poulain, Alexandre Valentian and Hervé Fanet

CEA-LETI, Minatec Campus, 17 Rue des Martyrs, Grenoble 38054, France;
E-Mails: christophe.poulain@cea.fr (C.P.); alexandre.valentian@cea.fr (A.V.);
herve.fanet@cea.fr (H.F.)

*   Author to whom correspondence should be addressed; E-Mail: samer.houri@cea.fr

Abstract: This paper qualitatively explores the performance limits, *i.e.*, energy *vs.* frequency, of adiabatic logic circuits based on nanoelectromechanical (NEM) switches. It is shown that the contact resistance and the electro-mechanical switching behavior of the NEM switches dictate the performance of such circuits. Simplified analytical expressions are derived based on a 1-dimensional reduced order model (ROM) of the switch; the results given by this simplified model are compared to classical CMOS-based, and sub-threshold CMOS-based adiabatic logic circuits. NEMS-based circuits and CMOS-based circuits show different optimum operating conditions, depending on the device parameters and circuit operating frequency.

Keywords: nanoelectromechanical switches (NEMS); adiabatic logic circuits

1. Introduction

Adiabatic logic circuits were introduced as a possible means to achieve ultra-low power circuits by taking advantage of the adiabatic charging principle [1–4] (the adiabatic charging principle states that if the voltage in a circuit changes slower than the electric time constant of the circuit, then resistive losses are reduced). However, while the adiabatic charging principle affords a possibly unlimited energy-performance compromise [3], at least in theory, the fact is that CMOS-based adiabatic circuits suffer from non-adiabatic residues and leakage loss components that deteriorate their performance.

On the other hand, electrostatic nanoelectromechanical (NEM) switches have already been suggested and demonstrated for use in classical logic circuits [5–8]. NEM switches offer the advantage of zero leakage current and therefore zero static power dissipation, which is an appealing property for low power low performance circuits. However, these switches require high operating voltages and suffer from low switching speeds when compared to MOSFETs, see for example [5] and [7] for a comprehensive review. These factors constitute serious obstacles to replace CMOS circuits by NEM relays for low power solution.

Nonetheless, NEM switches are ideal candidates to replace classical CMOS elements in adiabatic logic circuits [9], where the zero leakage current experienced in NEM switches allows the efficient operation of NEM-based adiabatic logic circuits at low frequencies without any static dissipation penalties, and at the same time the use of adiabatic charging makes it possible to offset the high energy dissipation that accompanies the voltages required to operate the nanoelectromechamical relays.

In this paper, the basic principles of adiabatic circuits will be presented, the dissipation of adiabatic circuits will be explicitly derived based on a simplified circuit model and compared for three different device technologies, which are: the classical CMOS-based circuit, the sub-threshold CMOS-based circuit and NEMS-based circuit respectively. It will be shown that replacing a MOSFET switch by an ideal electromechanical one does contribute in reducing significantly the power dissipation. It will also be demonstrated that the contact resistance of electromechanical switches may well be a limiting factor on the performance of NEM-based adiabatic logic circuits. This paper aims to present a modeling based results that gives to a good approximation the qualitative behavior of NEMS-based adiabatic logic circuits.

This paper starts by introducing the principle of adiabatic charging and how it translates to energy saving in logic circuits in Section 2. The energy saving terms will be derived for both CMOS, and sub-threshold CMOS in Section 3. Afterwards, electrostatic nanoelectromechanical relays will be introduced in Section 4, and attention will be given to the contact resistance of NEM switches, followed by an explicit derivation of the performance limits of NEM-based adiabatic circuits. A brief discussion of the results will be given in Section 5. Finally, the paper ends by a discussion regarding the implication of obtained results for circuit design and possible future trends in Section 6.

2. Adiabatic Charging of an RC Circuit

The operation of a logic circuit typically consists of charging a capacitor through a series resistance, where the capacitor is mainly that of the fan out interconnect and the resistance is dominated by the switching element series resistance.

In conventional circuits, the capacitor charging is done under constant voltage, which is the circuit operating voltage V_{dd}. This constant voltage charging results in dissipative energy losses of CV_{dd}^2 for a charge-discharge cycle. This remains the case regardless of the values of the series resistance or load capacitance.

In adiabatic charging, a slowly varying voltage source, usually a linearly ramped voltage, is used to charge and discharge the load capacitance [10]. This slowly varying voltage source results in a reduced dissipated energy E for the charge-discharge cycle given by:

$$E \cong 2\frac{RC}{T}CV_{dd}^2 \tag{1}$$

where R is the series resistance assumed to be dominated by the switching element, C is the load capacitance, and T is the ramp period. Equation (1) is accurate while $\left(\dfrac{RC}{T}\right) < 10$.

A simplified model of a typical adiabatic logic circuit is schematically represented in Figure 1, showing both the series resistance and the load capacitance. In the case of a NEMS switch, a variable capacitance C_S is added to represent the change in the NEMS device capacitance upon commutation (the origin of this change in capacitance will be detailed in Section 4). Also shown is the typical four-phase power clock (ϕ) used, where the ramp-up, hold, and the ramp-down times are shown each having a period T. In a typical adiabatic circuit, both the power clock and the input have the same waveform with a T phase shift between the two [4].

Figure 1. Schematic representation of an equivalent logic circuit showing the four phase power clock with equal length segments: F represents the block logic function; R_S is the series resistance dominated by the switch resistance; C_L represents the load capacitance that is mainly due to interconnect capacitance; and C_S represents the nanoelectromechanical switches (NEMS) variable capacitance. The current provided by the power clock is labeled i; the current going into the static interconnect capacitance (C_L) is labeled i_1; and that going into the variable capacitance of the NEMS switch (C_S) is labeled i_2.

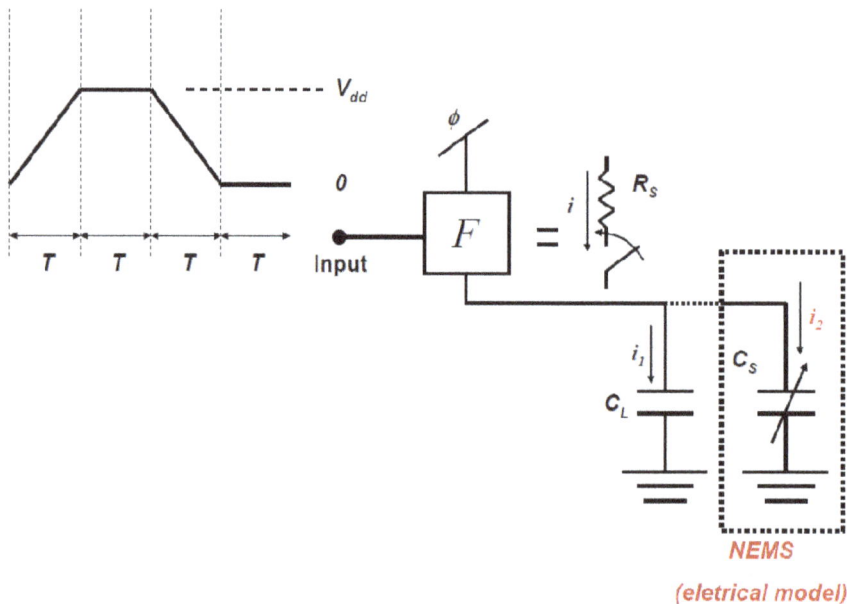

3. CMOS-Based Adiabatic Logic Circuits

The sources of dissipation in a CMOS-based adiabatic circuit can be attributed to either adiabatic or non-adiabatic residues. While it is possible to reduce the adiabatic residues by increasing the ramp-up time T, the non-adiabatic losses are not easily controlled, and depend on the device parameters.

The exact performance of CMOS-based adiabatic circuits depends on the exact circuit design; a derivation is presented based on the generic logic circuit diagram shown in Figure 2. Dissipation terms are derived for both classical CMOS circuits and sub-threshold CMOS circuits for the four phase power clock shown in Figure 1.

3.1. Classical CMOS Circuits

The energy dissipated in a complete charge-discharge clock cycle, including the mean non-adiabatic losses, is given by the following equation [4]:

$$E = 2\left(\frac{RC}{T}\right)CV_{dd}^2 + \frac{1}{2}CV_{th,p}^2 + V_{dd}\overline{I_{leak}}T \tag{2}$$

where R and C take the meanings defined previously; and V_{dd} is the voltage of the hold phase as shown in Figure 1.

In Equation (2), the first term represents the adiabatic residues of a charge-discharge cycle, while the second one represents the energy dissipation due to the sudden discharge of the residual output voltage [2]. Finally, the last term represents the power dissipation due to leakage currents, *i.e.*, the passive power dissipation, where in the above expression $\overline{I_{leak}}$ represents the average leakage current over a complete clock period.

Figure 2 shows an adiabatic circuit having an arbitrary logic function F: once the input is in its hold phase, the synchronized power clock ϕ will start its 4-phase period. During the time the power clock is applied, either the logic block or its complement will be passing, and therefore either the output or the inverted output will follow the clock voltage. Therefore, a leakage current will be experienced on one of the two latch nMOS transistors as shown in Figure 2.

Figure 2. Schematic representation of a CMOS implementation of an adiabatic logic function F, also identifying the leakage current. This architecture is known as the Positive Feedback Adiabatic Logic (PFAL) architecture [4]; however, the obtained results apply as a first order approximation to other architectures as well.

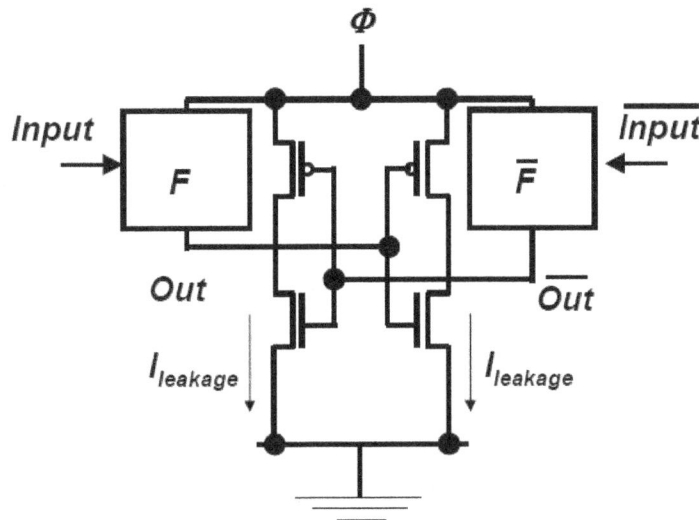

The main leakage current component in a MOSFET is due to the sub-threshold leakage current $I_{leakage}$ given by:

$$I_{leakage} = I_0 \left(1 - \exp\left(\frac{-V_{DS}}{Vt} \right) \right) \tag{3}$$

where I_0 is a function of the transistor size and parameters; V_t is the thermal voltage $\left(V_t = \frac{kT}{q} \cong 25mV \right)$; and V_{DS} is the source-drain voltage which is assumed to perfectly follow the power clock.

Considering the four phases of the power clock shown in Figure 1, the leakage energy term expressed in Equation (2) may be calculated by replacing the leakage current with Equation (3) and the voltage by the clock voltage and integrating their product over the clock period. Therefore, the passive dissipation is given by:

$$E_{leakage} = \int_0^T V_{dd} I_0 \left(1 - \exp\left(\frac{-tV_{dd}}{TV_t} \right) \right) \frac{t}{T} dt + \int_T^{2T} V_{DD} I_0 \left(1 - \exp\left(\frac{-V_{dd}}{V_t} \right) \right) dt$$

$$+ \int_{2T}^{3T} V_{dd} I_0 \left(1 - \exp\left(\frac{-V_{dd}\left(1 - \frac{t}{T}\right)}{TV_t} \right) \right) \left(3 - \frac{t}{T} \right) dt \tag{4}$$

$$\cong 2 I_0 V_{dd} T$$

where $E_{leakage}$ is the leakage energy dissipation, and the first, second and third integrals correspond to the rising phase, the hold phase, and the decreasing phase of the power clock respectively.

Furthermore, the series resistance R_S of a MOSFET is also a function of the transistor properties and the operating voltage, and is given by:

$$R_S = \frac{L^2}{\mu_n C_n \left(V_{dd} - 2V_{th} \right)} \tag{5}$$

where L, μ_n and C_n are device dependent parameters; and V_{th} is the transistor threshold voltage.

By replacing the right hand sides of Equations (4) and (5) into their respective terms in Equation (2), the energy dissipation of a CMOS-based adiabatic circuit may be expressed as:

$$E = 2 \frac{L^2 C^2}{\mu_n C_n T} \frac{V_{dd}^2}{\left(V_{dd} - 2V_{th} \right)} + \frac{1}{2} C V_{th,p}^2 + 2V_{dd} I_0 T \tag{6}$$

The energy dissipation as expressed in Equation (6) admits an optimum operating period $T_{optimum}$, that can be obtained by solving $dE/dT = 0$, and an optimum voltage equal to $3 V_{th}$. This optimum is visible in the energy-performance plot shown in Figure 3.

3.2. Sub-Threshold CMOS Circuits

Combining adiabatic logic and sub-threshold mode may provide interesting energetic performance if low frequency operation is allowed. Calculation of energy is more complex in the case of sub-threshold CMOS, but by using Equation (3), it is possible to obtain the following approximate equation for the 2N-2P adiabatic gate:

$$E = 2nCV_t^2 \cdot \frac{CV_{dd}}{I_0 T} + 2V_{dd}I_0 T \tag{7}$$

where n is the body effect factor and V_t the thermal voltage.

If low frequency mode is allowed (less than 1 MHz), comparison of Equations (6) and (7) defines conditions where sub-threshold mode is advantageous. Equation (7) is also plotted in Figure 3 for comparison using the same transistor parameters used to plot (6).

Figure 3. Comparison between the performance of adiabatic circuits using: conventional CMOS (solid red line) as given by Equation (2), and sub-threshold CMOS (dashed blue line) as given by Equation (7), both done for the same device parameters. The non-adiabatic residue in classical CMOS circuit is also shown for comparison (solid black line).

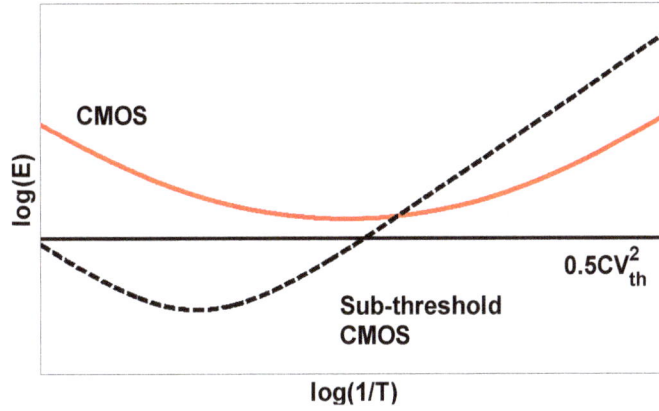

4. NEMS-Based Adiabatic Logic Circuits

Nanomechanical switches are devices that rely on a beam, cantilever, or a membrane to deform under the effect of electrostatic force in order to make and break electrical contact upon the application of an external voltage. NEMS switching elements offer the property of zero leakage current [5]. Although NEMS switches of varying design, dimensions and materials have been constructed and demonstrated, this paper relies on a 1-dimensional model as introduced in [11] that may be applied to all devices equally in order to obtain a generic formulation of NEMS-based adiabatic logic circuits.

A typical 1-dimensional reduced order model of a 3-terminal electrostatic NEMS switch, where the structure is modeled by a simple parallel plate capacitor, is schematically represented in Figure 4a: an electrostatic force is created between the gate electrode (G) and the suspended structure. When a bias voltage is applied, this force brings the structure which is connected to the drain (D), into contact with the source electrode (S). The source electrode is assumed to play no role in the electrostatic actuation of the structure. If the ratio of air gap d to actuation gap g is $\frac{d}{g} > \frac{1}{3}$, an instability known as pull-in takes place. The pull-in results in a hysteretic effect as shown for an ideal I-V response plotted in Figure 4b, in which case the voltage at which pull-in takes place, known as the pull-in voltage (V_{pi}), is larger than the voltage at which the structure breaks contact, known as the pull-out voltage V_{po}. In case the contact is established before the onset of pull-in, i.e., $\frac{d}{g} \le \frac{1}{3}$, then there is only a contact voltage

$V_{contact} = V_{pi} = V_{po}$.

The total energy dissipation E_{Total} in a NEMS-based adiabatic circuit may be expressed as:

$$E_{Total} = E_{Electrical} + E_{Mechanical} \qquad (8)$$

where $E_{Electrical}$ and $E_{Mechanical}$ are the energy dissipated by the electrical resistance and the energy dissipated through mechanical damping respectively. While the mechanical energy dissipation may depend on the ramp period T, it will be considered as a constant second order residue throughout this work.

Based on the equivalent circuit in Figure 1, the electrical dissipation $E_{Electrical}$ may be expressed as:

$$E_{Electrical} = \int_0^\infty R_S i^2 \, dt = \int_0^\infty R_S \left(i_1 + i_2 \right)^2 dt = \int_0^\infty R_S \left(i_1^2 + i_2^2 + 2i_1 i_2 \right) dt \qquad (9)$$

where i, i_1 and i_2 are the currents going through the series resistance R_S, the load capacitance C_L and NEMS capacitance C_S respectively, as shown in Figure 1. An expression for the current in each branch may be derived as follows:

$$
\begin{aligned}
i_1 &= \frac{dC_L V(t)}{dt} = C_L \frac{dV(t)}{dt} = C_L \frac{V_{dd}}{T} \\
i_2 &= \frac{dC_S V(t)}{dt} = C_S \frac{dV(t)}{dt} + V(t) \frac{dC_S}{dt}
\end{aligned}
\qquad (10)
$$

From Equations (9) and (10), one can remark that the NEMS parameters governing the dissipation in an adiabatic circuit are the series resistance R_S and the time-dependent capacitance value of the NEMS switch $\dfrac{dC_S}{dt}$. Furthermore, the minimum values of V_{dd} and T are set by the switch and therefore also dependent on the NEMS device.

Figure 4. Schematic illustration of a reduced order model of a nanoelectromechanical switch (**a**) showing the source (S), drain (D) and gate (G). Also visible are the actuation and contact gaps, g and d respectively; V and I represent the actuation voltage and the source-to-drain current respectively; and ϕ represents the power clock signal; (**b**) Typical I-V plots in a NEMS switch operating in the pull-in mode, where I_{sat} represents the saturation current of the device.

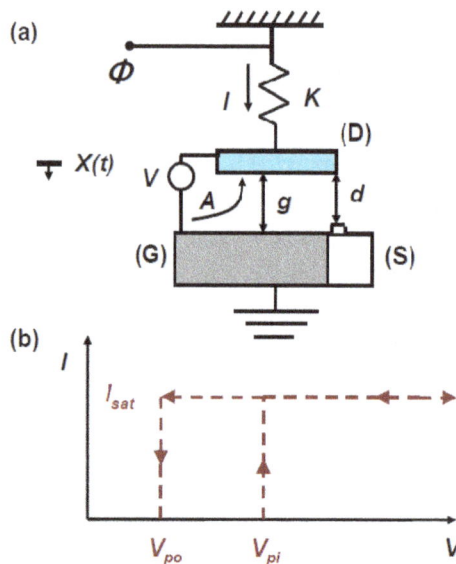

Therefore in order to obtain a comprehensive description of dissipation in NEMS-based adiabatic logic circuits, it is necessary to inject values for the switch series resistance and its transient behavior into Equations (9) and (10). These values are obtained from realistic contact mechanics models and dynamics NEMS models respectively, and are explicitly detailed below.

4.1. Contact Resistance in Nanomechanical Switches

The series resistance of an electromechanical switch is in fact dominated by the resistance at the contact between the movable and fixed electrodes [12]. Therefore, proper electromechanical modeling of the contact resistance is crucial when designing NEMS relays based circuits.

Recent literature review of electromechanical contact in nano- and microelectromechanical systems [13] indicate that to appropriately model the electromechanical contact, a multi-physics model that accounts for surface topography, elastic and plastic deformations, adhesion forces, surface contamination, and electron transport regime through the contacting areas needs to be employed. However, such complete modeling of contact aspects is beyond the scope of this paper, therefore a simplified contact model that considers two spherical asperities brought together under an applied force $F_{applied}$ will be used.

Several models exist to describe the resulting interplay between the mechanical deformation of the asperities and the electrical behavior of the contact. Here, two common contact mechanics models will be considered, the Hertz contact theory [14] and the JKR contact model [15]. In addition, two electrical resistance models will be considered, the Maxwell resistance, and the Sharvin ballistic resistance.

The Hertz contact assumes perfectly elastic deformations of two asperities, while at the same time neglecting any adhesion forces that may exist at the interface between the asperities: the radius r_H of the circular contact spot between two asperities shown schematically in Figure 5, is given by:

$$r_H = \sqrt[3]{\frac{3F_{applied}R}{4E^*}} \tag{11}$$

where R is the equivalent asperity radius given by $R = R_1R_2/(R_1 + R_2)$ where R_1 and R_2 are the radii of the first and second asperities respectively; and E^* is the effective elastic modulus given by $E^* = \left(\frac{1-v_1^2}{E_1} + \frac{1-v_2^2}{E_2}\right)^{-1}$, where E_1, v_1, E_2, v_2, are the Young moduli and Poisson ratios of the first and second asperities respectively.

In a reduced order model approximation, the force $F_{applied}$ applied to the contacting asperities may be expressed as:

$$F_{applied} = F_0 \frac{V_{dd}^2}{V_{pi}^2} - \alpha \tag{12}$$

where F_0 is the minimum electrostatic contact force when $V_{dd} = V_{pi}$, and α represents the restoring elastic force; the exact values of both F_0 and α depend on the device design and fabrication. Note that Equation (12) only applies once contact is established, i.e., $V_{dd} \geq V_{contact}$, regardless of whether the switch is operating in the pull-in or non-pull-in regimes.

The Johnson-Kendall-Roberts (JKR) contact model on the other hand accounts for possible adhesion forces while also considering a purely elastic contact. A first order approximation of the contact radius in the JKR model is given by:

$$r_{JKR} = \sqrt[3]{\frac{3F_{applied}R}{4E^*} + \frac{9\pi\Delta\gamma R^2}{4E^*}} = r_H \left(1 + \frac{3\pi\Delta\gamma R}{F_{applied}}\right)^{\frac{1}{3}} \tag{13}$$

where r_{JKR} is the contact radius for a JKR type contact, R and K take the same values as expressed previously, and $\Delta\gamma$ is the net adhesion energy between the two surfaces.

The electrical contact resistance given by Maxwell ($R_{Maxwell}$) in the diffusive transport regime, and Sharvin ($R_{Sharvin}$) in the ballistic transport regime are respectively expressed as:

$$R_{Maxwell} = \frac{\rho}{2r} \tag{14a}$$

$$R_{Sharvin} = \frac{4\rho\lambda_e}{3\pi r^2} \tag{14b}$$

where ρ is the electrical resistivity of the contacting asperities; r is the contact radius as defined above; and λ_e is the electron mean free path.

The expressions of the contact deformation and the contact resistance given by Equations (11) through (14), when combined together, give four electro-mechanical contact models. These contact models can be injected into Equation (9) to determine the effects of contact resistance on the total energy dissipation. The impact of these four models on the dissipation and performance of NEMS-based adiabatic logic circuits will be explored in the results and discussion section.

Figure 5. Schematic illustration of two asperities in contact, deformed under the effect of an applied load.

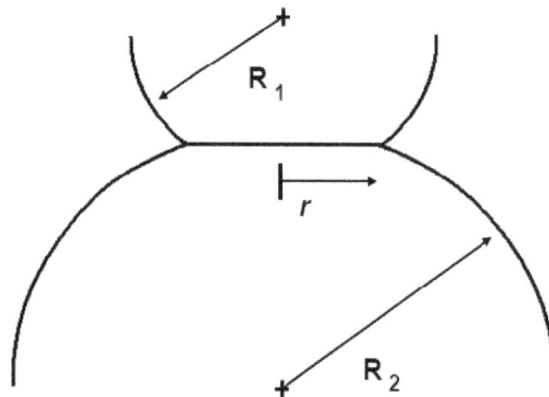

4.2. Switching Behavior of Nanomechanical Switches

The switching behavior of nanoelectromechanical switches will take on an important role in determining the losses in a NEMS-based circuit: this is due to the fact that the device capacitance, *i.e.*, C_S, is inversely proportional to the time-dependent position $X(t)$ of the movable mass, as given by the parallel plate capacitor value:

$$C_S = \frac{C_0}{1 - X(t)} \tag{15}$$

where we define C_0 as the NEMS switch capacitance in the initial position, *i.e.*, for $X = 0$.

The question of the time dependent response of a NEMS switch has been treated to some length in literature for the case of step voltage actuation, which is the type of actuation envisioned for classical circuits; see for example [6,16]. Furthermore, the dissipation resulting from the transient current that is generated upon commutation under a step voltage actuation was also obtained [17].

However, the voltage waveform in an adiabatic logic circuit is, by definition, required to be slowly varying. This constitutes the biggest difference between classical and adiabatic circuits with respect to NEMS dynamics. This fact results in an interplay between electrical and mechanical time constants, where the mechanical time constant is defined as the switching delay of the NEMS device. Therefore, two limiting cases will be considered separately, depending on the relations between the mechanical time constant (τ_{Mech}), the electrical time constant (τ_{Elec}), and the rise time of the clock signal (T).

First case ($\tau_{Mech} \gg T \gg \tau_{Elec}$): in this scenario, that we will also refer to as "*dynamic mode*", the electric time constant is considered to be several orders of magnitude smaller than the mechanical time constant. The rise time of the clock signal is also much smaller than the mechanical time constant, such that from a mechanical point of view, the voltage would be ramped up to its hold value of V_{dd}, before the mechanical structure even begins to move. Therefore, the structure responds in a manner that is similar to that when subjected to a step voltage. The net effect of this form of operation is to have a hold time that is significantly longer than the rise time, as shown in Figure 6.

The electrical dissipation, in this case, may be obtained by decoupling the electrical and mechanical time scales by rewriting Equation (9) as follows:

$$E_{Electrical} = \int_0^\infty R_S \left(i_1^2 + i_2^2 + 2i_1i_2\right)dt = R_S \left[\int_0^T \left(i_1^2 + i_2^2 + 2i_1i_2\right)dt + \int_T^\infty i_2^2 dt\right] \tag{16}$$

The first term on the right hand side of (16) represents the dissipation due to the ramped electric charging of the interconnect and the switch capacitance: because the charging takes place during a time T which is small compared to the τ_{Mech}, the switch capacitance is assumed to be a time-independent constant equal to C_0. While the second term represents the effect of a varying switch capacitance, which takes place upon mechanical commutation, under the effect of the now stable bias voltage V_{dd}. For each of these terms, the currents are defined differently, given by:

$$i_1 = C_L \frac{V_{dd}}{T}$$
$$\qquad \text{for } 0 < t < T \tag{17a}$$
$$i_2 = C_0 \frac{V_{dd}}{T}$$

$$i_1 = 0$$
$$\qquad \text{for } t > T \tag{17b}$$
$$i_2 = V_{dd} \frac{dC_S}{dt}$$

The transient current in Equation (17b) and the resulting dissipation were derived explicitly in [17], while the values in (17a) may be placed directly into Equation (16) to obtain a value for the electrical dissipation per switching operation. Furthermore, for the case of a switch response to an applied step voltage, the mechanically dissipated energy may be obtained by integrating the electrostatic force over the trajectory of the parallel plate [7], resulting in: $E_{Mech} = \frac{1}{2} \frac{d}{g-d} C_0 V_{dd}^2$, where g and d are the values of the actuation and contact gaps, respectively, as shown in Figure 4.

By combining Equations (16) and (17b), and including the mechanical losses, the total energy dissipated per switching cycle may be expressed as:

$$E_{Total} = E_{ramp} + E_{transient} + E_{Mech} \tag{18a}$$

where:

$$E_{ramp} = 2R_S \frac{V_{dd}^2}{T} \left[C_L^2 + C_0^2 + 2C_L C_0 \right] \tag{18b}$$

and:

$$E_{transient} = R_s C_0 \omega_0 C_0 V_{dd}^2 \int_0^{d/g} \frac{1}{(1-X)^4} \sqrt{\frac{X V_{dd}^2}{1-X} - X^2} \, dX \tag{18c}$$

where all symbols in Equation (18a–c) take their previously defined meaning, and ω_0 is the mechanical resonance frequency of the spring-mass system that models the NEMS switch. The *ramp* subscript is meant to represent the dissipation during the ramp-up and ramp-down phases of the power clock cycle.

Figure 6. Schematic representation of the necessary clock signal for NEMS-adiabatic circuits working in the limit of ($\tau_{Mech} \gg T \gg \tau_{Elec}$) with the hold phase lasting longer than the ramp-up and ramp-down phases. Also shown in the figure is the required synchronization between two phase clocks.

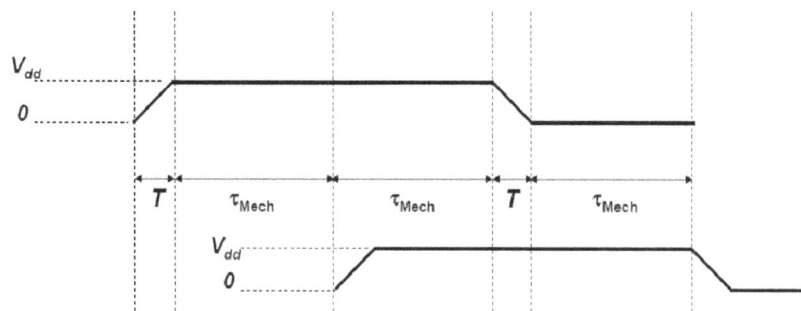

Second case ($T \gg \tau_{Mech} \gg \tau_{Elec}$): another limit case to consider is when the rise time of the clock signal is slow compared to the mechanical time constant: therefore, the mechanical structure is assumed to move slowly. We will refer to this case as the "*quasi-static mode*". Both the power clock rise time and the mechanical time constant are considered to be significantly longer than the electrical time constant.

It is important to note that, for this case, if the contact is to be established beyond pull-in, then a transient dissipation term will have to be introduced again into the dissipation equations, therefore negating the advantages that are sought to be obtained under this mode of operation. Therefore, for this case, only switches operating before pull-in will be considered.

To obtain the response of a NEMS switch, and hence the dissipation, it is necessary to solve the following governing normalized nonlinear differential equation [18]:

$$\ddot{X} + X = \frac{V^2(t)}{2(1-X)^2}$$ (19)

where $\ddot{X} = \frac{d^2X}{dt^2}$, and $V(t)$ is the time dependent ramp-up voltage; and $X(t)$ is the time dependent displacement as shown in Figure 4.

Equation (19) can only be solved numerically, and it is therefore difficult to provide a general solution for different values of ramp-up periods.

An approximate solution is therefore derived based on the hypothesis that the nanomechanical structure is moving slowly compared to its mechanical time constant (which is the underlying assumption for this case). More precisely, if $T >> Q/\omega_0$, where Q is the mechanical quality factor, a solution based on static equilibrium equations is considered to be a good approximation. Such a solution is explicitly derived in [19], and gives an expression for the total energy dissipation as:

$$E_{Total} = \frac{2R_S V_{dd}^2}{T}\left[C_L^2 + C_0^2\left(1 + 2\frac{d}{g}\right) + 2C_0 C_L\left(1 + \frac{d}{g}\right)\right]$$ (20)

In fact, two additional dissipation terms exist, one related to the mechanical dissipation, while the other is a non-adiabatic residue due to broken ramp-up ramp-down symmetry caused by adhesion forces. However, both these terms are small and are neglected in this analysis, as their derivation is beyond the scope of this publication.

5. Results and Discussion

From what is already introduced in the previous section, it is clear that the two most influential features of NEMS-based adiabatic circuits are: the contact resistance and the dynamic response of the switch.

The dynamic response of the switch and the way it affects power consumption are related to the mode of actuation of the switch, where a quasi-static actuation, i.e., "mechanically adiabatic", is theoretically possible if a long enough ramp-up time is used. This form of operation of NEMS-switches is however less poised to be a candidate for practical adiabatic circuits, due to the low commutation speeds required for that mode of operation. In the dynamic operation mode, on the other hand, the switch is operated near its mechanical switching frequency, and consequently the adiabatic circuit operates near the NEMS switch commutation frequency.

The difference between the two operation modes in terms of power dissipation as a function of frequency is plotted in Figure 7; the energy is normalized with respect to $C_0 V_{dd}^2$. The energy dissipation curves are calculated for $C_L = 10C_0$, a value chosen by industry standards [20]. Additionally, the contact gap to electrostatic gap ratio is taken to be $d/g = 2/3$. The horizontal axis of the graph in Figure 7 represents the operating frequency of the circuit normalized to the nominal resonance frequency of the NEMS structure.

Figure 7. Comparison of the performance of NEMS-based adiabatic circuits shown for: the dynamic mode of operation (black line) as given by Equation (18a–c), and the quasi-static mode of operation (red line) as given by Equation (20), both done for $d/g = 2/3$, and $C_L = 10C_0$.

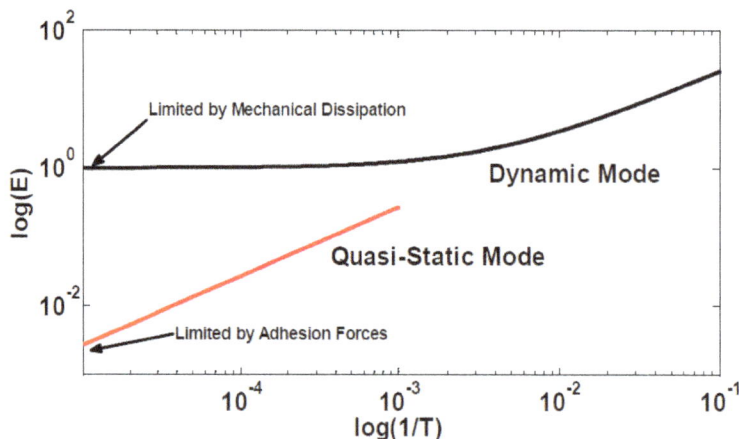

The first difference between the two operating modes shown in Figure 7 is the frequency of operation, which differs by almost two orders of magnitude. The second difference is the magnitude of dissipated energy per operation. A difference of approximately two orders of magnitudes lower is observed for quasi-static operation. This latter effect is due to the mechanical energy dissipated per charge-discharge in the dynamic mode; this energy remains constant independently from the circuit operating frequency. Finally it is worth mentioning that, in the case of the quasi-static mode, adhesion forces set the lower limit of dissipation. Indeed, according to Equation (20) the energy dissipation limit is directly proportional to the contact resistance. As mentioned before, the JKR contact model considers the adhesion energy or adhesion force between the contacting asperties. Equation (13) indicates that the more the adhesion force, the larger the contact spot radius, and as a consequence the lower the contact resistance, whatever the electronic transport regime. The energy dissipation limit is then strongly influenced by adhesion forces. However the exact determination of adhesion forces requires significant efforts and is beyond the scope of this paper.

The contact resistance is a critical feature of a NEMS switch as it dictates energy dissipation and plays a direct role as it sets the RC/T adiabatic gain factor of the circuit.

The energy dissipation of a NEMS-based adiabatic circuit operating in the dynamic mode is calculated and plotted in Figure 8 for the different contact resistance models presented in Section 4.1. These plots are based on a gold-gold contact with the following physical parameters: equivalent asperity $R = 4$ μm, surface energy $\Delta\gamma = 1.4$ J/m^2 [15], electrical resistivity and electron mean free path of gold are $\rho = 22 \times 10^{-9}$ Ω·m, and $\lambda_e = 38$ nm respectively.

Even though these classical contact models are used to simulate micrometer scale contacts, recent literature considers them to be inadequate to address nanometer scale contact, and expects the actual contact resistance to be more sophisticated to simulate and to present larger values than those actually given by the above models [8]. In addition, these models are more suited to describe metal-metal contact, and give poor results when applied to semiconductor-semiconductor contact, as these latter ones may be dominated by tunneling currents [8].

Therefore, contact resistance in nanomechanical relays remains a key parameter that dictates the adiabatic energy saving term in adiabatic logic circuits, and a parameter that needs further theoretical and experimental investigation.

Figure 8. Plots of the contact resistance, described by Equations (11–14b), given by the different models on the performance of NEMS-based adiabatic circuits operating in the dynamic mode. The quasi-static mode is similarly affected (not shown) by the value of contact resistance values.

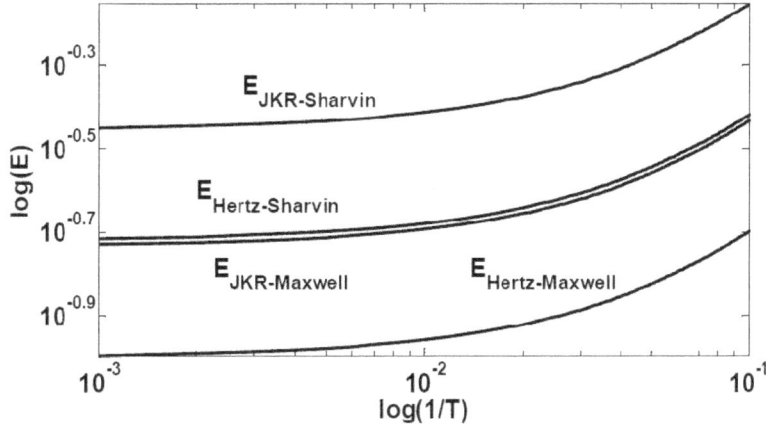

At this point it is worth mentioning that having a supply voltage that is higher than the pull-in voltage, *i.e.*, $V_{dd} > V_{pull-in}$, does not interfere with the proper operation of the circuit provided that the different phases of the power clock are well synchronized, and that is the case for both "dynamic mode" and "quasi-static mode". This is because if the second power clock phase does not commence until the first one has started its hold value, the switch will already be in the close or open position (depending on the state) regardless of the supply voltage (as long as that is at least equal to the pull-in voltage) which is the purpose in exploring these limit case scenarios.

In fact a certain amount of voltage overdrive should be expected in any nanoelectromechanical integrated circuit, since fabrication variability will undoubtedly results in dispersion of the value of pull-in voltage. Therefore, in any NEMS-based logic circuit certain devices will be slightly overdriven, as it is the case with CMOS logic circuits. While this voltage overdrive does not interfere with the circuit function, it does result in an increased energy dissipation as given by equations (18a–c) and (20) depending on the mode of operation of the circuit.

Finally, we would like to provide a numerical example that compares directly the expected performance of CMOS-based and NEMS-based adiabatic (operating in the "dynamic mode") logic circuits. In this example we compare the performance of 45 nm CMOS adiabatic circuit with a 1 fF load capacitance to that of a NEMS switch with the following properties: device capacitance = 0.1 fF, pull-in voltage = 1 V, switching time > 1 ns, feature size = 50 nm, contact resistance = 10 KΩ, and a load capacitance of 1 fF.

The NEMS device parameters listed above correspond to the best projected parameters as expected by the ITRS roadmap [20], as well as some parameters extracted from literature. Although a reliable NEMS switch with the above parameters have yet to be demonstrated, the following example gives an insight into the energy gain possible if the ITRS device targets are achieved. The results of this simulation are shown in Figure 9, where the NEMS switch based circuit performs visibly more favorably than both classical and sub-threshold CMOS circuits.

Figure 9. Plots comparing the performance of the classical (black) and sub-threshold (blue) CMOS adiabatic logic circuits (for 45 nm technology node and a 1fF load capacitance), and the performance of a NEMS-based adiabatic logic circuit (also with a 1fF load capacitance).

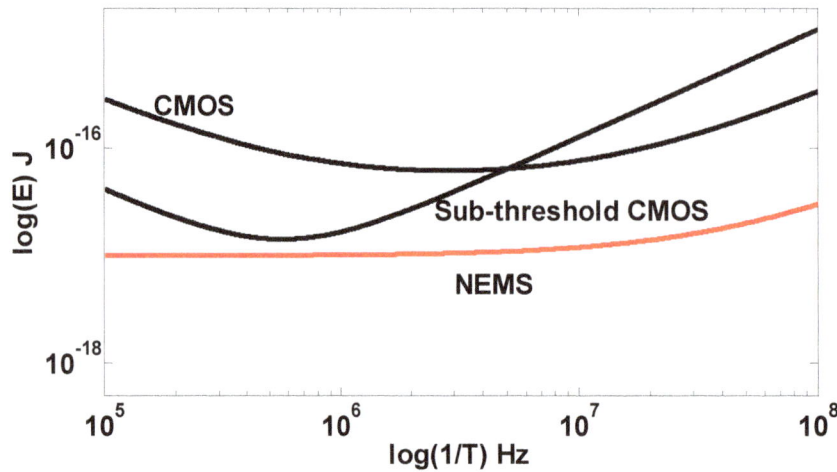

6. Conclusions

In summary, in this work, analytical expressions based on simplified circuit models were derived to obtain the performance limits of adiabatic circuits based on conventional CMOS, sub-threshold CMOS and NEMS based devices. Comparisons show that CMOS implementations are plagued by their leakage currents: when operating under a given frequency, the energy dissipation starts increasing again, and this remains the case for CMOS circuits designed and operated in the sub-threshold regime. It should be noted that the energy-optimum of sub-threshold CMOS is lower than that of a classical CMOS, but at high frequencies energy dissipation strongly increases.

It is also shown that for NEMS-based adiabatic circuits, an entire spectrum of possible operating regimes is available, from which two limiting cases were considered. In the first mode, the mechanical structure is assumed to be actuated similarly to a classical, *i.e.*, non-adiabatic, NEMS-based circuit; while on the other operating extreme, the mechanical switch is assumed to operate slowly, in a quasi-static mode. These two operating modes offer contrasting performance in which either the circuit operating frequency is maximized at the cost of higher dissipation (dynamic mode), or the circuit energy dissipation is minimized at the cost of a lower operating frequency (quasi-static mode).

Furthermore, this paper identified several critical nanomechanical parameters that affect circuit behavior. In particular, the mechanical energy injected into the system limits the performance of NEMS switches operating in the dynamic mode, while adhesion energy will play a limiting role on the performance of quasi-statically operated switches. In both cases, it was also observed that the contact resistance of the NEMS device sets the magnitude of the dissipated energy; the full impact of contact resistance should be subject to further investigation, especially as it is correlated with adhesion forces.

Finally, it is of interest to note that the energy dissipated by NEMS-based adiabatic circuits is expected to be significantly lower than is the case of CMOS-based circuit, especially if low to medium circuit operating frequency is desired, a frequency range where their CMOS counterpart performs poorly due to static power loss.

Acknowledgments

The authors would like to thank Mr. Alexis Peschot for his help with the experimental characterization of the electromechanical switches.

This work was funded by the action Carnot AdiaMéca.

Conflicts of Interest

The authors declare no conflict of interest.

References and Notes

1. Koller, J.G.; Athas, W.C. Adiabatic Switching, Low Energy Computing, and the Physics of Storing and Erasing Information. In Proceedings of the Workshop on Physics and Computation, 1992 (PhysComp '92), Dallas, TX, USA, 2–4 October 1992; pp. 267–270.

2. Athas, W.C.; Svensson, L.J. Reversible Logic Issues in Adiabatic CMOS. In Proceedings of the Workshop on Physics and Computation (PhysComp '94), Dallas, TX, USA, 17–20 November 1994; pp. 111–118.

3. Paul, S.; Schlaffer, A.M.; Nossek, J.A. Optimal charging of capacitors. *IEEE Trans. Circuits Syst. I* **2000**, *47*, 1009–1016.

4. Teichmann, P. *Adiabatic Logic: Future Trend and System Level Perspective*; Springer: Dordrecht, the Netherlands, 2012.

5. Loh, O.Y.; Espinosa, H.D. Nanoelectromechanical contact switches. *Nature Nanotechnol.* **2012**, *7*, 283–295.

6. Akarvardar, K.; Elata, D.; Parsa, R.; Wan, G.C.; Yoo, K.; Provine, J.; Peumans, P.; Howe, R.T.; Wong, H.-S.P. Design Considerations for Complementary Nanoelectromechanical Logic Gates. In Proceedings of the IEEE International Electron Devices Meeting, Washington, DC, USA, 10–12 December 2007; pp. 299–302.

7. Kam, H.; King Liu, T.-J.; Stojanovic, V.; Markovic, D.; Alon, E. Design, optimization, and scaling of MEM relays for ultra-low power digital logic. *IEEE Trans. Electron Devices* **2011**, *58*, 236–250.

8. Pawashe, C.; Lin, K.; Kuhn, K.J. Scaling limits of electrostatic nanorelays. *IEEE Trans. Electron Devices* **2013**, *60*, 2936–2942.

9. Fanet, H. Circuit logique à faible consommation et circuit intégré comportant au moins un tel circuit logique (in French). EP 2549654 A1, 22 July 2011.

10. Svensson, L.; Koller, J.G. Adiabatic Charging without Inductors. In Proceedings of the International Workshop on Low-Power Design, Napa, CA, USA, 24–27 April 1994; pp. 159–164.

11. Nathenson, H.C. The resonant gate transistor. *IEEE Trans. Electron Devices* **1967**, *14*, 117–133.

12. Majumder, S.; McGruer, N.E.; Adams, G.G.; Zavracky, P.M.; Morrison, R.H.; Krim, J. Study of contacts in an electrostatically actuated microswitch. *Sens. Actuators A: Phys.* **2001**, *93*, 19–26.

13. Toler, B.F.; Coutu, R.A., Jr; McVride, J.W. A review of micro-contact physics for microelectromechanical systems (MEMS) metal contact switches. *J. Micromech. Microeng.* **2013**, *23*, doi:10.1088/0960-1317/23/10/103001.

14. Slade, P.G. *Electrical Contacts: Principles and Applications*; Marcel Dekker: New York, NY, USA, 1999.

15. Johnson, K.L. *Contact Mechanics*; Cambridge University Press: Cambridge, UK. 1987.

16. Younis, M.I.; Abdel-Rahman, E.M.; Nayfeh, A. A reduced-order model for electrically actuated microbeam-based MEMS. *IEEE J. Microelectromech. Syst.* **2003**, *12*, 672–680.

17. Houri, S.; Valentian, A.; Fanet, H. Transient Dissipation in NEMS-Based Circuits. In Proceedings of the Third Berkeley Symposium on Energy Efficient Electronic Systems, Berkeley, CA, USA, 28–29 October 2013.

18. Leus, V.; Elata, D. On the dynamic response of electrostatic MEMS switches *IEEE J. Microelectromech. Syst.* **2008**, *17*, 236–243.

19. Houri, S.; Valentian, A.; Fanet, H. Comparing CMOS-based and NEMS-based adiabatic logic circuits. *Lect. Notes Comput. Sci.* **2013**, *7948*, pp. 36–45.

20. Emerging Research Devices. International Technology Roadmap for Semiconductors, 2009 Edition. Available online: http://www.itrs.net/Links/2009ITRS/Home2009.htm (accessed on 29 September 2013).

5

Mechanisms of Low-Energy Operation of XCT-SOI CMOS Devices—Prospect of Sub-20-nm Regime

Yasuhisa Omura [1,2,*] and Daiki Sato [2]

[1] Organization for Research and Development of Innovative Science and Technology (ORDIST), Kansai University, Yamate-cho, Suita 564-8680, Japan

[2] Graduate School of Science and Engineering, Kansai University, Yamate-cho, Suita 564-8680, Japan; E-Mail: k561544@kansai-u.ac.jp

* Author to whom correspondence should be addressed; E-Mail: omuray@kansai-u.ac.jp

Abstract: This paper describes the performance prospect of scaled cross-current tetrode (XCT) CMOS devices and demonstrates the outstanding low-energy aspects of sub-30-nm-long gate XCT-SOI CMOS by analyzing device operations. The energy efficiency improvement of such scaled XCT CMOS circuits (two orders higher) stems from the "*source potential floating effect*", which offers the dynamic reduction of effective gate capacitance. It is expected that this feature will be very important in many medical implant applications that demand a long device lifetime without recharging the battery.

Keywords: XCT-SOI MOSFET; quasi-static body floating effect; source potential floating effect; low energy; medical applications

1. Introduction

Since the 1990s, various families of fully-depleted SOI MOSFET have been proposed and extensively studied [1,2] due to their various merits in terms of device scaling (high drivability, steep swing, less short-channel effects, smaller foot print, *etc.*). One of the authors (Omura) proposed the cross-current tetrode SOI MOSFET (XCT-SOI MOSFET) for analog applications (see Figure 1), in 1986, on the basis of partially-depleted SOI MOSFET technology [3], and conducted experiments to evaluate the fundamental aspects of the device. While the scaling feasibility of XCT-like devices has been studied recently [4,5], we expect that XCT-SOI devices will yield new applications, such as medical implants, which demand low-energy operation with high noise margin. In order to discuss those applications in detail, device models have already been proposed to perform circuit simulations [6,7]. However, modeling an XCT-SOI device is not so easy due to the three-dimensionality of its operations [3,5]. Recently, our laboratory group examined the device model proposed in [6], and its usefulness was comprehensively revealed [8]. In addition, a mechanism analysis demonstrated the potential of scaled XCT CMOS devices for extremely-low-energy operation [9].

Figure 1. XCT-SOI MOSFET (bird's eye view and equivalent circuit models): (**a**) Bird's eye view of XCT-SOI device with parameter definitions and current flow. Broken arrows reveal the current flow; (**b**) Basic equivalent circuit model of n-XCT-SOI MOSFET [3]; and (**c**) AC circuit model of n-XCT-SOI MOSFET [9].

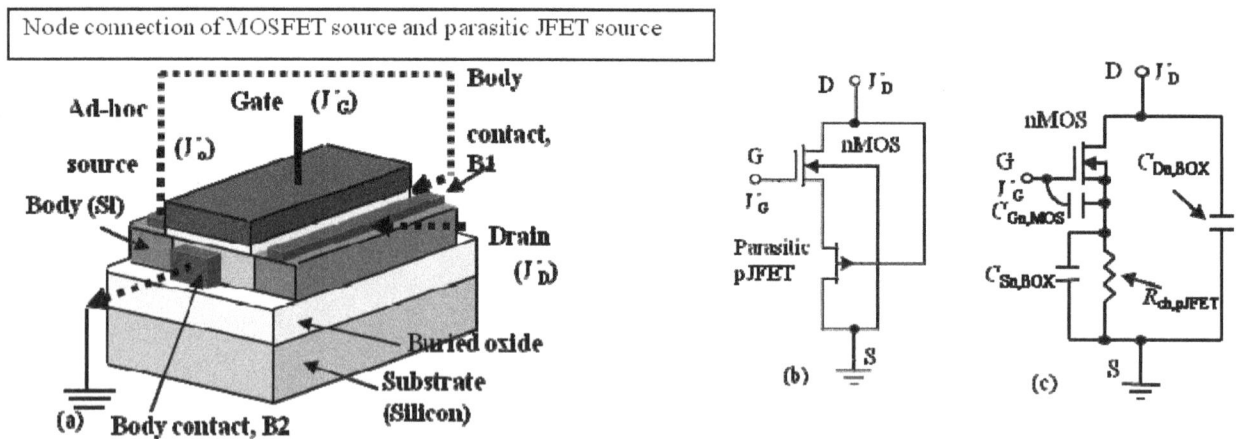

This paper considers the dynamic and standby power dissipation characteristics of the sub-30-nm-long gate XCT-SOI MOSFET. We start by analyzing the low-energy operation of XCT-SOI CMOS circuits; the model proposed here strongly suggests that the "*source potential floating effect* (SPFE)" substantially reduces the operation power consumption. The great advantages of the design methodology are also elucidated. In addition, this study addresses the reality of sub-20-nm-long gate XCT-SOI devices and a scaling scheme to suppress the standby power consumption for future low-energy applications.

2. Device Structure and Assumptions for Modeling

The schematic device structure is shown in Figure 1a. In an n-channel XCT-SOI device, the n-channel SOI MOSFET and p-channel JFET are self-merged and the electron current of nMOSFET is relayed to the hole current of pJFET in series; the broken arrows reveal the current flow by the node connection of the MOSFET source and the JFET source.

As an example, we show the fundamental I_D-V_D characteristics of a 0.1-μm-long gate partially-depleted nMOSFET without the parasitic pJFET, see Figure 2a, following 3-D device simulations [10]. Device parameters are summarized in Table 1. As the XCT-SOI device has two active body contacts (B1 and B2) of the SOI MOSFET, the conventional *"quasi-static body-floating effect"* [1] is eliminated automatically. Details of the device operation are described in a previous paper [8]. The nMOSFET shows slight short-channel effects due to channel length modulation. Figure 2b shows the I_D-V_D characteristics of an n-channel XCT-SOI MOSFET as determined from 3-D device simulations. It is seen that the XCT-SOI device has lower I_D than the MOSFET because of the series resistance of the parasitic pJFET. In addition, the XCT-SOI device shows the negative-differential conductance (NDC) in the saturation region of drain current [3,6,7]. On the other hand, the short-channel effects (SCEs) of the original MOSFET are sufficiently suppressed in the XCT device [8]; subthreshold swing is 76 mV/dec at $V_D = 1$ V. This is one of great advantages of XCT devices.

Figure 2. Simulated I_D-V_D characteristics. (a) Original 0.1-μm-gate SOI MOSFET without parasitic JFET; (**b**) nXCT-SOI MOSFET.

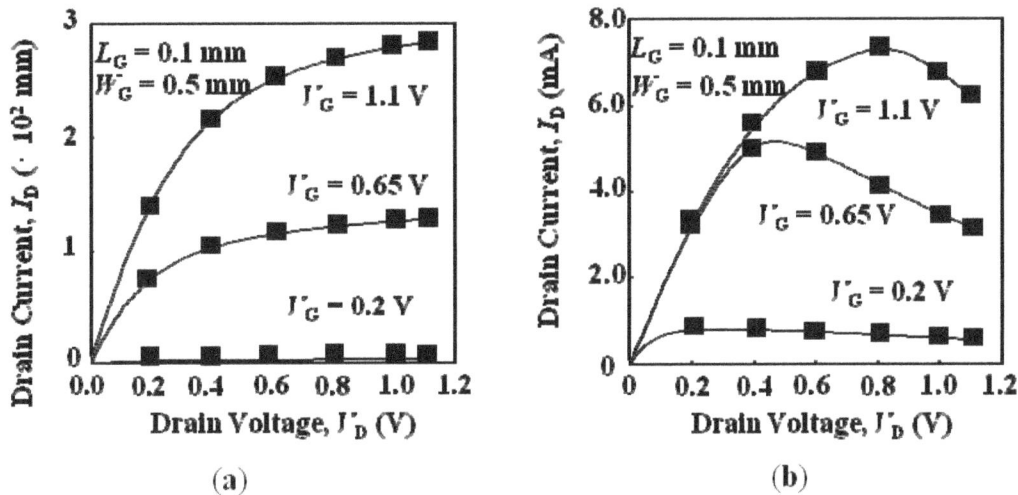

(a)   (b)

Table 1. Scaling scheme and device parameters for sub-100-nm-long gate XCT-SOI MOSFET.

Parameters	N_A (cm^{-3})	L_G (nm)	EOT (nm)	t_{SOI} (nm)	t_{BOX} (nm)	V_D (V)	V_{TH} (*) (V)
Scaling scheme	$k^{4/3}$	$1/k$	$4/3k$	$1/k^{1/3}$	$1/k^{1/3}$	$1/k^{1/2}$	-
$L_G = 100$ nm	2.0×10^{18}	100	2.0	129	111	1.00	0.20
$L_G = 75$ nm	4.0×10^{18}	75	1.5	117	100	0.97	0.16
$L_G = 20$ nm	2.3×10^{19}	20	0.4	75	65	0.50	−0.06
$L_G = 15$ nm	3.4×10^{19}	15	0.3	69	59	0.43	−0.10

(*) n^+-poly-Si gate is assumed.

In support of XCT-SOI MOSFET modeling, we have already proposed the equivalent circuit (for quasi-static analysis), shown in Figure 1b. The basic feasibility of this model has been examined by circuit simulations [5,7]. In addition, mechanisms of the low-energy operation were examined by a simplified analysis, based on the model shown in Figure 1c [9], where $C_{Gn,MOS}$, $C_{Sn,BOX}$, $C_{Dn,BOX}$ and $R_{ch,pJFET}$ denote the gate-to-source capacitance of SOI MOSFET, the source-to-substrate capacitance of SOI MOSFET, the drain-to-substrate capacitance of SOI MOSFET, and the source-to-drain resistance of the parasitic pJFET, respectively. In the case of p-channel XCT-SOI MOSFET, we label them as $C_{Gp,MOS}$, $C_{Sp,BOX}$, $C_{Dp,BOX}$, and $R_{ch,nJFET}$, respectively.

3. Circuit Simulation Results of SOI CMOS and XCT-SOI CMOS

Here, we advance the discussion to better understand XCT-CMOS EXOR circuit features. We assume the OR-NAND type EXOR circuit that consists of four CMOS inverters (standard layout). We concentrate the discussion on the energy ratio of CMOS-EXOR circuits, where energy ratio (*ER*) is defined as the energy dissipated by the XCT-CMOS EXOR over that of the comparable conventional SOI-CMOS EXOR. Calculation results, based on HSPICE [11] simulation results, are shown in Figure 3.

Figure 3. Simulation results of energy ratio (*ER*) of the energy dissipation of 1.0-μm-long gate CMOS EXOR and 100-nm-long gate CMOS EXOR. The energy ratio is defined by the energy dissipated by the XCT-EXOR divided by that of the conventional CMOS EXOR.

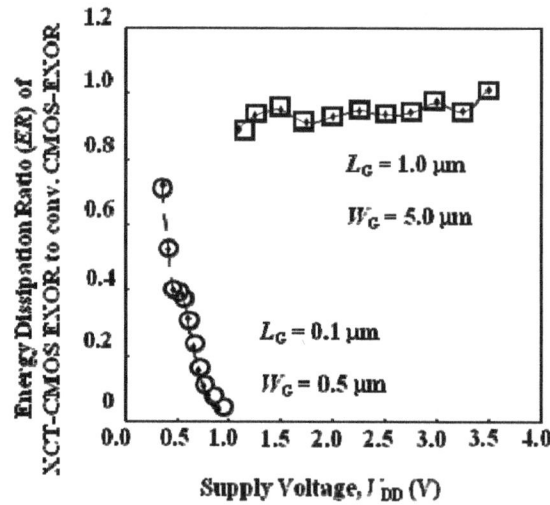

First, it is seen that the *ER* value of 1-μm-long gate devices is almost unity regardless of the V_{DD} value. This behavior is reasonable for the following reasons. The energy dissipation of conventional devices, evaluated by the P_d-t_d product, is not a function of V_{DD} due to the simple recognition of the MOS gate capacitor's charging and discharging operations. It is anticipated that the XCT-CMOS, with a 1-μm-long gate, follows this principle. In the case of 0.1-μm-long gate devices, on the other hand, it is seen that the *ER* value rapidly falls as V_{DD} rises. As this is a very interesting result and somewhat mysterious, we discuss below a possible mechanism based on physics.

Frequency-dependent energy ratio (*ER*) is defined by [9]:

$$ER(\omega) = \frac{C_{Gn,XCT} + C_{Gp,XCT} + C_{Dn,BOX} + C_{Dp,BOX}}{C_{Gn,MOS} + C_{Gp,MOS} + C_{Dn,BOX} + C_{Dp,BOX}} \tag{1}$$

where $C_{Gn,XCT}$ and $C_{Gp,XCT}$ denote the effective gate capacitance of *n*-XCT-SOI MOSFET and *p*-XCT-SOI MOSFET, respectively. Here, for simplicity, we do not take account of the depletion layer beneath the buried oxide layer. These capacitances are calculated using the equivalent circuit model shown in Figure 1c as:

$$C_{Gn,XCT} = \frac{C_{Gn,MOS}(1 + \omega^2 C_{Sn,BOX}^2 R_{ch,PJFET}^2)}{1 + \omega^2 C_{Sn,BOX} R_{ch,PJFET}^2 (C_{Gn,MOS} + C_{Sn,BOX})} \tag{2a}$$

$$C_{Gp,XCT} = \frac{C_{Gp,MOS}(1 + \omega^2 C_{Sp,BOX}^2 R_{ch,nJFET}^2)}{1 + \omega^2 C_{Sp,BOX} R_{ch,nJFET}^2 (C_{Gp,MOS} + C_{Sp,BOX})} \tag{2b}$$

At the low-frequency limit ($\omega \to 0$), $C_{Gn,XCT}$ and $C_{Gp,XCT}$ are reduced to $C_{Gn,MOS}$ and $C_{Gp,MOS}$, respectively, as expected. At the high-frequency limit, however, we have:

$$C_{Gn,XCT} = \frac{C_{Gn,MOS} C_{Sn,BOX}}{C_{Gn,MOS} + C_{Sn,BOX}} \tag{3a}$$

$$C_{Gp,XCT} = \frac{C_{Gp,MOS} C_{Sp,BOX}}{C_{Gp,MOS} + C_{Sp,BOX}} \tag{3b}$$

Generally speaking, reducing the gate length (L_G) raises the operation frequency at the same supply voltage. The *ER* rises when the scaling is enhanced; the scaling scheme assumed here [5,8] slightly reduces $C_{Sn,BOX}R_{ch,pJFET}$ and $C_{Sp,BOX}R_{ch,nJFET}$ as the scaling is advanced. For $f < 1/(C_{Sn,BOX}R_{ch,pJFET})$ and $f < 1/(C_{Sp,BOX}R_{ch,nJFET})$, the roles of $C_{Sn,BOX}$ and $C_{Sp,BOX}$ are lost; *i.e.*, we have $C_{Gn,XCT} \sim C_{Gn,MOS}$ and $C_{Gp,XCT} \sim C_{Gp,MOS}$. This is equivalent to the low-frequency limit. As a result, the *ER* value approaches unity; the intrinsic advantage of the XCT-SOI CMOS is lost. For $f > 1/(C_{Sn,BOX}R_{ch,pJFET})$ and $f > 1/(C_{Sp,BOX}R_{ch,nJFET})$, on the other hand, we have $C_{Gn,XCT} < C_{Gn,MOS}$ and $C_{Gp,XCT} < C_{Gp,MOS}$. Device operation approaches the high-frequency limit. In this case, the *ER* value decreases as the frequency rises. The *ER* value approaches zero as the supply voltage rises, shown in Figure 3. It is anticipated that the depletion layer beneath the buried oxide layer reduces the parasitic capacitance of source diffusion. In order to achieve a small *ER* value, therefore, we have to increase the effective channel resistance of the parasitic JFET when scaling is advanced.

Calculation results of the energy ratio (*ER*), defined by Equations (1) and (2), are shown in Figure 4. The model clearly predicts that high-frequency drive will drastically reduce XCT-SOI CMOS power consumption. This stems from the "*source potential floating effect* (SPFE)" obtained by the model shown in Figure 1c. Consequently, we can conclude that the SPFE of the source diffusion of the SOI MOSFET plays an important role in reducing the energy dissipated by XCT-SOI CMOS devices.

Figure 4. Dissipated energy ratio as a function of the scaling factor. It is assumed that the energy dissipation consists only of the charging and discharging processes of the gate capacitor. Scaling scheme is described in [5]. The dotted arrows reveal the lowest-to-highest range of supply voltage applicable to 1-μm-long gate CMOS (for $k = 1$) and 0.1-μm-long gate CMOS ($k = 20$).

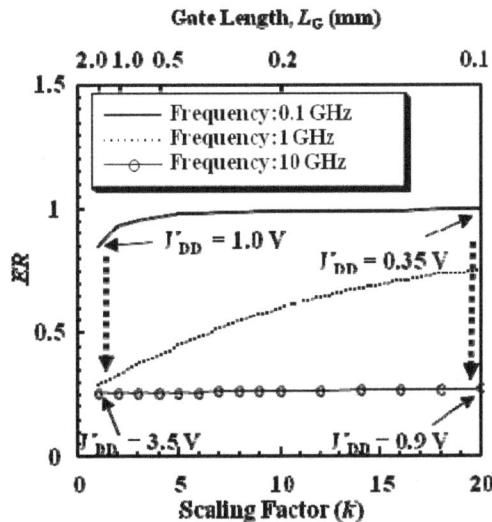

Figure 5 shows simulation results of the time-dependent through-current of the conventional SOI CMOS EXOR chain circuit and the XCT-SOI CMOS EXOR chain circuit, both under dynamic operation. It is seen that the standby power of the XCT-SOI CMOS EXOR chain circuit is about two orders lower than that of the conventional SOI CMOS EXOR chain circuit. This suggests that logic circuits composed of XCT-SOI CMOS can offer drastically lower standby-energy dissipation as well as lower switching-energy dissipation.

Figure 5. Simulation results of through-current of CMOS EXOR for $L_G = 100$ nm. Arrows show standby power reduction.

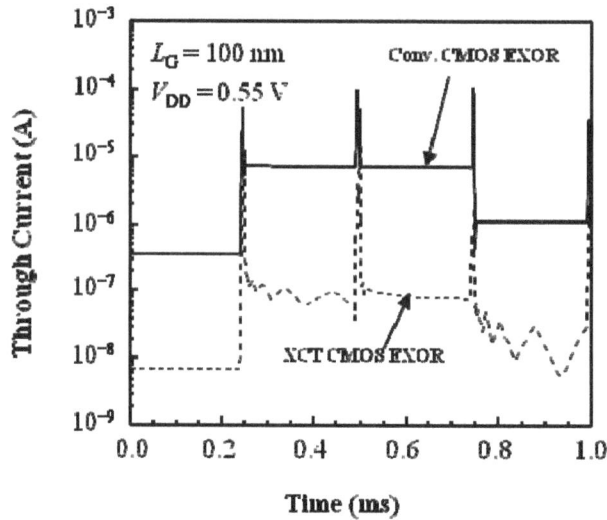

4. Further Scaling Potential of XCT-SOI MOSFET

We now investigate the scaling potential of the XCT-SOI CMOS. The fundamental scaling scheme of the XCT-SOI MOSFET has been already studied in [5,8] and a 100-nm-long gate XCT-SOI CMOS has been realized. However, body doping is apt to rise to 10^{19} cm^{-3} as the XCT-SOI MOSFET is inherently a partially-depleted SOI device. We investigated how well the XCT-SOI MOSFET can be scaled down to realize a 15-nm-long gate. Using 3D device simulations [10], we simulated the performance of 20-nm-long gate and 15-nm-long gate XCT-SOI MOSFETs under the scaled bias condition, where we assumed abrupt source and drain junctions, for simplicity, in the scaling scheme proposed recently [12]. As the previous scheme [5,8] cannot be applied to the sub-100-nm regime, we restructured the scheme (see Table 1) so that devices work well [12].

Here, simulated I_D-V_D characteristics of the 15-nm-long gate XCT-SOI MOSFET, with device parameters, shown in Table 1, shown in Figure 6; the device has the abrupt source and drain junctions. The device shows the negative differential conductance in the saturation region, as is expected.

Simulated subthreshold characteristics of the 15-nm-long gate XCT-SOI MOSFET and the 20-nm-long gate XCT-SOIMOSFET are compared in Figure 7; both devices have the abrupt junctions. It is seen in Figure 7 that the subthreshold swing (S) is 77 mV/dec, for the 20-nm-long gate device, and 81 mV/dec, for the 15-nm-long gate device. Short-channel effects are well suppressed. However, the gate-induced drain-leakage (GIDL) current level of the 15-nm-long gate XCT-SOI device is somewhat high (~10^{-9} A/μm) as the device simulation assumes the abrupt junction for simplicity.

Figure 6. Simulation results of I_D-V_D characteristics of 15-nm-long gate XCT-SOI MOSFET. L denotes the effective channel length.

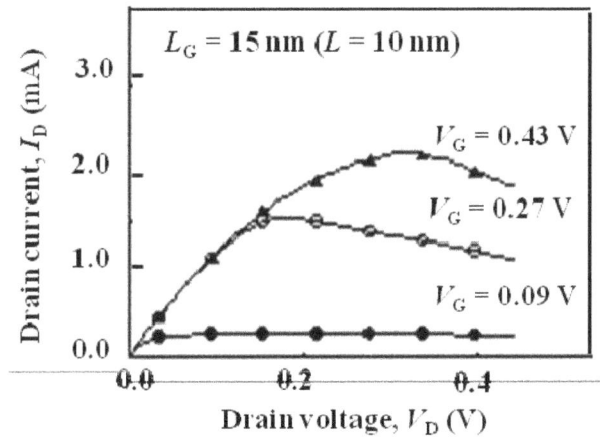

Figure 7. Simulation results of I_D-V_G characteristics of 20-nm-long and 15-nm-long gate XCT-SOI MOSFET. It is assumed that the devices have abrupt source and drain junctions.

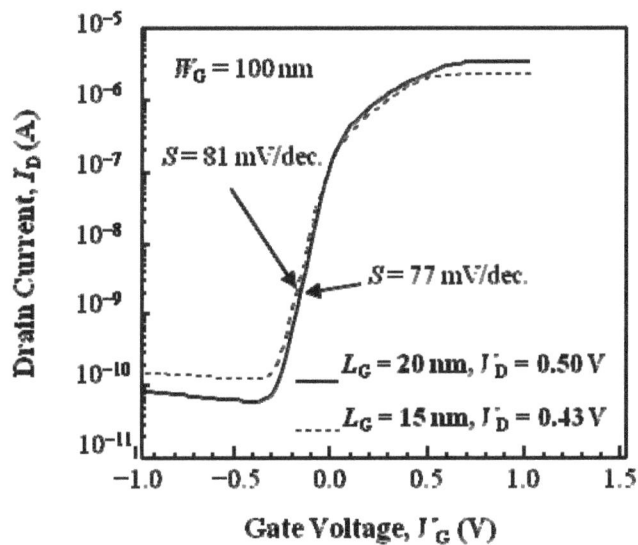

In response, we introduced a 10-nm-long graded doping region for source and drain junctions as a "new" scaling scheme, as shown in Figure 8. Simulated I_D-V_G characteristics of the 15-nm-long gate XCT-SOI MOSFET are shown in Figure 9. Drain current characteristics of 15-nm-long gate XCT-SOI devices with the abrupt and the graded junctions are compared. It should be noted that the subthreshold swing is 65 mV/dec for the 15-nm-long gate device with the graded junction. Short-channel effects are well suppressed in the 15-nm-long gate XCT-SOI MOSFET. The GIDL current level is lowered to 200 pA/μm. The most noticeable result is the improved drivability of the device with the graded junction; this is due to the remarkable improvement in carrier velocity over the whole device region, as shown in Figure 10 [13], and the reduction of the channel resistance of the parasitic pJFET [12].

Figure 8. Doping profiles from the source to drain. The doping profile with a graded junction is compared to that with an abrupt junction.

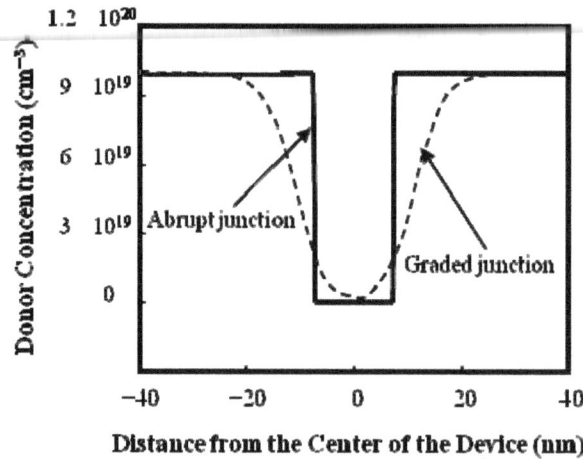

Figure 9. Simulation results of I_D-V_G characteristics of 15-nm-long gate XCT-SOI MOSFET. Impact of doping profile on I_D-V_G characteristics is compared.

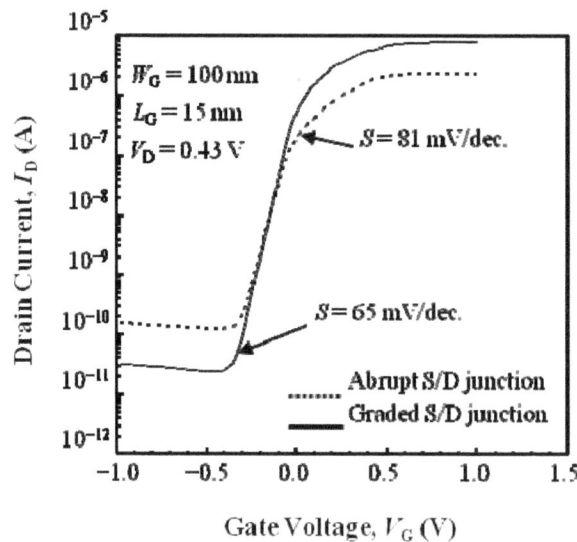

Figure 10. Simulation results of carrier velocity along the channel for the 15-nm-long gate XCT-SOI MOSFET. Impact of doping profile on velocity profile is compared.

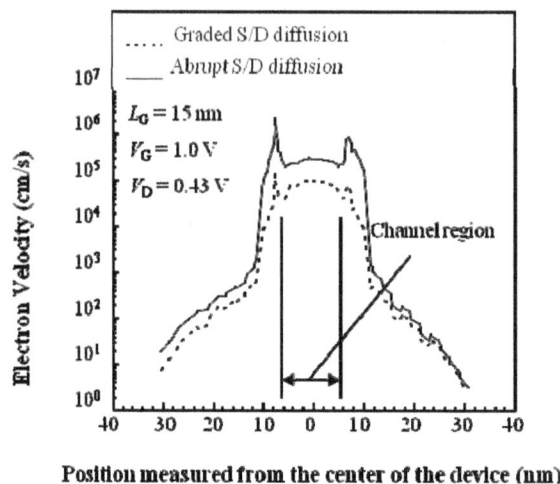

5. Performance Expected from the Scaled XCT-SOI MOSFET

This section comprehensively examines the potential of the scaled XCT-SOI MOSFET. Calculated values of the on-current (I_{on}), the off-current (I_{off}), the intrinsic switching time ($C_G V_D/I_{on}$), the switching energy (E_{sw}), and the standby power (P_{stby}), are summarized in Table 2, where it is assumed that the threshold voltage (V_{TH}) is 0.2 V and $W_G/L_G = 5$, regardless of scale. Here, we assume that $C_{G,XCT} = C_{G,MOS}$; in other words, SPFE is not assumed in Table 2. From Table 2, we can identify the following.

(i)   The graded source/drain junction improves the drivability;

(ii)   The graded source/drain junction does not always reduce the off-current;

(iii) The intrinsic switching time is degraded as the scaling is advanced if an abrupt junction is assumed. However, it can be drastically improved by using a graded junction due to the improved drivability;

(iv)   The switching energy will be further reduced when SPFE is taken into account.

Table 2. Performance perspectives of scaled XCT-SOI MOSFET.

L_G (nm)/W_G (nm)		I_{on} (A)	I_{off} (A)	$C_G V_D/I_{on}$ (s)*	E_{sw} (J)*	P_{stby} (W)*
100/500		5.4×10^{-6}	2.5×10^{-9}	1.7×10^{-10}	5.1×10^{-16}	2.8×10^{-9}
75/375		1.0×10^{-5}	1.3×10^{-9}	1.6×10^{-11}	3.0×10^{-16}	1.3×10^{-9}
20/100	abrupt S/D	1.7×10^{-6}	7.6×10^{-9}	4.9×10^{-11}	2.1×10^{-17}	3.8×10^{-9}
	graded S/D	2.2×10^{-6}	2.4×10^{-9}	3.8×10^{-11}	2.1×10^{-17}	1.9×10^{-9}
15/75	abrupt S/D	5.2×10^{-7}	2.1×10^{-10}	1.4×10^{-10}	1.2×10^{-17}	9.0×10^{-11}
	graded S/D	4.4×10^{-6}	1.9×10^{-9}	1.6×10^{-11}	1.2×10^{-17}	8.2×10^{-10}

*It is assumed that $C_{G,XCT} = C_{G,MOS}$.

In order to evaluate the above aspects of the scaled XCT-SOI MOSFET, we start by referring to the ITRS roadmap 2011 [14]. Figure 10 shows the projected switching performance suggested by the ITRS roadmap 2011; the calculation results of various XCT-SOI devices are also plotted. Simple calculation results reveal that 15-nm-long gate XCT-SOI devices without SPFE are identical to the non-volatile memory device group. However, it is worthwhile to note that the 15-nm-long gate XCT-SOI device is expected to lie in the zone below the non-volatile memory zone; SPFE should provide XCT-SOI devices with much better performance ("expected zone") than that shown in Figure 11.

Figure 11. Comprehensive examination of scaled XCT-SOI MOSFET based on the ITRS roadmap 2011. Expected zone is estimated by assuming the SPFE, based on the model shown in Figure 1c.

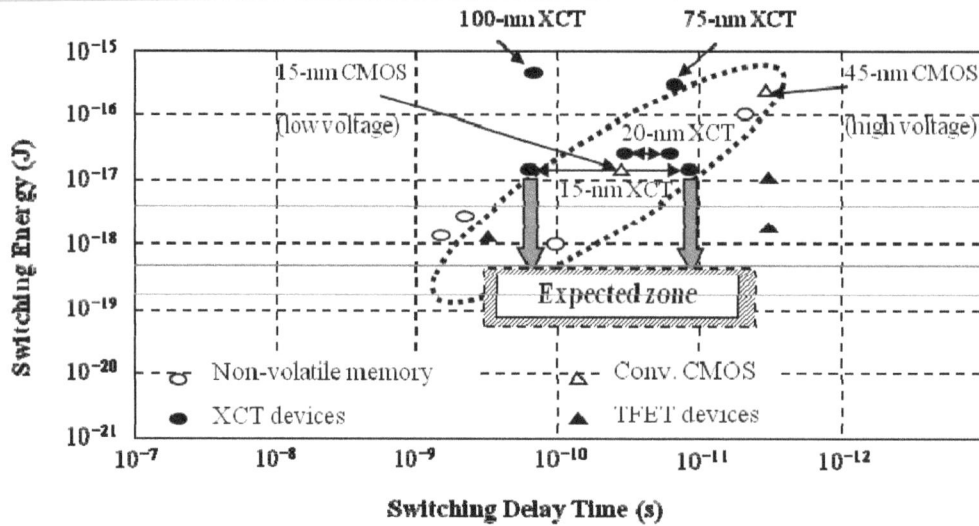

6. Conclusions

We demonstrated the low-energy operation of XCT-SOI CMOS devices scaled down to 15 nm and analyzed the key underlying mechanism. It was shown that the source-follower like operation of the XCT-SOI MOSFET dynamically reduces the effective input capacitance, and, thus, the energy dissipated by XCT-SOI devices. This operation should be called the "*source potential floating effect (SPFE)*". It is predicted, based on a physics-based model, that realizing SPFE in XCT-CMOS circuits will significantly suppress standby power dissipation.

It was also suggested that such aspects are still available in the sub-30-nm-long gate regime. Therefore, we can state that XCT-SOI CMOS devices are very promising for future extremely-low-energy circuits that suit medical implant applications.

Acknowledgments

A part of this study was supported by "Strategic Project to Support the Formation of Research Bases at Private Universities" in Japan: Matching Fund Subsidy from MEXT (Ministry of Education, Culture, Sports, Science and Technology) in Japan.

Conflicts of Interest

The authors declare no conflict of interest.

References

1.   Colinge, J.P. *Silicon-On-Insulator Technology: Materials to VLSI*, 3rd ed.; Kluwer Academic Publishers: Dordrecht, the Netherlands, 2004; pp. 151–277.

2.   *FinFETs and Other Multi-Gate Transistors*; Colinge, J.P., Ed.; Springer: New York, NY, USA, 2008; Chapter 1, pp. 4–28.

3. Omura, Y.; Izumi, K. High-Gain Cross-Current Tetrode MOSFET/SIMOX and Its Application. In Proceedings of the 18th International Conference of the Solid State Devices and Materials, Tokyo, Japan, 1986; pp. 715–716.

4. Blalock, B.J.; Cristoloveanu, S.; Dufrene, B.M.; Allibert, F.; Mojarradi, M. The multiple-gate MOSFET-JFET transistor. *Int. J. High Speed Electron. Syst.* **2002**, *12*, 511–520.

5. Omura, Y.; Fukuchi, K.; Ino, D.; Hayashi, O. Scaling scheme and performance perspective of cross-current tetrode (XCT) SOI MOSFET for future ultra-low power applications. *ECS Trans.* **2011**, *35*, 85–90.

6. Azuma, Y.; Yoshioka, Y.; Omura, Y. Cross-Current SOI MOSFET Model and Important Aspects of CMOS Operations. In Proceedings of the International Conference of the Solid State Devices and Materials, Tsukuba, Japan, 2007; pp. 460–461.

7. Omura, Y. Cross-current silicon-on-insulator metal-oxide semiconductor field-effect transistor and application to multiple voltage reference circuits. *Jpn. J. Appl. Phys.* **2009**, *48*, 04C07–04C11.

8. Omura, Y.; Azuma, Y.; Yoshioka, Y.; Fukuchi, K.; Ino, D. Proposal of preliminary device model and scaling scheme of cross-current tetrode silicon-on-insulator metal-oxide-semiconductor field-effect transistor aiming at low-energy circuit applications. *Solid-State Electron.* **2011**, *64*, 18–27.

9. Omura, Y.; Ino, D. Definite Feature of Low-Energy Operation of Scaled Cross-Current Tetrode (XCTY) SOI CMOS Circuit. In Proceedings of the 17th Workshop on Synthesis and System Integration of Mixed Information Technologies (SASIMI 2012), Beppu, Japan, 2012; pp. 355–360.

10. Synopsys Inc. *Sentaurus Operations Manual*; Synopsys: Mountain View, CA USA, 2008.

11. Synopsys Inc. *HSPICE Users Manual*; Synopsys: Mountain View, CA USA, 2008.

12. Sato, D.; Omura, Y. Scaling Scheme Prospect of XCT-SOI MOSFET Aiming at Medical Implant Applications Showing Long Lifetime with a Small Battery. In Proceedings of the IEEE International Meeting for Future of Electron Devices, Kansai (IMFEDK), Suita, Japan, 2013; pp. 38–39.

13. Omura, Y.; Nakano, S.; Hayashi, O. Gate-field engineering and source/drain diffusion engineering for high-performance Si wire GAA MOSFET and low-power strategy in sub-30-nm-channel regime. *IEEE Trans. Nanotechnol.* **2011**, *10*, 715–726.

14. ITRS Roadmap. http://www.itrs.net/Links/2011ITRS/Home2011.htm (accessed on 20 January 2012).

6

Ultralow-Power SOTB CMOS Technology Operating Down to 0.4 V [†]

Nobuyuki Sugii [1,*]**, Yoshiki Yamamoto** [1]**, Hideki Makiyama** [1]**, Tomohiro Yamashita** [1]**,
Hidekazu Oda** [1]**, Shiro Kamohara** [1]**, Yasuo Yamaguchi** [1]**, Koichiro Ishibashi** [2]**,
Tomoko Mizutani** [3] **and Toshiro Hiramoto** [3]

[1] Low-Power Electronics Association & Project, Tsukuba, Ibaraki 305-8569, Japan;
    E-Mails: yoshiki.yamamoto.pc@renesas.com (Yo.Ya.); hideki.makiyama.wx@renesas.com (H.M.);
    tomohiro.yamashita.ue@renesas.com (T.Y.); hidekazu.oda.yf@renesas.com (H.O.);
    shiro.kamohara.uh@renesas.com (S.K.); yasuo.yamaguchi.uf@renesas.com (Ya.Ya.)

[2] Department of Engineering Science, Graduate School of Informatics and Engineering Departments,
    The University of Electro-Communications, Chofu, Tokyo 182-8585, Japan;
    E-Mail: ishibashi@ee.uec.ac.jp

[3] Institute of Industrial Science, The University of Tokyo, Meguro, Tokyo 153-8505, Japan;
    E-Mails: mizutani@nano.iis.u-tokyo.ac.jp (T.M.); hiramoto@nano.iis.u-tokyo.ac.jp (T.H.)

[†] This article is based on "V_{min} = 0.4 V LSIs are the real with Silicon-on-Thin-Buried-Oxide
    (SOTB)—How is the application with 'Perpetuum-Mobile' micro-controller with SOTB?",
    which appeared in Proceedings of the IEEE 2013 SOI-3D-Subthreshold Microelectronics
    Technology Unified Conference (S3S) ©2013 IEEE.

[*] Author to whom correspondence should be addressed; E-Mail: n-sugii@ieee.org

Abstract: Ultralow-voltage (ULV) CMOS will be a core building block of highly energy
efficient electronics. Although the operation at the minimum energy point (MEP) is
effective for ULP CMOS circuits, its slow operation speed often means that it is not used
in many applications. The silicon-on-thin-buried-oxide (SOTB) CMOS is a strong
candidate for the ultralow-power (ULP) electronics because of its small variability and
back-bias control. Proper power and performance optimization with adaptive V_{th} control

taking advantage of SOTB's features can achieve the ULP operation with acceptably high speed and low leakage. This paper describes our results on the ULV operation of logic circuits (CPU, SRAM, ring oscillator and other logic circuits) and shows that the operation speed is now sufficiently high for many ULP applications. The "Perpetuum-Mobile" micro-controllers operating down to 0.4 V or lower are expected to be implemented in a huge number of electronic devices in the internet-of-things (IoT) era.

Keywords: ultralow power; ultralow voltage; CMOS; minimum energy point; variability; back bias; FDSOI; silicon-on-thin-buried-oxide (SOTB); thin BOX

---

1. Introduction: Issues for ULV Operation Possibly Staying on MEP Point

A huge number of small electronic devices composing big data are expected to be used across the globe as the "internet of things" (IoT). The CMOS integrated circuit is a core part of these devices. The energy efficiency of the CMOS circuits should therefore be greatly improved. It is well known that the operating voltage (V_{dd}) is a primarily important parameter for reducing the energy per operation cycle in the CMOS circuits. As shown in Figure 1, the energy is a sum of active (E_{ac}) and leakage (E_{leak}) energy as shown in Equation (1) in the simplified form.

$$E = E_{ac} + E_{leak} = C_{load}V_{dd}{}^2 + I_{leak}V_{dd}/af \tag{1}$$

where C_{load}, I_{leak}, a, and f denote load capacitance, leakage current, activity, and frequency, respectively. With decreasing V_{dd}, E_{ac} decreases since it is proportional to $C_{load}V_{dd}{}^2$. However, E_{leak} relatively increases as f decreases with decreasing V_{dd}. These two terms determine the minimum energy point (MEP). The energy efficiency of CMOS circuits has been greatly improved by the miniaturization of CMOS transistors. This improvement is mainly accomplished by E_{ac} reduction due to C_{load} reduction with the transistor scaling. However, most of the circuits do not operate at MEP and the improvement in terms of the efficiency has not been perfect so far. In recent generations, the scaling has increased the V_{dd} at MEP, as shown in Figure 2 [1,2]. This is because E_{leak} tends to increase with the miniaturization that has been seen in recent generations, especially for the performance-oriented applications. In the energy-efficiency conscious design like [1], the E_{min} has already an increasing trend (minimum point of E_{min} at 90-nm node), as shown in Figure 2.

The near- or sub-V_{th} operation is attractive to improve the energy efficiency. The operating speed of these circuits, however, is not high. The maximum frequency rapidly drops with decreasing V_{dd} in the conventional CMOS [3]. In the device design for ULP circuits, it is important to optimize both V_{dd} and V_{th}. With decreasing V_{dd}, V_{th} should increase to minimize the energy [4]. This drastically decreases the frequency and in many cases the MEP operation is a sub-V_{th} operation and its frequency is less than MHz. The variable V_{th} approach with adaptive back-bias control can mitigate the situation: optimizing frequency and decreasing energy as low as possible down to the MEP value. Both V_{dd} and V_{th} are controlled to minimize the energy, while satisfying the required workload: required frequency. In the dynamic voltage and frequency scaling (DVFS), only V_{dd} is controlled. In order to achieve higher energy efficiency, the control of V_{th} should be accompanied.

Figure 1. Schematic relationship between energy per operation E *versus* operating voltage V_{dd}.

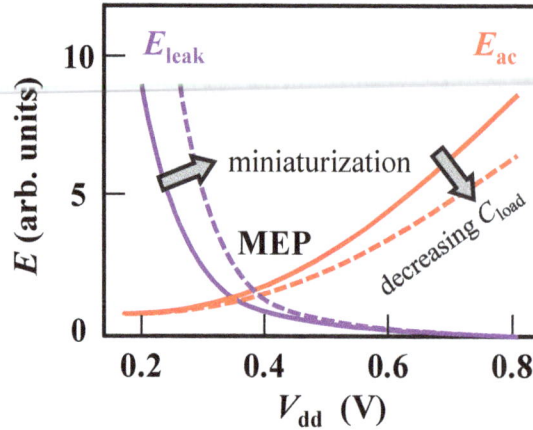

Figure 2. Minimum energy E_{min} and V_{dd} at MEP as a function of technology node number after [1,2].

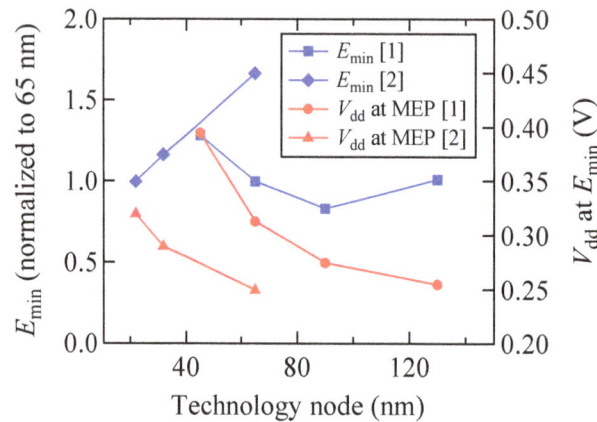

It is well known that the characteristic variability of transistors is recognized as a major obstacle for the performance/power tradeoff, especially at low V_{dd}. The increasing variability also increases V_{dd} at MEP [5] simply because of the increase in leakage current in a circuit that is a sum of transistor leakages [6]. Moreover, increasing the transistor variation causes delay variation, especially at low V_{dd} and causes a significant performance drop [7]. Design to cope with the increasing variability at low V_{dd} becomes more complex. The variability tolerant design prefers to increase the transistor width W; however, this directly increases the power [8,9]. Another variability tolerant logic design prefers a smaller number of pipeline stages and longer logic depth. However, these design strategies decrease the frequency [10,11] and increase E_{min}.

We hypothesize the main issues for the highly energy efficient CMOS circuits are adaptive V_{th} control and small variability, as described in this section. In order to solve these issues, we are developing the silicon on thin buried oxide (SOTB) [12–16]. In this article, we show SOTB's low voltage capability, including small variability and back-bias control through device and circuit results.

2. SOTB Device Technology

Schematic cross section of the SOTB is shown in Figure 3. There are four major factors: (i) small local variability due to low-impurity fully depleted SOI channel; (ii) high back-bias coefficient due to thin BOX layer and doped ground plane (n GP and p GP) just below the BOX layer; (iii) flexible V_{th} tuning by impurity density control of the ground plane and (iv) high design compatibility with the conventional CMOS due to mostly identical planar layout to the bulk and a hybrid bulk integration for I/O. Details of the SOTB fabrication process are reported elsewhere [15,16].

Figure 3. Schematic cross section of silicon on thin buried oxide (SOTB) and hybrid bulk transistors. V_{bp} and V_{bn} denote back-bias terminal for p- and n-type SOTB, respectively.

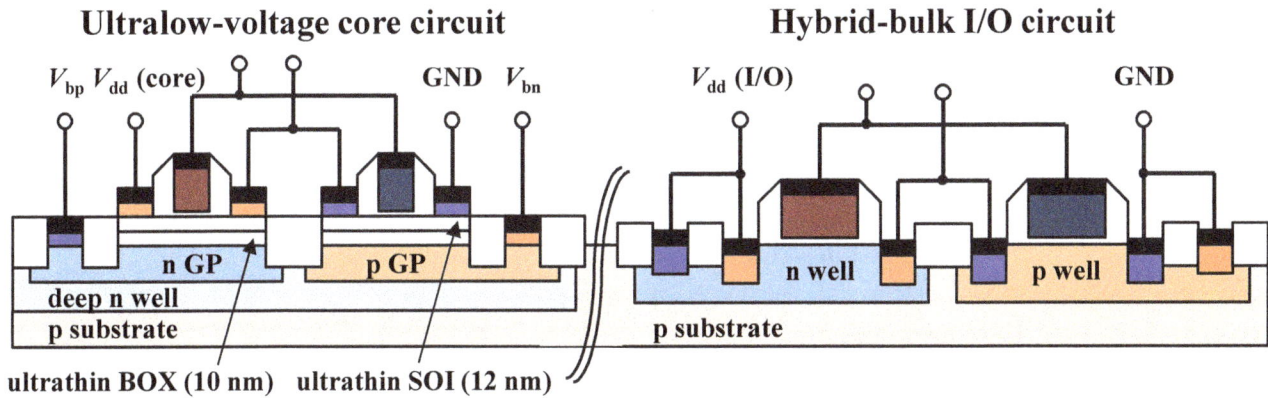

V_{th} control optimized for the ULV operation has been an important process issue. In this optimization, we controlled the V_{th} values at around 0.2 V, which is by about 0.2 V lower than that for the low-standby-power (LSTP) application, with the multiple V_{th} option utilizing a poly-silicon and high-k gate-stack technology and a proper doping profile control of the ground plane [15]. As shown in Figure 4, we utilize a small amount of high-k (Hf and Al) oxide mixed with the conventional SiON gate dielectric and control proper effective work function (EWF) both for NMOS and PMOS. Typical I_d-V_g characteristics of triple V_{th} option are shown in Figure 5. LVT, SVT, and HVT denote low, standard, and high V_{th} option, respectively. Two-orders-of-magnitude off-leakage-current variation can be done only by changing the doping density below the BOX layer with the same gate stack.

Figure 4. Effective work function (EWF) control with high-k/SiON gate stack [15]. Circles and diamonds represent EMF of P-type and N-type gate stacks for PMOS and NMOS, respectively.

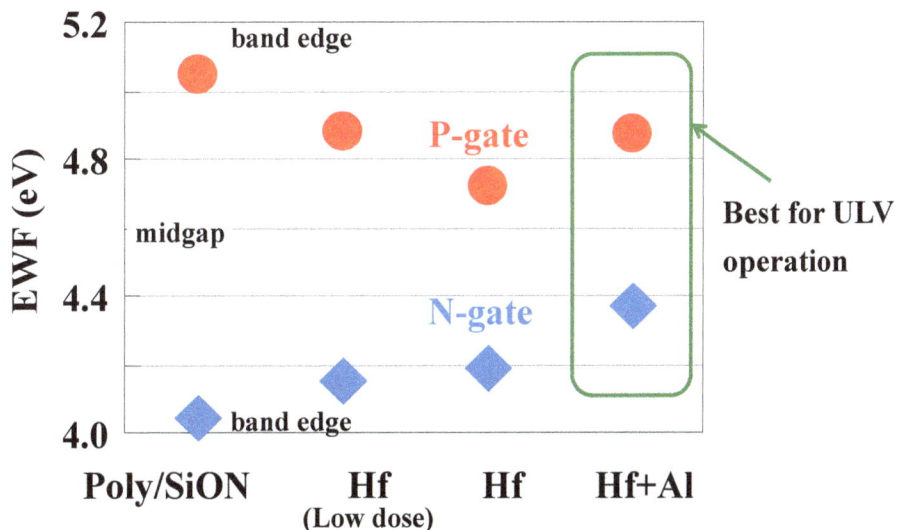

Figure 5. I_d-V_g characteristics of triple V_{th} option controlled by ground plane doping [15].

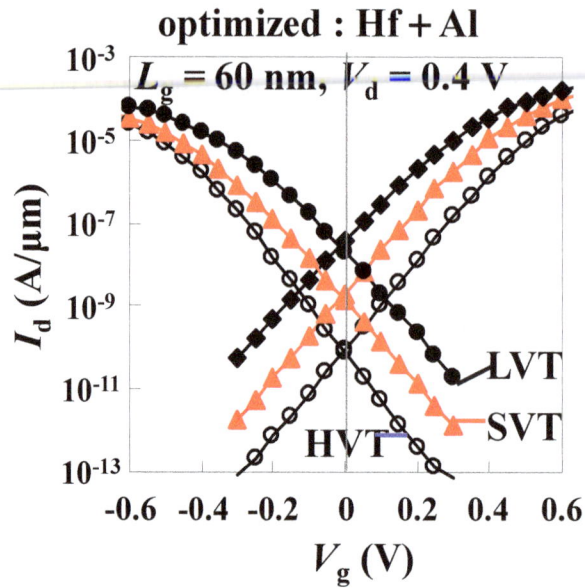

We have demonstrated significant reduction of the V_{th} variability. The Pelgrom coefficient (A_{VT}) of the SOTB was about 1.2–1.3 mVμm [16], which is less than half of the bulk. Moreover, we measured the V_{th} variation of one-million transistors (gate length and width: 0.06 and 0.14 μm, respectively) [16] as shown in Figure 6 and confirmed regular distribution without dropout transistors. Variation of on-state current I_{on} is important and must be decreased since this directly affects the delay variation of circuits. The I_{on} variation was demonstrated to be less than half of bulk [17] and we confirmed a significant reduction of the I_{on} variability for one-million transistors as shown in Figure 7 [16]. The lowest V_{th} values of SOTB and bulk are the same as shown in Figure 6. This means the highest leakage current among one-million transistors is the same. Besides, the highest V_{th} transistor determines the delay. As shown in Figure 7, the smallest I_{on} value for SOTB is about twice as high as the bulk's worst value. This is a strong advantage of SOTB's small variability in terms of the circuit performance.

Figure 6. V_{th} distribution of 1 M transistors. [16] Vertical-axis value shows deviation from V_{th} median value. Positive or negative σ values indicate that the corresponding V_{th} value is upper or below median, respectively.

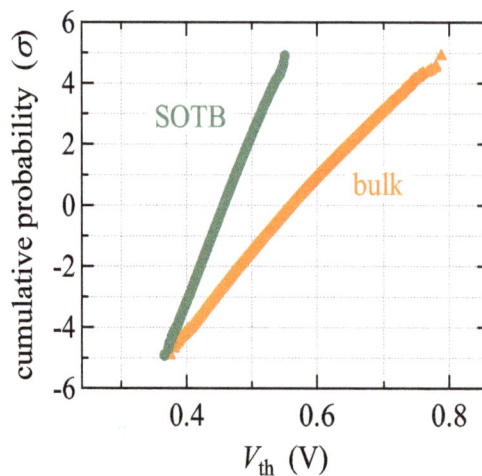

Figure 7. I_{on} distribution of 1 M transistors [16].

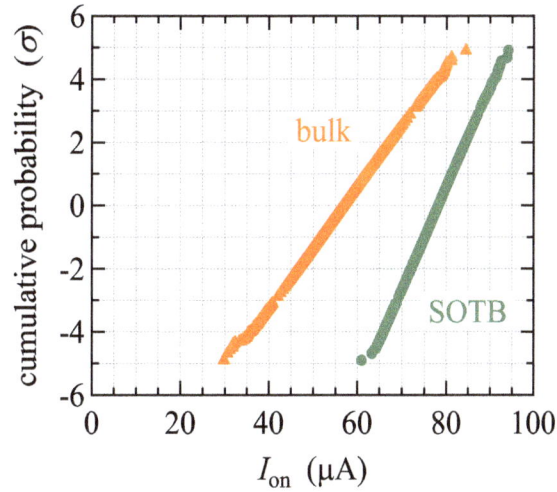

3. ULV Operation of SOTB Circuits

3.1. V_{min} Reduction of 6T-SRAM and Leakage Control by Back-Bias

Thanks to the significant reduction of the variability, as shown in Figures 6 and 7, we successfully demonstrated 2-Mbit SRAM operation at V_{min} = 0.37 V [16] of the standard six-transistor (6T) layout and without assist circuits as shown in Figure 8. Due to the V_{th} optimization mentioned in the previous section, very small access time (5.5 ps at V_{dd} = 0.4 V) was demonstrated. This enables a circuit operation with SRAM at several tens of MHz. The standby leakage decreased more than two orders of magnitude by a reverse back biasing and achieved 1.2 pA/cell without destroying the data. Moreover, the above V_{min} value can be kept at lower at elevated temperatures with a proper back-bias control. In Figure 9, the V_{min} value at room temperature was the same as the value shown in Figure 8. At 80 °C, the V_{min} value increased to 0.46 V. This is because the V_{th} values of NMOS and PMOS transistors differently shifted from the room-temperature values. Although these V_{th} values were smaller than the room-temperature values (this increased the leakage current about two orders of magnitude higher than that at room temperature), balance of V_{th}s (current drivabilities) between the NMOS and PMOS transistors untuned. This deteriorated the SRAM cell stability and increased the V_{min}. By applying proper back-bias voltages for both transistors, the V_{min} value again reduced to less than 0.4 V as shown in Figure 9. Leakage current also can be minimized by the back-bias control regardless of temperature, which is roughly the same as the room-temperature value.

3.2. Ring Oscillator Circuit Results

We have developed a standard logic cell library of the SOTB technology with a hybrid bulk I/O library. The delay characteristics of the cells were evaluated through the RO measurements [18]. Figure 10 shows propagation delay t_{pd} of a 101-stage inverter RO for SOTB and bulk. The delay of SOTB was by 42% smaller than bulk at V_{dd} = 0.4 V. Note that V_{th}s of SOTB and bulk were the same at V_{dd} = 0.4 V. The speed gain of SOTB was higher at lower V_{dd} because of better I_{eff}/I_{off} and smaller DIBL. The delay variability was then investigated. The standard deviation of t_{pd} for SOTB exhibited a very weak $1/\sqrt{N}$ dependence. The result means that the local delay variability of SOTB is very small. We also succeeded in the significant reduction of die-to-die delay variability by a proper back-biasing [19]. The logic circuits contain various types of logic cells such as inverter, NAND, NOR, *etc.* We found that back-biasing considering a drivability balance of NMOS and PMOS transistors is essential for an effective suppression of die-to-die delay variability for various types of the cells.

Figure 8. Fail-bit count of 2-M bit SRAM array as a function of V_{dd} [16].

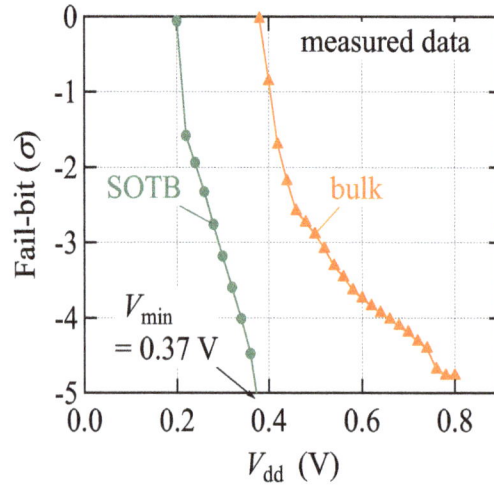

Figure 9. V_{min} of 2-M bit SRAM array as a function of back-bias voltage $|V_b|$: absolute values of back-bias voltage for NMOS and PMOS, $|V_b| = |V_{bn}| = |V_{bp}|$.

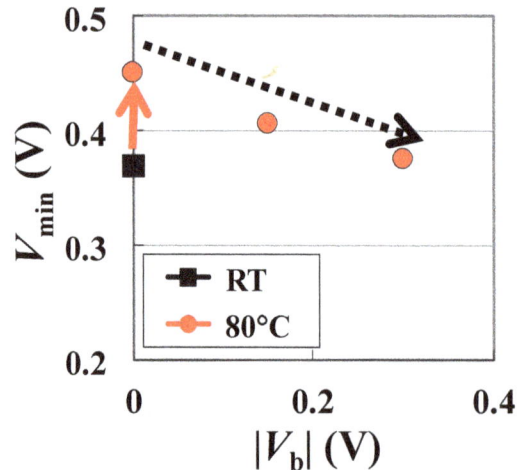

The minimum energy consumption of SOTB logic circuits of 50 kgates was estimated based on the RO results by optimizing the back-bias voltage [20]. At the same energy per operation, SOTB operates about ×10 higher than bulk. The power consumption of 44 µW at 10 MHz (4.4 pJ/cycle) is expected at $V_{dd} = 0.33$ V whereas bulk operates at 1 MHz with the same energy per cycle as SOTB.

Figure 10. Inverter delay t_{pd} as a function of V_{dd} [18].

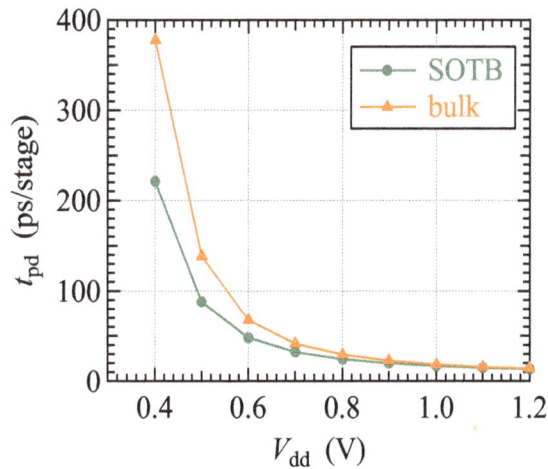

3.3. Demonstration of ULV and ULP Operation of Logic Circuits

The design flow for the SOTB integrated circuits is basically the same as the conventional one. Using our newly developed design flow with the SOTB/bulk hybrid library, several ULV circuits were designed. Significant power reduction was demonstrated by the post-layout timing and power analysis. The reconfigurable accelerator named cool mega array (CMA) was designed and silicon results were obtained [21]. The bulk CMA operates at 0.8–1.2 V (with dynamic voltage scaling) and 210 MHz, and the SOTB version operates at 0.4 V (with back-biasing) and 65 MHz. The energy efficiency executing the Alpha blender test program was 38 and 65 MOPS/mW for bulk and SOTB, respectively.

The back-bias control offers a strong advantage for the FPGA circuits. The flex-power FPGA of the SOTB version was firstly implemented and silicon results were obtained [22]. After the FPGA configuration, the back-bias control enables that the only critical-path logic elements are set to low V_{th}. This significantly reduces the leakage power with no operation speed penalty.

High-efficiency generator of back-bias voltage is important for the SOTB technology since standby leakage current is kept low by applying reverse back-bias voltage. A superior point of the SOTB technology is that current load of the back-bias generator is very small because back-gate region of the SOTB transistor is electrically isolated by the BOX layer. This leads to a significant reduction of current consumption of the back-bias generator itself. We designed the generator circuit using the standard Dickson's charge pump for the SOTB and bulk hybrid platform and silicon results were obtained [23]. The generator operates at V_{dd} = 0.1 V and higher, and generates sufficient back-bias voltages for NMOS and PMOS of 0.85 and −1.5 V, respectively, at V_{dd} = 0.4 V with a current consumption of only 13 μA. By applying these back-bias voltages, leakage current of a 500 kgate logic circuit reduced to 2 μA corresponding to 4 pA/gate. There are still several points of the optimization and the generator current consumption is still higher than our target specification. The optimization is now under way.

We have confirmed a successful operation [24] of the proto-type ULV micro-controller chip as shown in Figure 11. This chip is composed of 32-bit RISC CPU with five-stage pipeline, 144 kByte SRAM, and interfaces (ROM, UART, SPI, and GP) and can be connected with sensors and rf modules for the sensor-network node. The micro-controller chip operates at $V_{dd} = 0.35$ V and consumes only $E = 13.4$ pJ ($f = 14$ MHz) as shown in Figure 12. Note that the E values for SOTB and bulk at $V_{dd} \geq 0.8$ V are identical because it is determined by E_{ac} (same C_{load} of the same 65-nm technology) in this region. Sleep current is only 0.14 μA. By taking advantage of the ULP capability of our SOTB micro-controller chip, named "Perpetuum-Mobile", the sensor node is expected to operate with a sufficient frequency (>10 MHz) for a long period with a single battery or further longer operation with an energy harvester.

Figure 11. Block diagram of "Perpetuum-Mobile" micro-controller chip for a sensor-node application. (PLL: phase locked loop; ADC: analog to digital converter; SPI: serial peripheral interface; UART: universal asynchronous receiver transmitter; GP: general purpose).

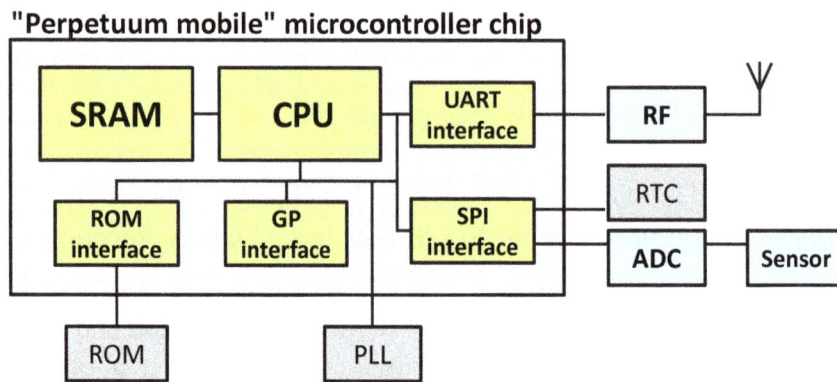

Figure 12. Energy per cycle E as a function of operating voltage V_{dd} for the "Perpetuum-Mobile" micro-controller chip [24].

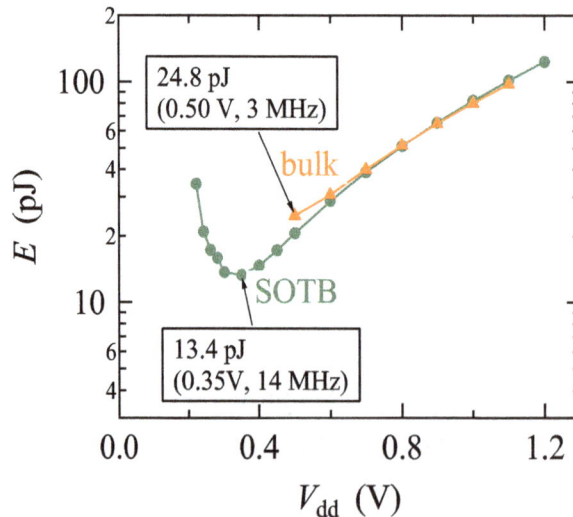

4. Conclusions

Silicon on thin buried oxide (SOTB) is suitable for the ULV operation thanks to its small variability and back-gate bias controllability. We have demonstrated significant variability reduction, 0.4-V operation of SRAM, and reducing power consumption of logic circuits including a micro-controller chip with a significant speed gain even at ULV. Many ULP applications are expected with SOTB chips. The "Perpetuum-Mobile" micro-controller chips will work as a core electronics parts in various types of electronic apparatuses of the "internet of things".

Acknowledgments

This work was performed as "Ultra-Low Voltage Device Project" funded and supported by the Ministry of Economy, Trade and Industry (METI) and the New Energy and Industrial Technology Development Organization (NEDO). The authors thank the staff of Renesas Electronics Corporation for the SOTB process integration.

Author Contributions

Nobuyuki Sugii coordinated and discussed the overall research and prepared the manuscript. Yoshiki Yamamoto and Hideki Makiyama designed the wafer fabrication process, prepared the device wafer, and measured their electrical characteristics. Tomohiro Yamashita, Hidekazu Oda, and Yasuo Yamaguchi designed the fabrication process and discussed the electrical characteristics. Shiro Kamohara and Koichiro Ishibashi designed and characterized the test chips. Tomoko Mizutani and Toshiro Hiramoto measured the electrical characteristics of the device wafers and discussed the results.

Conflicts of Interest

The authors declare no conflict of interest.

References

1. Bol, D.; Kamel, D.; Flandre, D.; Legat, J.-D. Nanometer MOSFET Effects on the Minimum-Energy Point of 45nm Subthreshold Logic. In Proceedings of the 14th IEEE/ACM International Symposium on Low-Power Electronics and Design, San Francisco, CA, USA, 19–21 August 2009; pp. 3–8.
2. Chandrakasan, A.P.; Daly, D.C.; Finchelstein, D.C.; Kwong, J.; Ramadass, Y.K.; Sinangil, M.E.; Sze, V.; Vermaet, N. Technologies for ultradynamic voltage scaling. *Proc. IEEE* **2010**, *98*, 191–214.
3. Zhai, B.; Nazhandali, L.; Olson, J.; Reeves, A.; Minuth, M.; Helfand, R.; Pant, S.; Blaauw, D.; Austin, T. A 2.60 pJ/Inst Subthreshold Sensor Processor for Optimal Energy Efficiency. In Proceedings of the 2006 Symposium on VLSI Circuits, Digest of Technical Papers, Honolulu, HI, USA, 15–17 June 2006; pp. 154–155.
4. Wang, A.; Chandrakasan, A. A 180-mV subthreshold FFT processor using a minimum energy design methodology. *IEEE J. Solid-State Circuits* **2005**, *40*, 310–319.
5. Slimani, M.; Silveira, F.; Matherat, P. Variability-Speed-Consumption Trade-off in Near Threshold Operation. In Proceedings of the 21st The International Workshop on Power and Timing Modeling, Optimization and Simulation (PATMOS), Madrid, Spain, 26–29 September 2011; pp. 308–316. doi:10.1007/978-3-642-24154-3_31.

6. Sugii, N.; Tsuchiya, R.; Ishigaki, T.; Morita, Y.; Yoshimoto, H.; Iwamatsu, T.; Oda, H.; Inoue, Y.; Hiramoto, T.; Kimura, S. Evaluation of threshold-voltage variation in silicon on thin buried oxide complementary metal–oxide–semiconductor and its impact on decreasing standby leakage current. *Jpn. J. Appl. Phys.* **2009**, *48*, 04C043.

7. Seo, S.; Dreslinski, R.G.; Woh, M.; Park, J.; Charkrabari, C.; Mahlke, S.; Blaauw, D.; Mudge, T. Process Variation in Near-Threshold Wide SIMD Architectures. In Proceedings of the 49th ACM/EDAC/IEEE Design Automation Conference (DAC), San Francisco, CA, USA, 3–7 June 2012; pp. 980–987.

8. Kwong, J.; Chandrakasan, A.P. Variation-Driven Device Sizing for Minimum Energy Sub-Threshold Circuits. In Proceedings of the 2006 International Symposium on Low Power Electronics and Design (ISLPED'06), Tegernsee, Germany, 4–6 October 2006; pp. 8–13.

9. Blaauw, D.; Zhai, B. Energy Efficient Design for Subthreshold Supply Voltage Operation. In Proceedings of the 2006 IEEE International Symposium on Circuits and Systems (ISCAS) 2006, Island of Kos, Greece, 21–24 May 2006; p. 32.

10. Bowman, K.A.; Duvall, S.G.; Meindl, J.D. Impact of Die-to-Die and within-Die Parameter Fluctuations on the Maximum Clock Frequency Distribution. In Proceedings of the 2001 IEEE International Solid-State Circuits Conference (ISSCC), Digest of Technical Papers 2001, San Francisco, CA, USA, 7 February 2001; pp. 278–279.

11. Datta, A.; Bhunia, S.; Mukhopadhyay, S.; Roy, K. Delay modeling and statistical design of pipelined circuit under process variation. *IEEE Trans. Comput. Aided Des. Integr. Circuits Syst.* **2006**, *25*, 2427–2436.

12. Tsuchiya, R.; Horiuchi, M.; Kimura, S.; Yamaoka, M.; Kawahara, T.; Maegawa, S.; Ipposhi, T.; Ohji, Y.; Matsuoka, H. Silicon on Thin BOX: A New Paradigm of the CMOSFET for Low-Power High-Performance Application Featuring Wide-Range Back-Bias Control. In Proceedings of the IEEE International Electron Devices Meeting, IEDM Technical Digest, San Francisco, CA, USA, 13–15 December 2004; pp. 631–634.

13. Morita, Y.; Tsuchiya, R.; Ishigaki, T.; Sugii, N.; Iwamatsu, T.; Ipposhi, T.; Oda, H.; Inoue, Y.; Torii, K.; Kimura, S. Smallest V_{th} Variability Achieved by Intrinsic Silicon on Thin BOX (SOTB) CMOS with Single Metal Gate. In Proceedings of the 2008 Symposium on VLSI Technology, Honolulu, HI, USA, 17–19 June 2008; pp. 166–167.

14. Sugii, N.; Tsuchiya, R.; Ishigaki, T.; Morita, Y.; Yoshimoto, H.; Kimura, S. Local V_{th} variability and scalability in Silicon-on-Thin-BOX (SOTB) CMOS with small random-dopant fluctuation. *IEEE Trans. Electron Devices* **2010**, *57*, 835–845.

15. Yamamoto, Y.; Makiyama, H.; Tsunomura, T.; Iwamatsu, T.; Oda, H.; Sugii, N.; Yamaguchi, Y.; Mizutani, T.; Hiramoto, T. Poly/high-k/SiON Gate Stack and Novel Profile Engineering Dedicated for Ultralow-Voltage Silicon-on-Thin-BOX (SOTB) CMOS Operation. In Proceedings of the 2012 Symposium on VLSI Technology, Honolulu, HI, USA, 12–14 June 2012; pp. 109–110.

16. Yamamoto, Y.; Makiyama, H.; Shinohara, H.; Iwamatsu, T.; Oda, H.; Kamohara, S.; Sugii, N.; Yamaguchi, Y.; Mizutani, T.; Hiramoto, T. Ultralow-Voltage Operation of Silicon-on-Thin-BOX (SOTB) 2 Mbit SRAM Down to 0.37 V Utilizing Adaptive Back Bias. In Proceedings of the 2013 Symposium on VLSI Technology, Kyoto, Japan, 11–13 June 2013; pp. T212–T213.

17. Mizutani, T.; Yamamoto, Y.; Makiyama, H.; Tsunomura, T.; Iwamatsu, T.; Oda, H.; Sugii, N.; Hiramoto, T. Reduced Drain Current Variability in Fully Depleted Silicon-on-Thin-BOX (SOTB) MOSFETs. In Proceedings of the 2012 IEEE Silicon Nanoelectronics Workshop (SNW), Honolulu, HI, USA, 10–11 June 2012; pp. 1–2.

18. Makiyama, H.; Yamamoto, Y.; Shinohara, H.; Iwamatsu, T.; Oda, H.; Sugii, N.; Ishibashi, K.; Yamaguchi, Y. Speed enhancement at V_{dd} = 0.4 V and random τ_{pd} variability reduction and analyisis of τ_{pd} variability of silicon on thin buried oxide circuits. *Jpn. J. Appl. Phys.* **2014**, *53*, doi:10.7567/JJAP.53.04EC07.

19. Makiyama, H.; Yamamoto, Y.; Shinohara, H.; Iwamatsu, T.; Oda, H.; Sugii, N.; Ishibashi, K.; Mizutani, T.; Hiramoto, T.; Yamaguchi, Y. Suppression of Die-to-Die Delay Variability of Silicon on Thin Buried Oxide (SOTB) CMOS Circuits by Balanced P/N Drivability Control with Back-Bias for Ultralow-Voltage (0.4 V) Operation. In Proceedings of the 2013 IEEE International Electron Devices Meeting (IEDM), Washington, DC, USA, 9–11 December 2013; pp. 822–825.

20. Morohashi, S.; Sugii, N.; Iwamatsu, T.; Kamohara, S.; Kato, Y.; Pham, C.-K.; Ishibashi, K. A 44 µW/10 MHz Minimum Power Operation of 50 K Logic Gate Using 65 nm SOTB Devices with Back Gate Control. In Proceedings of the 2013 IEEE SOI-3D-Subthreshold Microelectronics Technology Unified Conference (S3S), Monterey, CA, USA, 7–10 October 2013; pp. 165–166.

21. Su, H.; Amano, H. *Real Chip Evaluation of a Low Power Reconfigurable Accelerator with SOTB Technology*; Technical Report of The Institute of Electronics, Information and Communication Engineers (IEICE), RECONF2013-52; IEICE: Tokyo, Japan 2013; Volume 113, pp. 71–76 (In Japanese).

22. Hioki, M.; Ma, C.; Kawanami, T.; Ogasahara, Y.; Nakagawa, T.; Sekigawa, T.; Tsutsumi, T.; Koike, H. The First SOTB Implementation of Flex Power FPGA. *J. Low Power Electron. Appl.* Submitted.

23. Nagatomi, H.; Le, D.-H.; Pham, C.-K.; Sugii, N.; Kamohara, S.; Iwamatsu, T.; Ishibashi, K. A 4 pA/Gate Sleep Current 65 nm SOTB Logic Gates Using On-chip VBB Generator for Energy Harvesting Sensor Network Systems. In Proceedings of the 2013 International Conference on Integrated Circuits, Design, and Verification (ICDV 2013), Ho Chi Minh City, Vietnam, 15–16 November 2013.

24. Ishibashi, K.; Sugii, N.; Usami, K.; Amano, H.; Kobayashi, K.; Pham, C.-K.; Makiyama, H.; Yamamoto, Y.; Shinohara, H.; Iwamatsu, T.; *et al.* A Perpetuum Mobile 32 bit CPU with 13.4 pJ/cycle, 0.14 µA Sleep Current using Reverse Body Bias Assisted 65 nm SOTB CMOS Technology. In Proceedings of the COOL Chips XVII, Yokohama, Kanagawa, Japan, 14–16 April 2014.

7

An Ultra-Low Energy Subthreshold SRAM Bitcell for Energy Constrained Biomedical Applications †

Arijit Banerjee * and Benton H. Calhoun

The Charles L. Brown Department of Electrical and Computer Engineering, University of Virginia, Charlottesville, VA 22904, USA; E-Mail: bhc2b@virginia.edu

† The original of this paper had been presented in IEEE S3S Conference 2013.

* Author to whom correspondence should be addressed; E-Mail: ab9ca@virginia.edu

Abstract: Energy consumption is a key issue in portable biomedical devices that require uninterrupted biomedical data processing. As the battery life is critical for the user, these devices impose stringent energy constraints on SRAMs and other system on chip (SoC) components. Prior work shows that operating CMOS circuits at subthreshold supply voltages minimizes energy per operation. However, at subthreshold voltages, SRAM bitcells are sensitive to device variations, and conventional 6T SRAM bitcell is highly vulnerable to readability related errors in subthreshold operation due to lower read static noise margin (RSNM) and half-select issue problems. There are many robust subthreshold bitcells proposed in the literature that have some improvements in RSNM, write static noise margin (WSNM), leakage current, dynamic energy, and other metrics. In this paper, we compare our proposed bitcell with the state of the art subthreshold bitcells across various SRAM design knobs and show their trade-offs in a column mux scenario from the energy and delay metrics and the energy per operation metric standpoint. Our 9T half-select-free subthreshold bitcell has 2.05× lower mean read energy, 1.12× lower mean write energy, and 1.28× lower mean leakage current than conventional 8T bitcells at the TT_0.4V_27C corner. Our bitcell also supports the bitline interleaving technique that can cope with soft errors.

Keywords: subthreshold; SRAM; half-select; half-select-free; 9T; bitcell; ultra-low-energy; biomedical; minimum energy point; energy per operation

---

1. Introduction

Portable biomedical devices requiring long-term data processing have stringent energy requirements. This includes portable electrocardiograms (ECG), electromyograms (EMG), and electroencephalograms (EEG) type devices that can process critical disease related data at operating frequency ranging from a few hundred kHz to a few MHz [1,2]. These devices impose energy constraints on biomedical system on chip (SoC) components and SRAM design. Due to the square law dependency of energy with supply voltage, scaling down the supply voltage reduces energy in logic and SRAMs in SoCs. In a CMOS process, reducing the supply voltage below the threshold voltage (V_T) of the MOSFET makes it enter into the subthreshold region. Prior works have shown that operating both logic and memory in subthreshold supply voltages reduces energy dissipation and minimizes energy per operation [3,4]. Although voltage-scaling increases delay in logic and SRAMs, subthreshold logic and SRAMs provide enough performance to meet the throughput requirements for the biomedical devices.

On the other hand, due to device variations in subthreshold SRAMs, the conventional 6T bitcell has poor read static noise margin (RSNM) [5] and is unreliable for subthreshold operation. There are many proposed subthreshold bitcells [6–9] present in the literature having some improvement in write-ability and read stability related design metrics by trading-off other metrics. However, subthreshold bitcells such as the 8T [10] bitcell, face half-select [8] problems in a (column) mux scenario, which can cause read-disturb and unnecessary energy drainage during a write operation. This imposes further constraints on usage of write assists such as the boosted wordline [11,12] due to degraded read stability in half-selected [12] bitcell.

In order to avoid this half-select problem, we either can implement read-before-write operation [7,13] instead of normal write in SRAMs or we can design half-select-free SRAM bitcells [7–9,14] that decouple read and write operations. However, implementing read-before-write SRAM architectures in subthreshold supply voltages can be a more complex and time-consuming task than designing a simple column mux based SRAM design. On the other hand, in subthreshold memories, soft error disturbs (SED) are critical [15], and bitline interleaving in memory architecture [7] uses column multiplexing to improve on SED.

Given all these subthreshold bitcells, their design trade-offs and various architecture related issues, we compare our subthreshold bitcell with available subthreshold bitcells in a column mux scenario. In this work, we assume that applying appropriate peripheral read and write assist methods [11,12] can solve read stability [12] and write-ability [11] related known issues in subthreshold SRAM bitcells with less penalty in energy per operation and area standpoint in an SoC. In addition, we also assume that we can trade-off SRAM area for better energy efficiency, since it is of less importance in biomedical applications. In this simulation-based paper, we compare our bitcell to the state of the art subthreshold SRAM bitcells from various SRAM design knob perspectives across a set of design metrics for biomedical applications. The rest of this paper is divided into seven sections. In Section 2, we introduce the state of the art subthreshold SRAM bitcell topologies. Section 3 talks about the limitations of the available subthreshold bitcells including the half-select issue. In Section 4, we introduce our half-select-free 9T subthreshold SRAM bitcell. Section 5 describes the concept of minimum energy per operation and read-write weighted energy per operation. In Section 6, we describe the experimental setup for comparison of the subthreshold SRAM bitcells. Section 7 presents the comparison results for various SRAM design knobs, and we conclude in Section 8.

2. Subthreshold Bitcells Topologies

The conventional 6T SRAM bitcell shown in Figure 1a is the most used bitcell topology in SRAMs. It has two back-to-back inverters, which act as a latch for storing logic "1" in one side, and "0" in the other side. There are two access transistors in the bitcell for both reading and writing. However, due to the poor read static noise margin [5] (RSNM) and half-select [12] issue, the 6T bitcell is not robust in subthreshold supply voltages. Almost all other SRAM bitcells, including subthreshold bitcells, are modified versions of the 6T. The most common subthreshold SRAM bitcell derived from the 6T is the conventional 8T subthreshold bitcell [10] with a 2T read buffer as shown in Figure 1b. The 2T read buffer senses the information stored in the bitcell in read operation. The conventional 8T allows decoupled read and write operations, which enable us to size the read and write path differently. This adds another knob for energy efficient design exploration with the 8T bitcell. Another subthreshold bitcell with reportedly lower minimum operating voltage (V_{MIN}) is the Schmitt-trigger based bitcell [6] (Figure 1c). This bitcell uses the hysteresis property of a Schmitt-trigger to strengthen the read operation still allowing lower V_{MIN}. Although the conventional subthreshold 8T and Schmitt-trigger based bitcells are robust in read and write operations, they are costly from an energy standpoint if used in a column mux scenario. This is due to the inherent half-select problem [12] in the 8T and Schmitt-trigger based bitcells in a write operation. There are many half-select-free bitcells available in the literature like Chang's 10T [7] (Figure 1d), Feki's 10T [8] (Figure 2a), and Chiu's 8T [9] shown in Figure 2b. Although Yang's 8T [14] (Figure 2c) is not mentioned as a subthreshold bitcell, due to structural symmetry with Chiu's bitcell, we include this bitcell for comparison in this paper. In most of the cases, these half-select-free bitcells have two separate wordlines for read and write. This allows us to size the read and write path independently such as Feki's bitcell. On the other hand, Chang's, Yang's and Chiu's bitcell has common read or write nodes, which prevents sizing their read and write paths independently. All of this work has shown some improvements in read stability, write-ability, V_{MIN}, and leakage metrics. In this paper, we show how our bitcell compare with state of the art subthreshold SRAM bitcells from the energy and delay metrics and the energy per operation metric perspective across various SRAM design knobs. As subthreshold designs are better suited to be designed with a lower leakage technology, we prefer an older technology for subthreshold design. We compare the bitcell using a commercial 130 nm technology in a typical typical corner (TT) as the 130 nm technology is very stable nowadays and it is available to us. As the applications targeted in this work are biomedical applications for Body Area Sensor Node (BASN) [1] applications, we use room temperature conditions of 27 °C for the comparion simulations in this paper.

Figure 1. (**a**) 6T SRAM bitcell; (**b**) Conventional 8T SRAM subthreshold bitcell; (**c**) Kulkarni's schmitt-trigger based subthreshold SRAM bitcell; (**d**) Chang's 10T subthreshold bitcell.

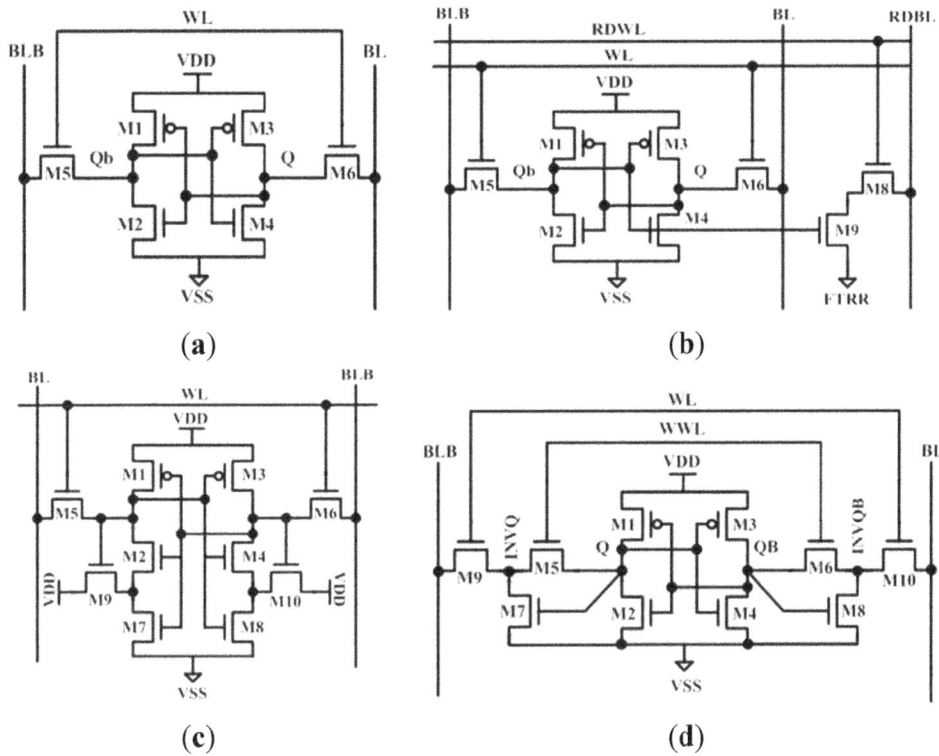

Figure 2. (**a**) Feki's 10T SRAM subthreshold bitcell; (**b**) Chiu's 8T subthreshold bitcell; (**c**) Yang's 8T SRAM bitcell.

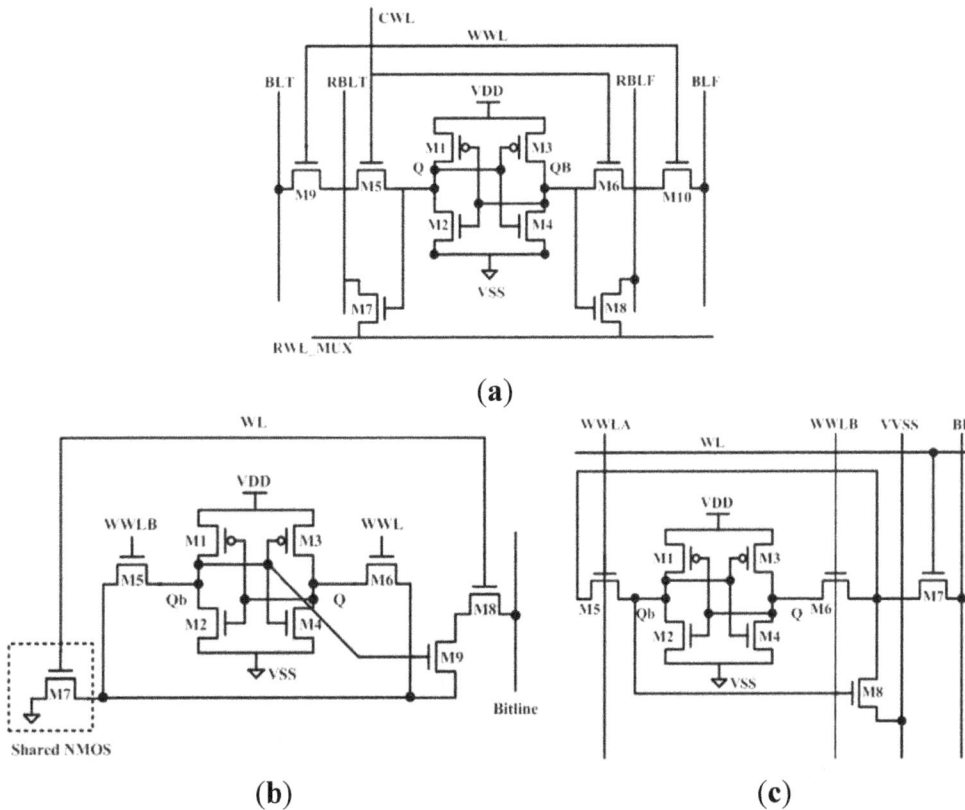

3. Limitations of Available Bitcells

Bitcells discussed in the previous section are not free from drawbacks. Although, Kulkarni's bitcell has the lowest reported V_{MIN}, it can consume more dynamic and leakage energy due to its Schmitt-trigger based feedback structure. The Schmitt-trigger based feedback structure uses the additional transistors M9 and M10 (Figure 2c) to strengthen the internal storage node resulting in higher dynamic energy dissipation and creates a greater number of source or sink paths causing more leakage current. Secondly, Kulkarni, Feki and Chang's bitcells have a 10T structure that inherently should burn more dynamic energy as those bitcells have more transistors than the 8T, Chiu and Yang's bitcells. This is due to the assumption that we size all the bitcells with respect to a set of common reference design metrics and thus extra transistors in bitcells will add to the increase of dynamic energy. Moreover, we can see that Chang's bitcell adds more leakage paths by introducing transistors M7 and M8 creating two additional leakage paths from bitline to ground, such as paths BLB-M9-M7-VSS, BL-M10-M8-VSS. In addition, assuming that the bitcell back-to-back inverter sizes are the same and each of the control signals such as wordlines have the same activity factor, bitcells having multiple wordlines as control signals—such as Chang's, Feki's, and Yang's bitcells—should drain more dynamic energy than bitcells having fewer wordline control signals triggered per read or write operation. Thirdly, bitcells those use the same path for read and write operations—such as Kulkarni's, Chiu's, and Chang's bitcells—should experience energy consumption due to precharging bitlines after the end of both read and write cycles. Moreover, in the column mux scenario, unselected bitcells in the same row are half-selected and they experience read stress in the write operation. Not only Kulkarni's bitcell, but also conventional subthreshold 8T bitcells such as Chiu and Yang's suffer from the half-select problem in the write operation. Hence, in order to capture all the aforementioned potential sources of energy dissipation, we need to simulate all these bitcells in a column mux scenario where all of these effects are taken into consideration.

3.1. SRAM Half-Select-Issue in Write Operation

Figure 3 shows the half-select problem in the presence of a column mux (CM) 4 in SRAM write operation. Here, our assumption is that the SRAM has multiple banks. Each of the SRAM banks has the same sized core array comprised of subthreshold bitcells. It shows that in column mux 4 scenarios, every four-bitcell columns constitute a single I/O column, which has precharge logic, read and write column muxes, write driver, and read logic. When a user asserts an SRAM address, it selects a word in an SRAM row by selecting one of the bank's physical rows and multiple physical columns. For example, if the user selects the first word in the row, then it selects only the first physical bitcell column of every four physical bitcell columns. Other bitcells in the same row being row wise selected but column wise unselected are half-selected bitcells. In write operation, these half-selected bitcells undergo read stress as if they are in a read operation, and it causes unnecessary energy drainage. Another potential issue with the half-select problem is that using wordline boost type write-assist [11,12] for write-ability improvement can cause the half-selected bitcells to have destructive read. In other words, applying a wordline-boost type write assist can cause the half-selected bitcells to flip. However, it is easy to implement column mux based SRAM architectures as the complexity of this type of designs are much less compared to read-before-write [7,13] subthreshold SRAM architectures. Another way of avoiding the half-select issue is to use a half-select-free [9] subthreshold SRAM bitcell. With this type of bitcells, column mux based designs are easy to implement. However, proposed half-select-free bitcells have more devices and control signals, which can cause unnecessary dynamic and leakage energy drainage. Moreover, half-select-free subthreshold bitcells have shared nodes in read and write paths which can cause sizing issues.

Figure 3. Half-select problem in SRAM bitcells in column mux (CM) 4 scenario.

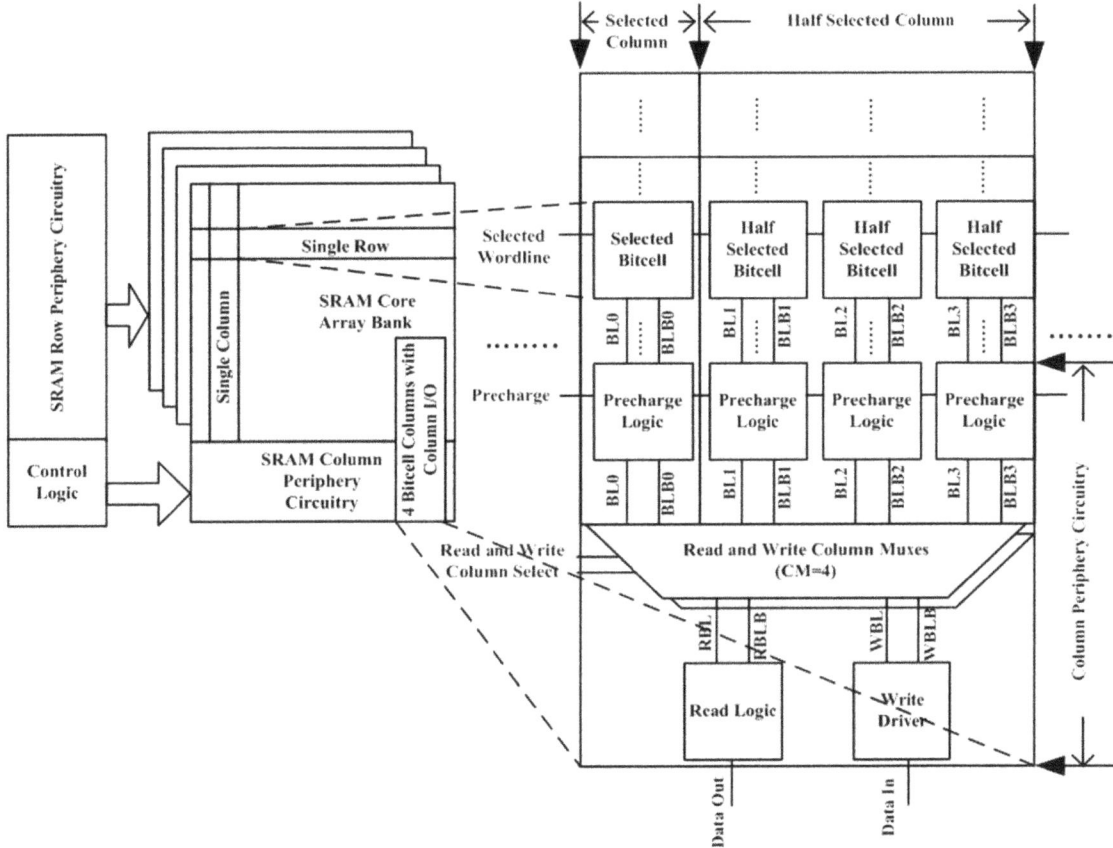

4. Proposed 9T Bitcell

The proposed bitcell [16] is shown in Figure 4a ($W_{M1, M3} = 0.4u$, $L_{M1, M3, M5, M6, M7} = 0.22u$, $W_{M2, M4} = 0.28u$, $L_{M2, M4, M8, M9} = 0.15u$, $W_{M5, M6, M7} = 0.45u$, and $W_{M8, M9} = 0.36u$). We choose the bitcell's back-to-back inverter (M1, M2, M3 and M4) and two transistor read-buffer transistor (M8, M9) sizes as per a reference conventional subthreshold 8T bitcell size used in subthreshold SRAMs in a Body Area Sensor Node Chip (BASN) chip [1] at University of Virginia. We run Monte Carlo simulations for write margin, HSNM, read time, *etc.* design metrics to choose the transistor gate widths and lengths as non-minimum to cope with the subthreshold process variations. The bitcell consists of a set of back-to-back inverters like 6T, and a differential amplifier-like structure used for write access. For reading, we use a two transistor read buffer like the 8T read buffer [10]. During the write operation, only one of the write bitlines WBL or WBLB goes high (Figure 4a), while the other remains low. Meanwhile, the write-wordline WWL becomes high, and if the corresponding internal node was storing "1", it is discharged through the write path. In write operation, pulling down WBL and WBLB nets of half-selected bitcells in the same SRAM row prevents the half-select issue. As our bitcell does not require any precharge operation of the write bitlines, it consumes less dynamic energy. In case of a read operation, we precharge the read bitline RBL to V_{DD}, and trigger the read wordline RWL to evaluate the read (Figure 4a). Here, node Qb is the reference node in the read operation. If the content of the bitcell is such that Qb holds logic "1", the RBL discharges, and this denotes a read "1" operation. Otherwise, if Qb holds logic "0", RBL stays at V_{DD}, which denotes a read "0" operation. In order to reduce standby leakage, the signals FTRR and FTRW only go to V_{DD} while in standby mode, and they do not toggle in normal read or write operations. Hence, the default states of FTRR and FTRW are logic "0" s. In this 130 nm technology, we report the leakage improvement by pulling the FTRR and FTRW to V_{DD} compared to leaving it at V_{SS} is 34% at TT_0.4V_27C corner. This technique of pulling the footer to V_{DD} can save significant leakage energy in lower technologies [10] too. Figure 4b shows the waveforms for read and write operations.

Figure 4. (**a**) Schematic of the proposed 9T bitcell; (**b**) Read (Rd)/Write (Wr) waveforms of proposed 9T bitcell.

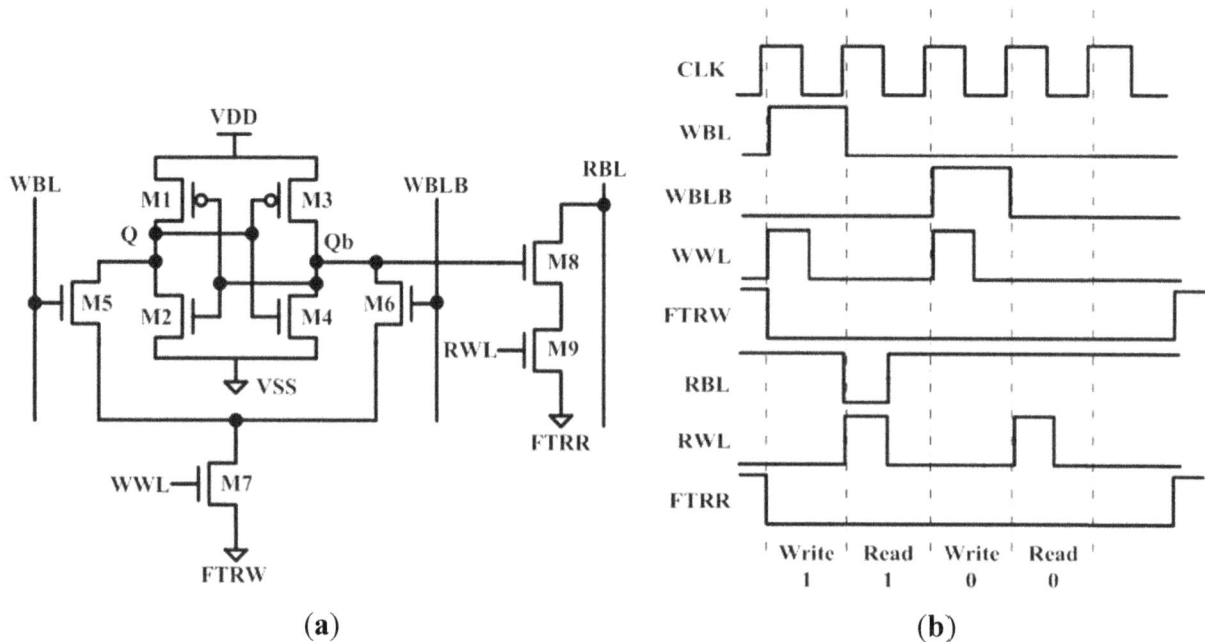

(**a**)   (**b**)

5. Minimum Energy per Operation of Subthreshold SRAMs

In subthreshold supply voltages, delay increases exponentially with decreasing supply voltage. Due to this fact, in subthreshold SRAMs, the leakage energy per operation also increases exponentially with decreasing supply voltage. On the other hand, the SRAM dynamic energy per operation decreases with supply voltage scaling in SRAMs as shown in Figure 5. As a resulting effect, the total energy of the SRAM has a minimum energy point (Figure 5) which lies within the subthreshold supply voltage region. In fairly bigger subthreshold SRAMs, the core array contributes to most of the leakage energy per operation. As the SRAM core array bitlines have higher capacitance due to the presence of multiple bitcells in a column, a significant amount of dynamic energy can come from the array itself. In order to lower the SRAM minimum energy supply voltage point, we can either increase the dynamic energy per operation of the SRAM keeping the leakage energy per operation fixed. Otherwise, we can lower the leakage energy per operation keeping the dynamic energy fixed. In the first case, the minimum energy point (MEP) will shift corresponding to a lower supply voltage, but the energy per operation will increase. However, if we lower leakage energy per operation, we can get two-fold benefit of lowering MEP as well as lowering MEP supply voltage. On the other hand, if we lower the leakage and dynamic energy per operation at the same rate, the MEP supply voltage can remain the same; however, it reduces the MEP itself.

Figure 5. Minimum energy point of SRAMs.

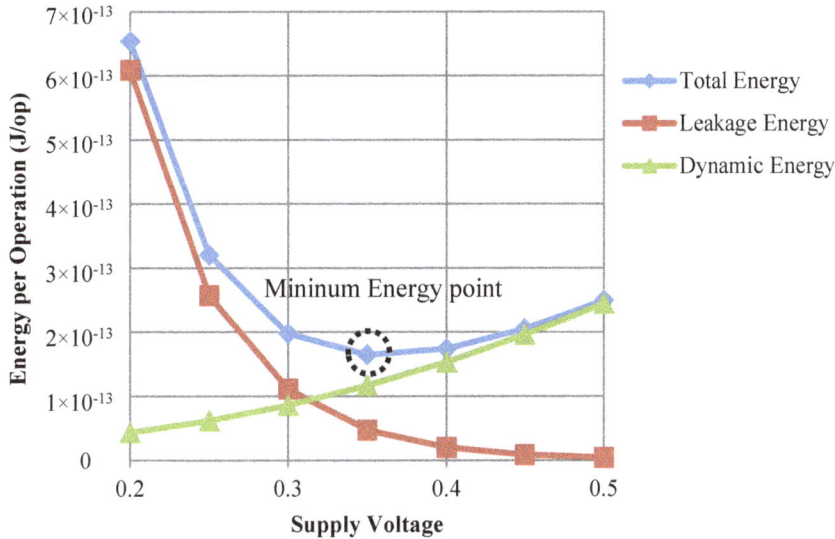

5.1. Read-Write Weighted Energy per Operation and Fraction of Read and Write

In SRAMs, usually we do more read operations than write. In order to get an equivalent minimum energy point (MEP), we have to weigh the read and write energy per operations accordingly to get the read-write weighted energy per operation. We express this weighted average energy per operation as Equation (1).

$$E_{\text{avgop}} = E_{\text{wr}} \times (1 - F_{\text{rdwr}}) + E_{\text{rd}} \times F_{\text{rdwr}} \tag{1}$$

Here, the parameter E_{avgop} denotes read-write weighted energy per operation; E_{wr} and E_{rd} are the write and read energy per operation, respectively. In Equation (1), the parameter F_{rdwr} is the fraction of read and write that denotes how many read operations on average are there out of total number of read-write operations. It is noticeable that if the E_{rd} is lower than the E_{wr}, increasing the F_{rdwr} parameter decreases the weighted energy per operation.

6. Experimental Setup

We do all our experiments in a commercial 130 nm technology at the TT_27C corner using Cadence's Spectre simulator. For the mismatch analysis, we run 1000 Monte Carlo simulations for each comparison at $V_{\text{DD}} = 0.4$ V. We perform two sets of experiments: one based on the experimental setup shown in Figure 6, where except from the actual drivers, we use voltage sources as input waveforms for comparisons of the energy and delay numbers. On the other hand, for comparison of the energy per operation metric, and to get the minimum energy point (MEP) data, we use the experimental setup shown in Figure 6a,b. Here, "WL" stands for wordline, "PREB" stands for precharge bar, and "WRITE_EN" stands for write enable. We use extracted inverter netlist from a standard cell library for the use of drivers for wordline, bitline and write enable, *etc.* signals. This will ensure that the rise and fall time of the buffer and inverter outputs to the wordlines, precharge enable, write enable, *etc.* signals are realistic in the subthreshold supply voltages. In each case of write and read setup shown in Figure 6a,b, we have two columns: the leftmost column represents the actual column for write or read setup, which has total modeled bitline load shown as rows per bank (RPB) times a single bitcell bitline load. On the other hand, the second column models the wordline load, which is column mux factor (CM) times a single bitcell wordline load. Overall, the setup models "RPB × CM" number of bitcells per set of physical bitcell-columns associated with a single column mux of an SRAM bank for dynamic energy measurement. For example, in an SRAM bank with CM = 4, we would have a set of four physical bitcell columns associated with each column mux 4 and with RPB = 16. In this case, our setup models the dynamic energy consumption of the set of four bitcell columns for 16 × 4 = 64 bitcells per column mux 4. For generating different energy, delay and MEP values, we simulate multiple instances with RPB = 4, 8, 16, 32 and 64 values. In order to generate dynamic energy values for different word-widths, we multiply set of columns' (consisting of multiple bitline columns as per column mux) dynamic energy values across SRAM word-widths of 2, 4, 8, 16, and 32. For extracting the leakage numbers, we use single bitcells' netlist with single voltage source for each circuit. We use these bitcell leakage values to generate the corresponding leakage values of the full SRAM core arrays. Finally, we added the modeled dynamic and leakage energy values for each SRAM macro to get the total energy values for minimum energy point calculations. We limit the observation of memory sizes from 2 to 32 KB range since prior works reported a similar range around 5–46 KB [1,2,17,18] of memory usage in biomedical SoCs.

Figure 6. (a) Experimental setup for dynamic write energy measurement for subthreshold SRAM bitcells in a column mux scenario; (b) Experimental setup for dynamic read energy measurement for subthreshold SRAM bitcells in a column mux scenario.

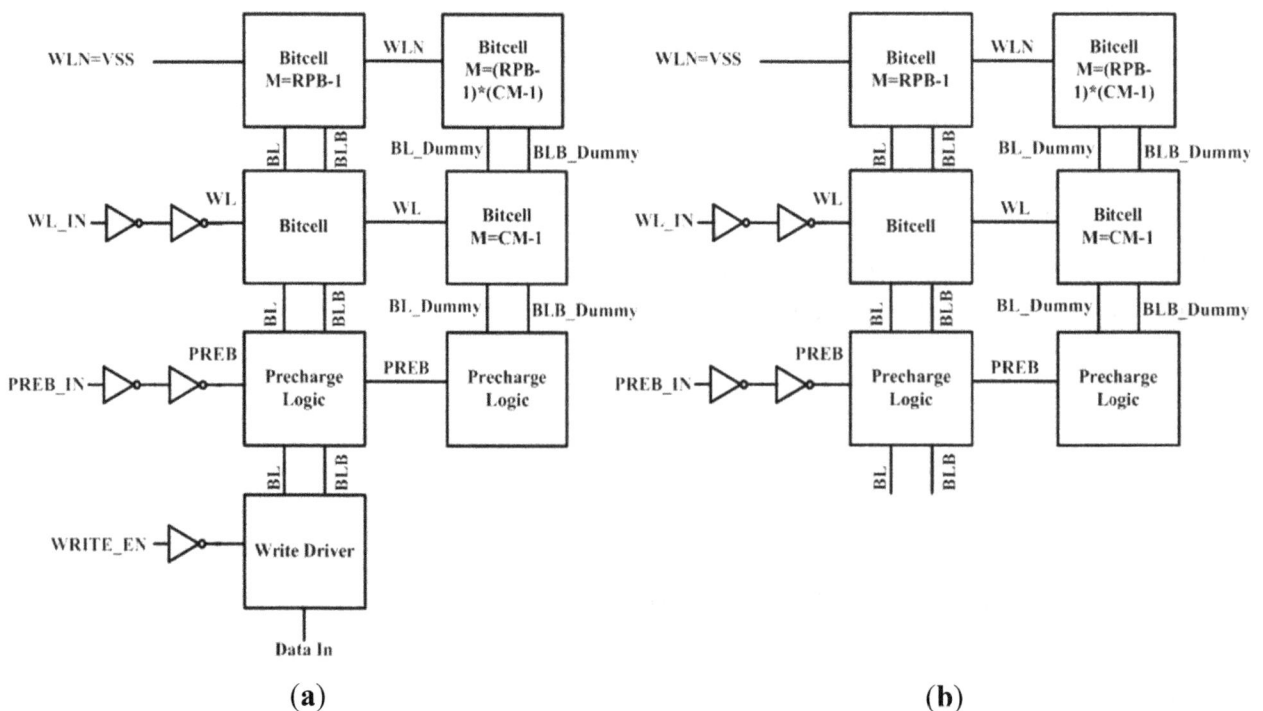

(a)   (b)

In order to determine which bitcell is more energy efficient we need to quantify the total energy per operation and the minimum energy point metrics with some assumptions. In a realistic scenario, we not only have bitcell arrays in SRAM, but also we have periphery drivers for wordline and bitline, precharge logic, and control logic *etc.* circuits. Hence, in order to make a fair estimate of the bitcells' minimum energy points we consider the bitcells having some drivers and periphery circuits that would be switching. We use the same driver stages for wordlines across all the bitcells and same driver stages for bitlines for most of the bitcells. The bitcells requiring a pull-down type write driver have the same write driver circuits. On the other hand, for pull-up type write driver for this work, we incorporate comparable strength buffers. In case of bitcells requiring precharge cycles, we include a precharge circuit (Figure 6a,b). It is obvious that the bitcells that require multiple wordlines for read or write operation or an extra precharge operation will consume higher dynamic energy due to overhead in peripheral circuits. However, the core arrays may have more leakage energy than the periphery. Hence, we repeat our experiment to get the dynamic energy per operation and leakage energy per operation as well as the total energy per operation for each bitcell array with the assumed periphery.

6.1. Experimental Assumptions

In this paper, we assume that all the read operations are full swing. Hence, we do not use sense amplifier in read operation for these experiments. Our model (Figure 6a,b) in a column mux scenario considers the energy consumption in bitlines and wordlines for a set of bitcell columns. We assume that the core array is sufficiently bigger, and its minimum energy point (MEP) will contribute most to the MEP in this experimental setup of modeled SRAM macros. Further inclusion of the actual control logic, pre-decoder and wordline drivers with the core array in a real SRAM scenario will affect the MEP trends accordingly as per the periphery energy consumption. However, in this paper we are interested in comparing the core MEP trend of all the bitcells' modeled SRAM macros assuming that the periphery and its MEP are same for all the cases.

7. Results and Comparisons

In this section, initially we discuss and compare the results of the energy and delay numbers of the bitcells, and later we move on to the comparisons from energy per operation perspective. In order to do a fair comparison, we size the 6T structures (back-to-back inverters: M1, M2, M3, M4 (Figures 1 and 2) and two NMOS pass transistors: M5 and M6 (Figures 1 and 2) which are the same in all the aforementioned bitcells ($W_{M1, M3}$ = 0.4u, $L_{M1, M3, M5, M6, M7}$ = 0.22u, $W_{M2, M4}$ = 0.28u, $L_{M2, M4, M8, M9}$ = 0.15u, $W_{M5, M6, M7}$ = 0.45u). Due to this reason, for all the bitcells under local and global variations we make the μ data retention voltage (DRV) nearly 74 mV, and the μ hold static noise margin (HSNM) roughly equal to 154 mV at the TT_0.4V_27C corner. As the bitcells have different read and write paths, it is hard to size them same with respect to multiple design metrics. However, we tried to make the bitcells' read and write paths similar. Apart from the M1–M6 being sized the same for all the bitcells, we size the $W_{M8, M9}$ = 0.36u and $L_{M8, M9}$ = 0.15u for conventional 8T, this work, and Chiu's bitcell, $W_{M7, M8}$ = 0.36u and $L_{M7, M8}$ = 0.15u for Chang's, Feki's and Yang's bitcell, $W_{M9, M10}$ = 0.45u and $L_{M9, M10}$ = 0.22u for Chang's and Feki's bitcell, W_{M7} = 0.45u and L_{M7} = 0.22u for Chiu's bitcell and $W_{M7, M8, M9, M10}$ = 0.16u and $L_{M7, M8, M9, M10}$ = 0.12u for Kulkarni's bitcell. For capturing unnecessary energy drainage, we constructed a 4 × 4 modeled array (Figure 6a,b) without the drivers for wordline, *etc.*) using RPB = 4 and column mux factor (CM) = 4. This model is similar to a 4 × 4 array in the presence of 4:1 column mux, which reveals the dynamic energy loss due to the effect of half-select problem and signal toggling. Comparing with the half-select-free bitcells, the mean read energy of this work is 3.18× lower than Chang's [7], 2.52× lower than Feki's [8], 2.05× lower than 8T [10], and 5.6% lower than Yang's [14]. On the other hand, the mean write energy of this work is 348× lesser than Chang's [7], 149× lower than Yang's [14], 1.12× lesser than 8T [10], and 2.4% lower than Feki's [8] at the TT_0.4V_27C corner with a column mux (CM) 4 in the worst case scenario. We report that the mean leakage current at the same corner is 1.28× lower than the 8T [10] bitcell (Table 1). However, our bitcell has 50% higher read time, and 7× higher write time compared to the conventional 6T at the same corner. Figure 7a–c and Table 1 show the comparison of the bitcells across voltages (0.2–0.5 V), and in the presence of statistical variations at the TT_0.4V_27C corner, respectively.

Figure 7. (**a**) Bitcell read time and total read energy (semi-log scale) *vs.* supply voltage at TT_27C corner; (**b**) Bitcell write time and total write energy (semi-log scale) *vs.* supply voltage at TT_27C corner; (**c**) Bitcell standby leakage current *vs.* supply voltage at TT_27C corner.

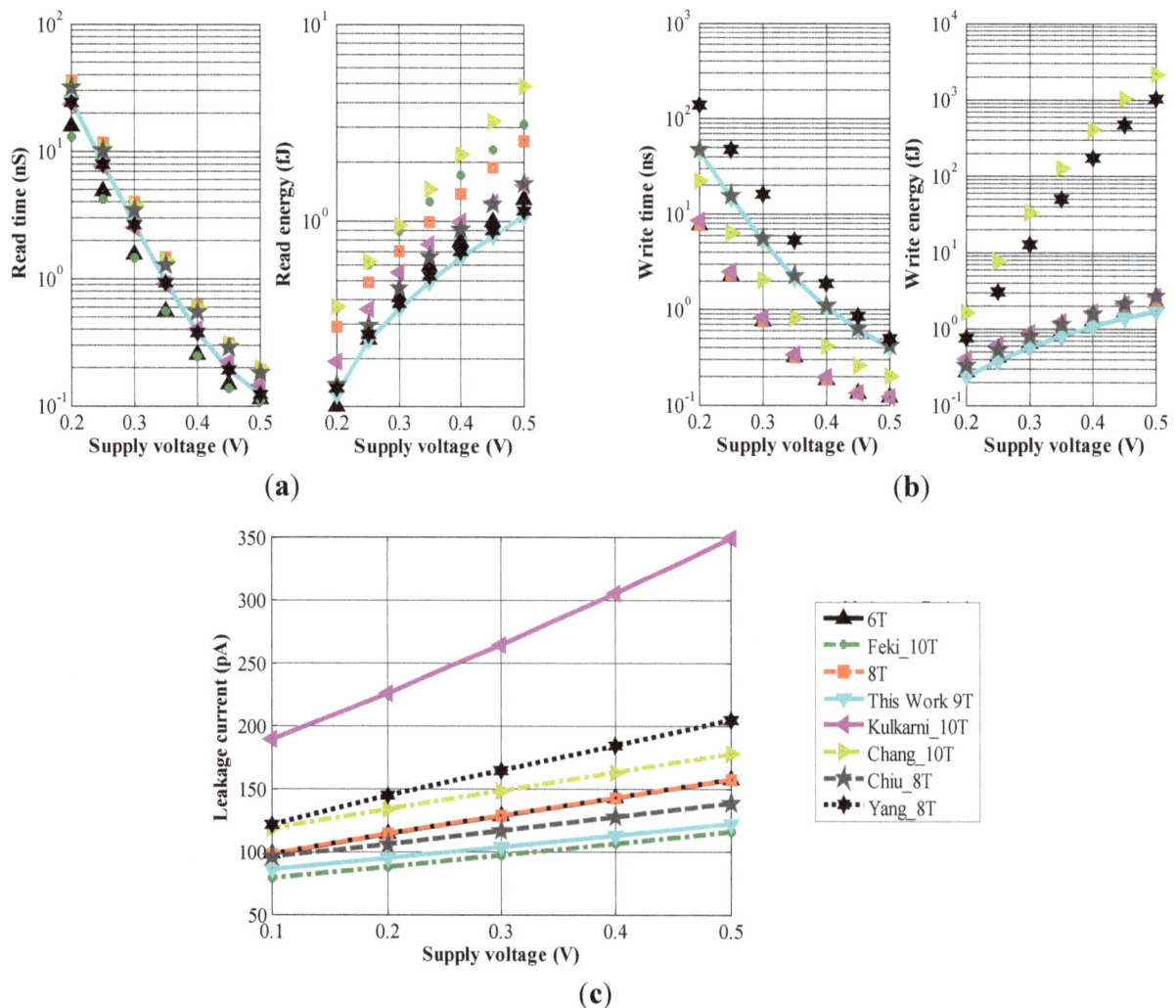

Table 1. Monte Carlo data comparison of bitcell design metrics at TT_0.4V_27C corner (energy in fJ, time in ns and current in pA units).

Metrics	6T	8T [10]	10T [8]	10T [6]	This work	10T [7]	8T [9]	8T [14]
Read time (μ)	0.30	0.73	0.28	0.48	0.45	0.69	0.65	0.45
Read energy (μ)	0.82	1.46	1.79	1.19	0.71	2.26	0.96	0.75
Write time (μ)	0.19	0.20	0.47	0.26	1.33	0.46	1.39	3.24
Write energy (μ)	1.35	1.36	1.24	1.98	1.21	421.71	1.69	180.67
Leakage current (μ)	187.8	188.2	136.1	468.4	146.1	211.8	161.9	245.3

7.1. Comparison of Total Energy per Operation

Figure 8a,b show total energy *vs.* supply voltage plots and minimum energy points (MEP) for the bitcells with column mux (CM) = 4 and RPB = 16. We generate this plot using the assumption that per four read-write operations, we have three reads and one write, which means that our value of fraction of read and write (F_{rdwr}) is 0.75. We can see that for most of the 8 KB SRAMs, the MEP supply voltage is around 0.3 V, and for most of the 32 KB SRAMs, this MEP supply voltage is around 0.35 V. There are two exceptions to this fact: Chang's bitcell does not have a minimum energy point within 0.2–0.5 V range in both the cases. This is because Chang's bitcell has much higher dynamic energy per operation in the subthreshold region compare to the leakage energy per operation than other bitcells (Figure 7a,b). We report 0.2 V as the MEP point for Chang's bitcell for bigger SRAM macros since it does not have an MEP within the 0.2–0.5 V region. On the other hand, although, Yang's bitcell has much higher MEP compare to other bitcells, its MEP supply voltage (V_{DD}) is around 0.25 V which is 16.66% lower than most of the bitcells' MEP V_{DD} (Figure 8a) in 8 KB SRAM and 28.57% lower than most of the bitcells' MEP V_{DD} (Figure 8b) in 32 KB SRAM.

Figure 8. (**a**) Total energy *vs.* supply voltage of 8 KB SRAMs (CM = 4, RPB = 16); (**b**) Total energy *vs.* supply voltage of 32 KB SRAMs (CM = 4, RPB = 16).

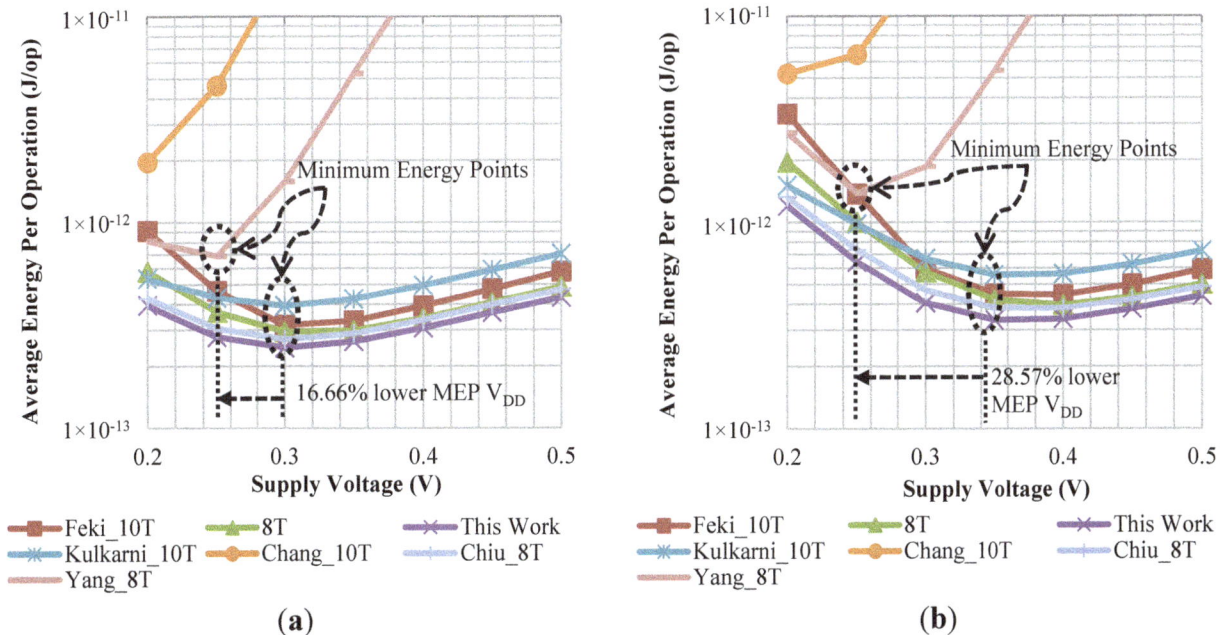

(**a**)   (**b**)

7.2. MEP vs. Fraction of Read and Write and Comparison Results

In order to observe the effect of F_{rdwr} on minimum energy point, we vary the value of F_{rdwr} in Equation (1) and plot the MEP *vs.* F_{rdwr} and MEP supply voltage *vs.* F_{rdwr} in Figure 9a,b with CM = 4. We can see that from Figure 9a that increasing the F_{rdwr} results in a decrease in weighted minimum energy points in all bitcells for 32 KB SRAMs with 16 rows per bank (RPB). It is also noticeable that with the increase of F_{rdwr} the slope of the MEP *vs.* F_{rdwr} changes more or less the same except for Chang's bitcell, which has much slower slope changes than other bitcells. We report a 49.5% decrease in MEP for this work (Figure 9a) as the F_{rdwr} increases from 0.5 to 0.9. This is because the read energy per operation of this work is much lower than the write energy per operation and weighing more in read energy per operation lowers the weighted MEP point. There is no clear trend observable from the MEP supply voltage *vs.* F_{rdwr} plot among the bitcells (Figure 9b). However, for Chiu's and our bitcell, the MEP supply voltage remains constant from F_{rdwr} = 0.6–0.8 at 0.45 V. On the other hand, Yang's and Chang's bitcell also shows constant MEP supply voltages across F_{rdwr} = 0.6–0.9. On the contrary, Feki's bitcell shows a linearly 20% decrease in MEP supply voltage from F_{rdwr} = 0.6–0.8. We also report that Chang's bitcell has 16.66% lower MEP supply voltage than Yang's bitcell from F_{rdwr} = 0.6–0.9. From Figure 9a,b, we can say that although Chang's and Yang's bitcell has much higher MEP, due to lower MEP supply voltages, it is suitable for bigger subthreshold SoCs having comparable energy per operation with a higher number of logic cells.

Figure 9. (**a**) Minimum energy point *vs.* fraction of read and write (F_{rdwr}) for 32 KB SRAM (CM = 4, RPB = 16); (**b**) MEP supply voltage *vs.* fraction of read and write (F_{rdwr}) for 32 KB SRAM (CM = 4, RPB = 16).

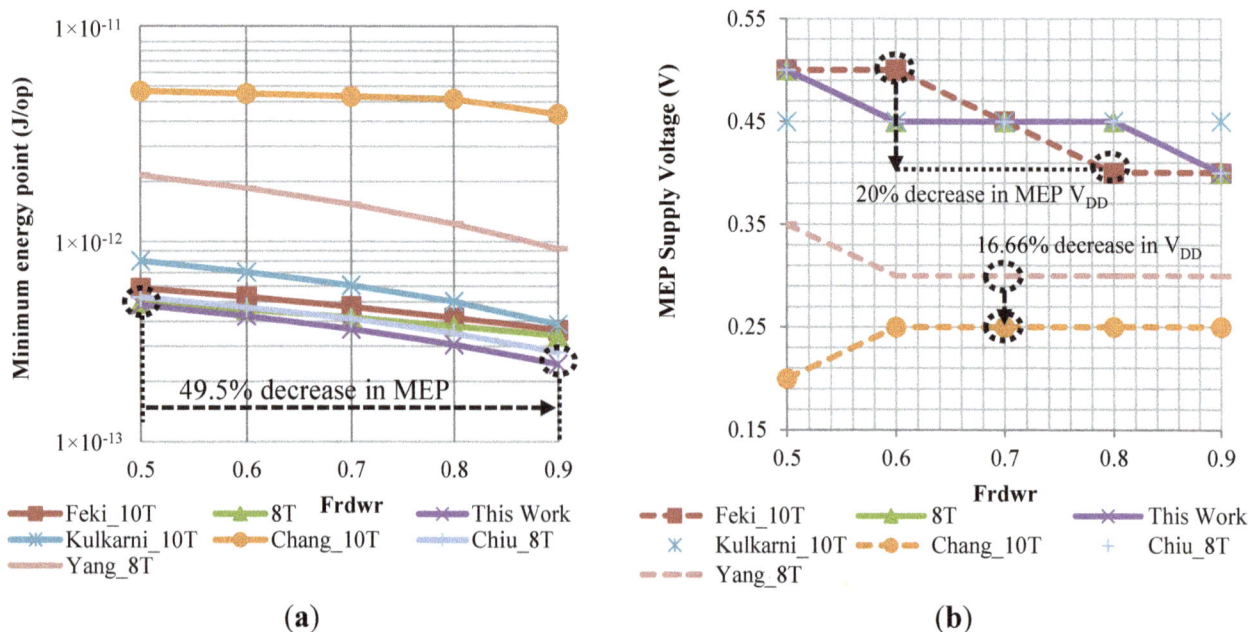

7.3. MEP vs. Number of Bitcell Rows per Bank Comparison Results

Figure 10a shows the variation of MEP with the number of bitcell rows per bank (RPB) for 32 KB SRAMs with CM = 4. This experiment uses a fixed SRAM macro size of 32 KB with word-width being fixed at word-width = 32 in a column mux 4 configuration. In order to keep the SRAM macro size fixed at 32 KB, the bank size and number of banks vary with RPB in this experiment. For the fixed size of 32 KB of SRAM macro size in this experiment, with the increase of RPB, the bank size increases and the number of banks decreases. We can see that all the modeled bitcell macros show a very similar trend of increasing MEP nonlinearly. This work shows minimum MEP variation across RPB = 4 to RPB = 64. However, from RPB = 32 to RPB = 64, Chiu's bitcell MEP variation is comparable to this work. Within RPB = 16–32, conventional subthreshold 8T and Chiu's bitcell MEPs are comparable too. We report Feki's bitcell has 1.46×, 8T has 1.24×, Kulkarni's bitcell has 1.65×, Chang's bitcell has 6.05×, Chiu's has 2.8%, and Yang's bitcell has 1.9× higher MEP at RPB = 32 for 32 KB SRAM. The modeled macro with our bitcell shows 4.48× and 1.78× increase in MEP for increasing the RPB 8× from RPB = 4–32, and 2X from RPB = 32–64, respectively. We can see a trend in the MEP supply voltage *vs.* RPB plot shown in Figure 10b for 32 KB SRAM. All the bitcells show constant MEP supply voltage from RPB = 32–64. From RPB = 16–32, Feki's, Kulkarni's and Chang's bitcell maintain their same constant MEP supply voltages as from RPB = 32–64. If we compare the MEP supply voltages of various bitcells above RPB = 32, we can see that Chang's bitcell has 33.33% lower MEP supply voltage (V_{DD}) than Yang's bitcell, Yang's has 14.28% lower MEP V_{DD} than Kulkarni's, this work and Chiu's bitcell. On the other hand, our bitcell has 12.5% lower MEP V_{DD} than Feki's bitcell.

Figure 10. (**a**) Minimum energy point (MEP) *vs.* number of bitcell rows per bank (RPB) for 32 KB SRAMs (CM = 4); (**b**) MEP Supply voltage *vs.* RPB of 32 KB SRAMs (CM = 4).

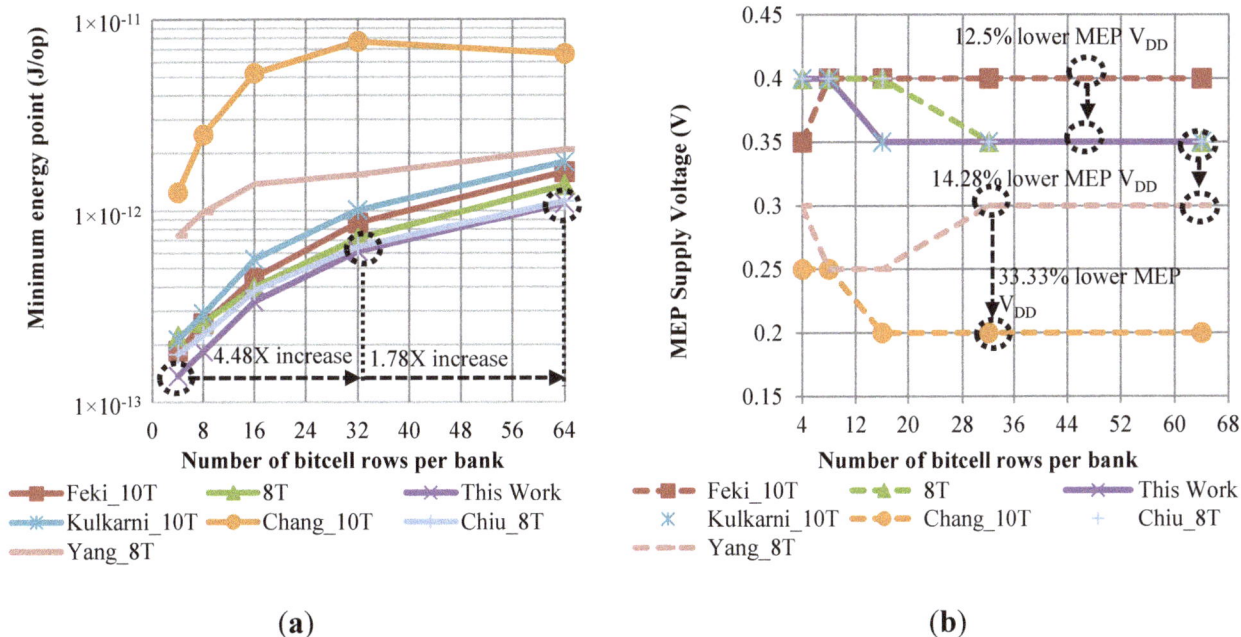

(**a**) (**b**)

7.4. MEP vs. Word-Width Comparison Results

Figure 11a shows the plot for MEP *vs.* number of SRAM bits in a word (word-width) for 32 KB SRAMs with CM = 4. We vary the word-width, and RPB at the same time, keeping the size of the banks fixed at 512 bits. Hence, the number of banks remains fixed at 512 for this experiment. In order to keep the bank size constant, the RPB decreases in a bank with the increase in word-width. As RPB and word-width both varies in this experiment with fixed bank size, we see a second order effect in MEP *vs.* word-width plot (Figure 11a): In almost all the bitcells (except Chang's and Yang's), the MEP first decreases and reaches a minimum point at some word-width then again it starts to increase. These minimum MEP points are at word-width = 8 for the 8T and Chiu's bitcell, and at word-width = 16 for Kulkarni's and Feki's bitcells, and this work. It is also, noticeable that our bitcell MEP varies much less than the Chiu's bitcell with increasing word-width. We report Feki's bitcell has 1.35×, subthreshold 8T has 1.62×, Kulkarni's bitcell has 1.55×, Chang's bitcell has 9.14×, Chiu's bitcell has 1.3×, and Yang's bitcell has 5.42× higher MEP than this work for 32 KB SRAMs with word-width = 32 (Figure 11a). Hence, with bigger memory macros, the combination of higher word-width and lower RPB is favorable for subthreshold SRAMs designed with our bitcell. Figure 11b shows the variation of MEP supply voltage *vs.* word-width. We can see a trend of decreasing MEP V_{DD} for all the bitcells except Chang's and Yang's bitcell. For the word-width increase of 4× from word-width = 8–32, Feki's bitcell shows 22.22% reduction in MEP V_{DD}. On the other hand, Chiu's and our bitcell show a 11.11% reduction in MEP V_{DD} for a 2× increase in word-width from word-width = 16–32.

Figure 11. (**a**) Minimum energy point (MEP) *vs.* word-width (bank size and number of banks kept fixed) for 32 KB SRAMs; (**b**) MEP supply voltage *vs.* word-width for 32 KB SRAMs.

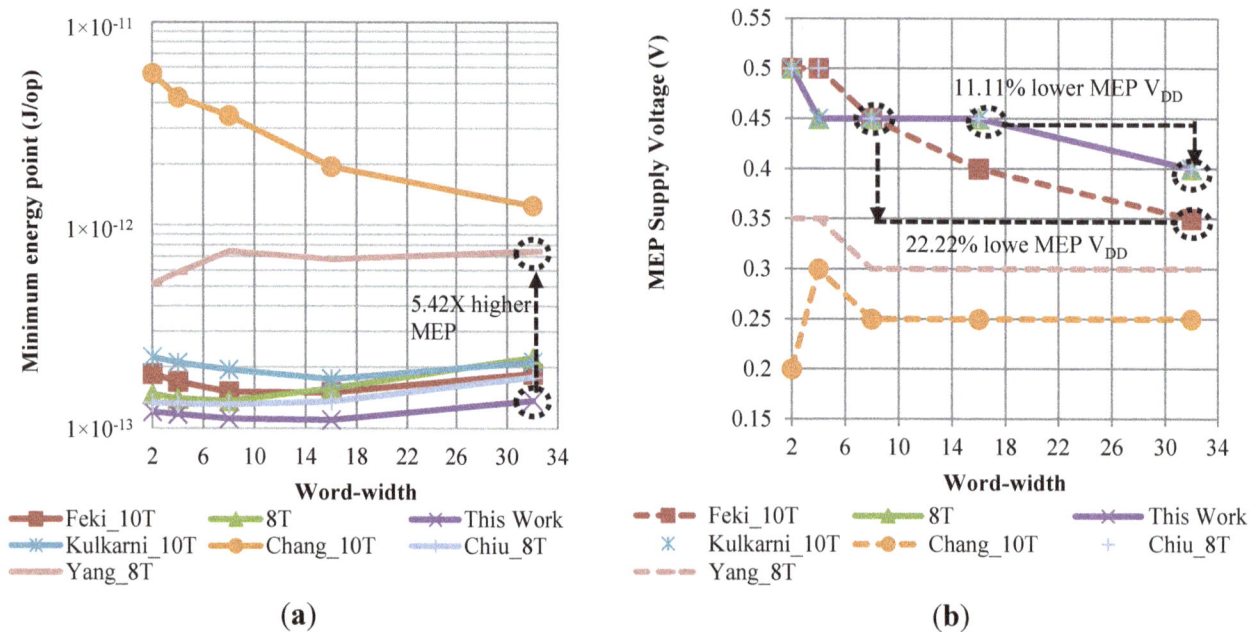

(**a**)   (**b**)

7.5. MEP vs. Column Mux Comparison Results

Figure 12a shows how the MEP varies with increasing column mux. For this experiment the RPB remains fixed at RPB = 64, the word-width at word-width = 32 and the size of the memory at 32 KB. In order to make the size of the memory constant, with the increase in column mux, the bank size increases and the number of banks decreases. We can see a linear trend of increasing MEP with column mux (CM). However, Kulkarni's and Chang's bitcells deviate from this trend in different parts in this plot. From CM = 2–16, although our bitcell MEP is comparable to Chiu's bitcell MEP, our bitcell MEP gets 9.3% lower than Chiu's bitcell MEP at CM = 32. We report that Feki's bitcell has 1.32×, 8T has 1.22×, Kulkarni's bitcell has 9.8%, Chang's bitcell has 1.53×, and Yang's bitcell has 17.36% higher MEP than our bitcell with CM = 32 for 32 KB SRAM macros. In addition, our bitcell shows the lowest MEP over all column mux configurations. For CM = 16, we report that Kulkarni's bitcell has 1.53× higher MEP than our bitcell as shown in Figure 12a. Figure 12b shows that with increasing column mux factor, the MEP supply voltage decreases with all the bitcell except Chang's bitcell. As Chang's bitcell in this memory configuration has lower MEP supply voltage below 0.2 V, we report 0.2 V as its MEP V_{DD}. We report that increasing the mux factor by 8× from CM = 4 to CM = 32, MEP supply voltage decreases by 25% for Feki's bitcell and 28.57% for conventional 8T as shown in Figure 12b.

Figure 12. (**a**) Minimum energy point (MEP) *vs.* column mux (words per row) for 32 KB SRAM; (**b**) MEP Supply voltage *vs.* column mux of 32 KB SRAMs.

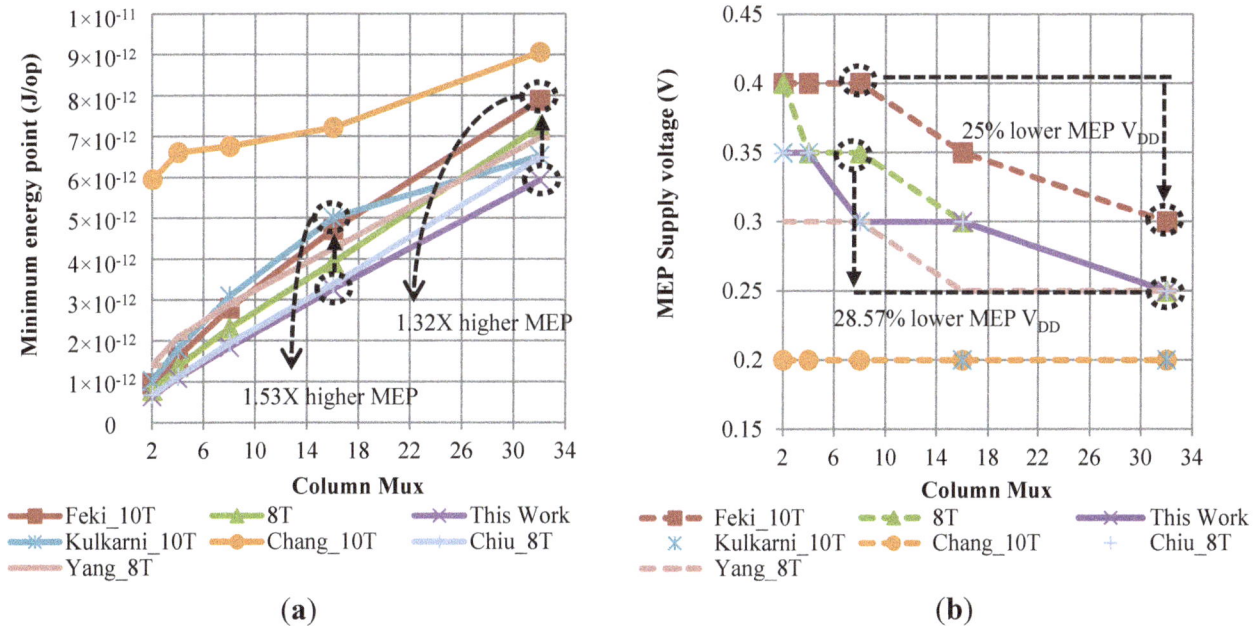

(**a**)   (**b**)

7.6. MEP vs. SRAM Size Comparison Results

Figure 13a shows the variation of MEP with increasing SRAM size with CM = 4. We conduct this experiment with the fixed bank size of 1024 bits per bank, RPB = 8 and word-width = 32 in a column mux 4 scenario. As the size of the SRAM banks remains fixed, the number of banks increases with the increase in memory size. We can see that the MEP of all bitcells increase with increasing SRAM memory size (Figure 13a). This is an expected trend as for a fixed word-width, increasing the SRAM size increases the leakage energy per operation and hence, the MEP shifts to a higher value. However, for this work, it has the lowest MEP across 2–32 KB SRAM memory sizes with RPB = 8. This is consistent with the results of this work's lower dynamic energy and leakage current data that keeps the MEP for this work lower compare to other bitcell macros. We report that for the SRAM size of 8 KB, Feki's bitcell has 1.31×, 8T has 1.39×, Kulkarni's bitcell has 1.51×, Chang's bitcell has 6.75×, Chiu's bitcell has 17.54%, and Yang's bitcell has 3.08× higher MEP than this work. Increasing the SRAM memory size 16× from 2 to 32 KB increases the MEP by only 1.89× for this work, but the other bitcells' MEP numbers increase by 2.04× for Feki's bitcell, 1.98× for Kulkarni's and 8T bitcell, 5.77× for Chang's bitcell, 2.03× for Chiu's bitcell, and 4.43× for Yang's bitcell. Figure 13b shows the variation of MEP supply voltage *vs.* SRAM macro size. We observe that with the increase in SRAM size, the MEP supply voltage increases for almost all the bitcells. We report a 33.33% increase in MEP supply voltage for Feki's, Chiu's, 8T and our bitcell. On the contrary, it is interesting to can see that from 4–32 KB, Yang's bitcell has a constant MEP supply voltage. Thus, even though Yang's bitcell has much higher MEP across different SRAM sizes, it can be suitable for bigger subthreshold SoCs having comparable logic energy per operation. However, for smaller low energy biomedical SoCs, our SRAM bitcell shows promising MEP numbers.

Figure 13. (**a**) Minimum energy point (MEP) *vs.* SRAM memory size (KB); (**b**) MEP supply voltage *vs.* SRAM memory size (KB).

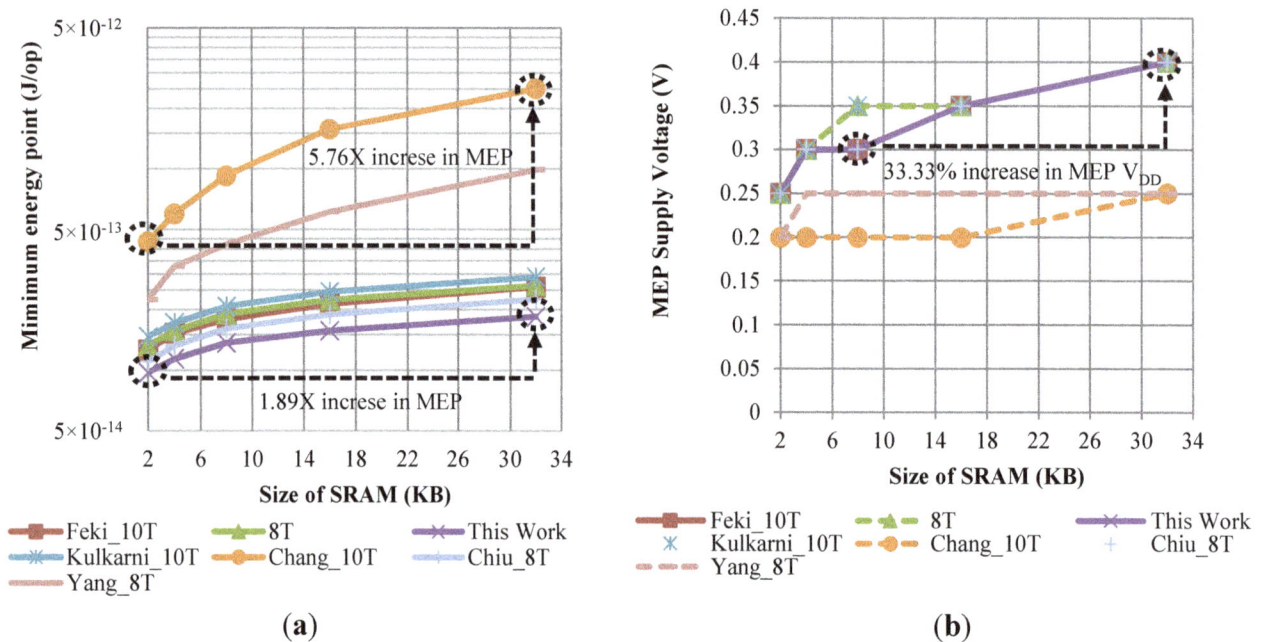

(**a**)   (**b**)

8. Conclusions

Across voltages of 0.25–0.5 V, our bitcell [16] has the lowest read energy among [6–16] and the conventional 6T. It has the lowest write energy among the bitcells across the voltages 0.35–0.5 V and second lowest leakage current in the 0.1–0.5 V range. Though our bitcell has lower numbers in energy and leakage current in subthreshold voltages, it suffers from a timing penalty. This work has demonstrated the lowest minimum energy point (MEP) across F_{rdwr} = 0.5–0.9 for 32 KB SRAMs. Our bitcell also provides the lowest MEP variation for 32 KB SRAMs across various rows per bank (RPB) ranging from RPB = 4–64; however, after RPB = 32, Chiu's bitcell has comparable MEP values for 32 KB SRAMs. This work shows that with varying word-width and fixed bank sizes and number of banks, most of the bitcell has a minima in the MEP curve around word-width = 8 and 16. This is due to a second order effect of varying two of the design knobs word-width and RPB simultaneously. In addition, our bitcell shows the lowest MEP values across word-width = 2–32. However, this work does not compare physical layout area of our bitcell with other bitcells, and therefore, it may have higher area penalty. MEP *vs.* column mux plots show a linear trend for most of the bitcells, and this work has the lowest MEP values across a mux factor from 2 to 32. Additionally, with RPB = 8, our bitcell has the lowest values of MEP across various SRAM sizes. However, for larger subthreshold SoCs with comparable logic energy per operation, Yang's and Chang's bitcells have lower MEP supply voltages, and those may be the best fit from the minimum energy per operation metric standpoint. We conclude that for energy constrained biomedical SoCs, where battery life is critical, operating in the frequency range of a few hundred kHz to several MHz, our 9T half-select-free SRAM bitcell offers lower energy numbers in read and write operations and the lowest MEP values across various subthreshold SRAM design knobs.

Acknowledgments

This project was supported in part by NVIDIA through the DARPA PERFECT program and by the NSF NERC ASSIST Center (EEC-1160483).

Author Contributions

In this research work, author Arijit Banerjee contributed to the literature search and coming up with the idea of the new bitcell that can be more energy efficient from MEP standpoint. He was responsible for simulating all the bitcells for comparison, collecting data from simulations and plotting them in meaningful figures. Arijit also wrote the initial draft version of the paper. Author Dr. Benton H. Calhoun guided Arijit for this research work to choose the research questions and to follow a predefined path for executing this research through technical discussions. He has contributed in this paper by reviewing the trends of the results presented in the paper and proofreading it from technical writing and formatting aspects.

Conflicts of Interest

The authors declare no conflict of interest.

References

1. Zhang, Y.Q.; Zhang, F.; Shakhsheer, Y.; Silver, J.D.; Klinefelter, A.; Nagaraju, M.; Boley, J.; Pandey, J.; Shrivastava, A.; Carlson, E.J.; *et al.* A batteryless 19 μW MICS/ISM-band energy harvesting body sensor node SoC for ExG applications. *IEEE J. Solid State Circuits* **2013**, *48*, 199–213.

2. Chen, G.; Fojtik, M.; Kim, D.; Fick, D.; Park, J.; Seok, M.; Chen, M.-T.; Foo, Z.Y.; Sylvester, D.; Blaauw, D. Millimeter-scale nearly perpetual sensor system with stacked battery and solar cells. In Proceedings of the 2010 IEEE International Solid-State Circuits Conference Digest of Technical Papers (ISSCC), San Francisco, CA, USA, 7–11 February 2010; pp. 288–289.

3. Wang, A.; Chandrakasan, A.P.; Kosonocky, S.V. Optimal supply and threshold scaling for subthreshold CMOS circuits. In Proceedings of the IEEE Computer Society Annual Symposium on VLSI, Pittsburgh, PA, USA, 25–26 April 2002; pp. 5–9.

4. Wang, A.; Chandrakasan, A. A 180-mV subthreshold FFT processor using a minimum energy design methodology. *IEEE J. Solid State Circuits* **2005**, *40*, 310–319.

5. Seevinck, E.; List, F.J.; Lohstroh, J. Static-noise margin analysis of MOS SRAM cells. *IEEE J. Solid State Circuits* **1987**, *22*, 748–754.

6. Kulkarni, J.P.; Kim, K.; Roy, K. A 160 mV robust schmitt trigger based subthreshold SRAM. *IEEE J. Solid State Circuits* **2007**, *42*, 2303–2313.

7. Chang, I.-J.; Kim, J.J.; Park, S.P.; Roy, K. A 32 kb 10T sub-threshold SRAM array with bit-interleaving and differential read scheme in 90 nm CMOS. *IEEE J. Solid State Circuits* **2009**, *44*, 650–658.

8. Feki, A.; Allard, B.; Turgis, D.; Lafont, J.; Ciampolini, L. Proposal of a new ultra low leakage 10T sub threshold SRAM bitcell. In Proceedings of the 2012 International SoC Design Conference (ISOCC), Jeju Island, Korea, 4–7 November 2012; pp. 470–474.

9. Chiu, Y.-W.; Lin, J.-Y.; Tu, M.-H.; Jou, S.-J.; Chuang, C.-Y. 8T Single-ended sub-threshold SRAM with cross-point data-aware write operation. In Proceedings of the 2011 International Symposium on Low Power Electronics and Design (ISLPED), Fukuoka, Japan, 1–3 August 2011; pp. 169–174.

10. Verma, N.; Chandrakasan, A.P. A 256 kb 65 nm 8T subthreshold SRAM employing sense-amplifier redundancy. *IEEE J. Solid State Circuits* **2008**, *43*, 141–149.

11. Chandra, V.; Pietrzyk, C.; Aitken, R. On the efficacy of write-assist techniques in low voltage nanoscale SRAMs. In Proceedings of the Design, Automation & Test in Europe Conference & Exhibition (DATE), Dresden, Germany, 8–12 March 2010; pp. 345–350.

12. Mann, R.W.; Nalam, S.; Wang, J.J.; Calhoun, B.H. Limits of bias based assist methods in nano-scale 6T SRAM. In Proceedings of the 2010 11th International Symposium on Quality Electronic Design (ISQED), San Jose, CA, USA, 22–24 March 2010; pp. 1–8.

13. Kim, T.; Liu, J.; Keane, J.; Kim, C.H. A high-density subthreshold SRAM with data-independent bitline leakage and virtual ground replica scheme. *IEEE J. Solid State Circuits* **2008**, *43*, 518–529.

14. Yang, H.-I.; Yang, S.-C.; Hsia, M.-C.; Lin, Y.-W.; Chen, C.-C.; Chang, C.-S.; Lin, G.-C.; Chen, Y.-N.; Chuang, C.-T.; Hwang, W.; *et al.* A high-performance low VMIN 55 nm 512 Kb disturb-free 8T SRAM with adaptive VVSS control. In Proceedings of the 2011 IEEE International SOC Conference (SOCC), Taipei, Taiwan, 26–28 September 2011; pp. 197–200.

15. Slayman, C. Soft errors—Past history and recent discoveries. In Proceedings of the 2010 IEEE International Integrated Reliability Workshop Final Report (IRW), Stanford Sierra, CA, USA, 17–21 October 2010; pp. 25–30.

16. Banerjee, A.; Calhoun, B.H. An ultra low energy 9T half-select-free subthreshold SRAM bitcell. In Proceedings of the 2013 IEEE SOI-3D-Subthreshold Microelectronics Technology Unified Conference (S3S), Monterey, CA, USA, 7–10 October 2013; pp.1–2.

17. Kim, H.; Kim, S.; van Helleputte, N.; Artes, A.; Konijnenburg, M.; Huisken, J.; van Hoof, C.; Yazicioglu, R.F. A configurable and low-power mixed signal SoC for portable ECG monitoring applications. *IEEE Trans. Biomed. Circuits Syst.* **2013**, *8*, 257–267.

18. Yan, L.; Bae, J.; Lee, S.; Roh, T.; Song, K.; Yoo, H.-J. A 3.9 mW 25-electrode reconfigured sensor for wearable cardiac monitoring system. *IEEE J. Solid State Circuits* **2011**, *46*, 353–364.

8

Study of Back Biasing Schemes for ULV Logic from the Gate Level to the IP Level

Guerric de Streel * and David Bol

ICTEAM institute, Université catholique de Louvain, Place du Levant 3, 1348 Louvain-la-Neuve, Belgium; E-Mail: david.bol@uclouvain.be

* Author to whom correspondence should be addressed; E-Mail: guerric.destreel@uclouvain.be

Abstract: Minimum energy per operation is typically achieved in the subthreshold region where low speed and low robustness are two challenging problems. This paper studies the impact of back biasing (BB) schemes on these features for 28 nm FDSOI technology at three levels of abstraction: gate, library and IP. We show that forward BB (FBB) can help cover a wider design space in terms of the optimal frequency of operation while keeping minimum energy. Asymmetric BB between NMOS and PMOS can mitigate the effect of systematic mismatch on the minimum energy point (MEP) and robustness. With optimal asymmetric BB, we achieve either a MEP reduction up to 18% or a $36\times$ speedup at the MEP. At the IP level, we confirm the MEP configurability with BB with synthesis results of microcontrollers at 0.35 V. We show that the use of a mix of overdrive FBB voltages further improves the energy efficiency. Compared to bulk 65 nm CMOS, we were able in 28 nm FDSOI to reduce the energy per cycle by 64% or to increase the frequency of operation by $7\times$, while maintaining energy per operation below 3 μW/MHz over a wide frequency range.

Keywords: digital CMOS circuits; ultra-low power; subthreshold logic; FDSOI; back gate biasing;variability; leakage currents; yield

1. Introduction

With the accelerating expansion of ultra-low-power computing and energy-autonomous systems requested by the vision of the Internet-of-Things, the need for compact battery-less wireless sensor nodes with consequent embedded data processing abilities is getting stronger [1]. In these systems, energy sobriety supersedes the traditional hunt for speed performances, and the Ultra Low Voltage operation region has became a hot topic in low-power research. As introduced in [2] and modeled in [3], energy minimization due to the tradeoff between leakage energy and switching energy is achieved in the subthreshold domain, as shown in Figure 1. Operating with such an aggressively scaled down supply voltage (V_{DD}) can lead to energy savings up to $10\times$ [4]. As shown in [5], subthreshold FDSOI circuits are emerging and can lead to very high energy efficiency. However, this scaling is limited by two factors: the target frequency constraint and the robustness constraint [6]. In order to meet the target frequency of the application, the V_{DD} may need to be raised above the optimal energy point. Below the threshold voltage (V_T), the I_{ON}/I_{OFF} ratio of MOSFET exponentially exacerbates the sensitivity to random variation and increasing functional failure probability [7]. A thorough examination of the design techniques that can be used in ULV to mitigate such constraints can be found in [8].

Figure 1. Optimal E_{cycle} and minimal V_{DD} to achieve a target clock frequency in 28 nm FDSOI regular V_T (RVT) and low V_T (LVT) devices from Figure 2. Simulation results for an eight-bit multiplier using the ELDO simulator from Mentor and 28-nm FDSOI CMOS models from STMicroelectronics. We estimated the 3σ worst-case delay obtained though 1 k Monte-Carlo (MC) simulations, and the robustness limit is extracted from 80 k MC simulations following the framework detailed in [6].

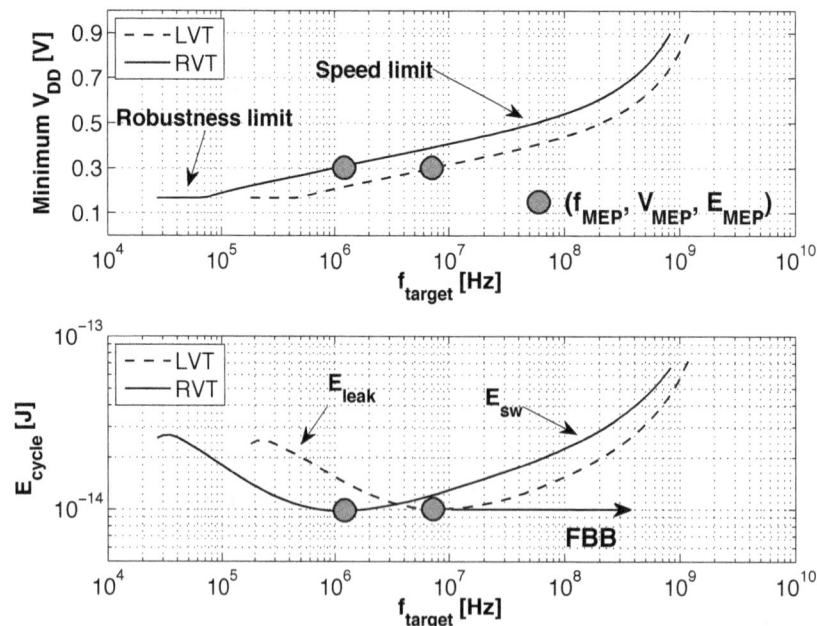

In this work, we explore the potential of back biasing technique in FDSOI technology to mitigate both constraints. Back biasing (BB) voltage controls the V_T through the back gate and can be used to dynamically adapt V_T in operation. The impact of back gate voltage biasing on V_T is detailed in Section 2. As the minimum energy point and the frequency of operation of the minimum energy point

(MEP) (f_{MEP}) are linked to the threshold voltage [9], dynamic V_T can help keep a circuit with variating computational workload near an energetic optimum. Figure 1 shows how the MEP is shifted toward higher operating frequencies between regular V_T (RVT) and low V_T (LVT) devices in the 28-nm FDSOI high-κ/metal-gate CMOS technology from STMicroelectronics [10].

Figure 2. Schematic view of RVT and LVT devices in 28-nm Ultra Thin Body and BOX FDSOI technology from STMicroelectronics. Back biasing voltages for: (**top**) RVT devices that are similar to bulk technologies; (**bottom**) LVT devices implemented with a flip well.

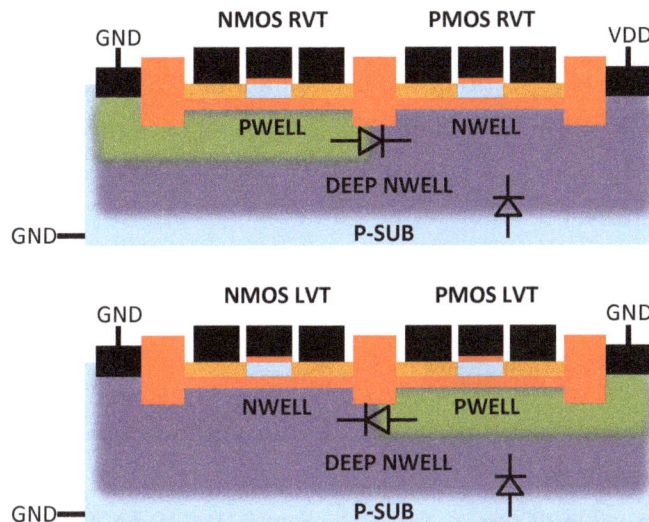

We study the impact of V_T modulation through BB on three abstraction levels:

- At the gate level, we evaluate the impact of back gate biasing and several schemes of back gate biasing on key factor of merit of a test bench circuit simulated at gate level. Technology scaling of bulk CMOS leads to increased variability, drain-induced barrier lowering (DIBL) and gate leakage that are harmful to the minimum energy level [9]. MEP reduction with technology scaling is limited by short channel effects in advanced bulk nodes, calling for FDSOI technologies to keep reducing MEP, while improving the corresponding f_{MEP}. In this work, the 28-nm FDSOI technology MEP is compared to 130-nm to 28-nm bulk technology MEP to quantify the high potential of FDSOI for ultra-low power and ultra-low voltage anticipated in [11–13]. Section 3.1 illustrates the effect of scaling and back biasing in the ULV domain. In addition to its use for trading the performance for energy efficiency, body biasing has already been proposed on bulk technology to mitigate random V_T mismatch under process variation (PV) in [14]. In [15], the authors proposed an analytical framework to analyze the PMOS/NMOS ratio variation with supply voltage and an adaptive scheme to optimize this ratio while compensating for PV. In this work, we use BB to control systematic PMOS/NMOS mismatch over the V_{DD} range without resorting to sizing modifications, allowing the use of standard cell libraries sized at nominal voltage. Systematic mismatch cancellation results in both energy and robustness improvement. We show that such adaptive BB can save 18% of energy per cycle at MEP and improve the gate count for a 95% functional die yield by a factor of six. Mismatch compensation schemes are described in Section 3.2, and energy efficiency and robustness results are shown in Sections 3.3 and 3.4;

- At the library level, we recharacterized a standard cell library when applying different BB voltages and investigate the obtained performances, as shown in Section 4;
- At the IP level, we study the scaling perspective of ULV microcontrollers cores towards a 28-nm implementation by using the recharacterized libraries. We validate the conclusions made at the gate and library levels based on the synthesis results of two microcontroller cores at 0.35 V in 28 nm FDSOI, as compared to the latest best-in-class results in 65-nm CMOS bulk [4]. Energy efficiency depending on BB use at the synthesis or during the operation of the microcontroller is discussed in Section 5.

This work consists of a collection of previously published results on forward BB (FBB) impact at the gate level [16] and at the IP level [17]. Sections 2 and 4 contain unpublished results, from which we are able to extract a coherent view of BB in FDSOI circuits from the device to synthesized IPs.

2. Back Biasing at the Device Level

The V_T modulation through back gate biasing can be modeled as follows [18]:

$$V_T = V_{T0} - \gamma \times V_{BS} \tag{1}$$

with γ, the back gate effect taking into account the electrostatic control of the back gate over the channel, and V_{BS}, the back gate biasing voltage.

Figure 3 shows the absolute value of the V_T for NMOS and PMOS LVT devices implemented using flipped well structures (*cf.* Figure 2). The V_T value drops with the forward BB down to 0.2 V at FBB = 3 V. Increasing the length of the devices increases the V_T for all FBB voltage, due to DIBL and V_T roll-off mitigation at a long channel length.

Figure 3. The impact of back biasing (BB) on V_T for a gate length from 30 to 500 nm. (**Left**) NMOS LVT device; (**right**) PMOS LVT device.

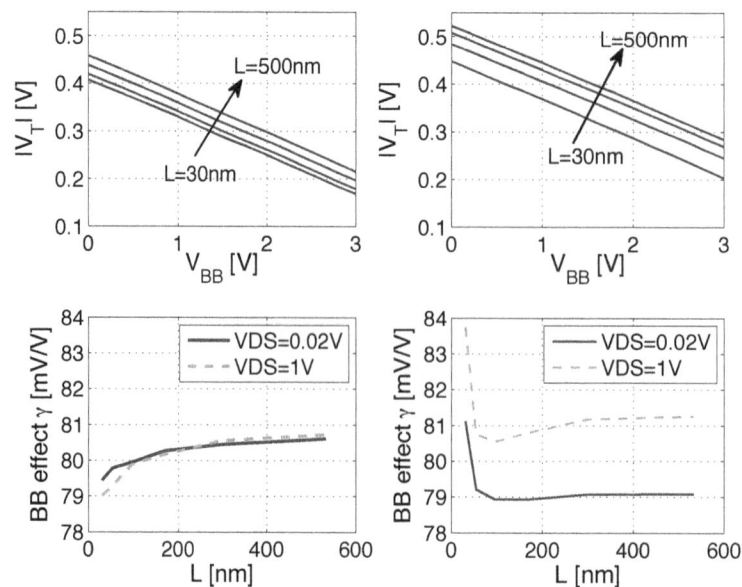

The γ parameter evolution with the length is also featured in Figure 3. For NMOS, γ is mostly independent of L_g, but the PMOS device shows a slightly higher variation with L, but also with V_{DS}, especially for a short gate length. This suggests that even for a long channel length, the DIBL of the PMOS is significant. We can reduce the V_T and improve γ for the PMOS by using a minimum gate length device. Lower V_T is interesting for high-speed ULV circuits in order to improve the I_{ON} current, as shown in Section 3.2, and a higher γ improves the efficiency of all of the back biasing schemes for performance tuning presented in this work. In particular, short channel PMOS presents a lower systematic V_T mismatch with the NMOS, hence improving the speed and robustness, as shown in Sections 3.3 and 3.4.

Figure 4 shows that the V_T is maximum at $V_{DS} = 0.1$ V. In ULV circuits with V_{DD} down to ≈ 0.3 V, the high V_T results in lower I_{ON}. The γ coefficient variation with V_{DS} is higher for the PMOS due to the higher DIBL.

Figure 4. V_T and γ coefficient for three FBB voltages depending on V_{DS}. (**Left**) NMOS LVT device; (**right**) PMOS LVT device.

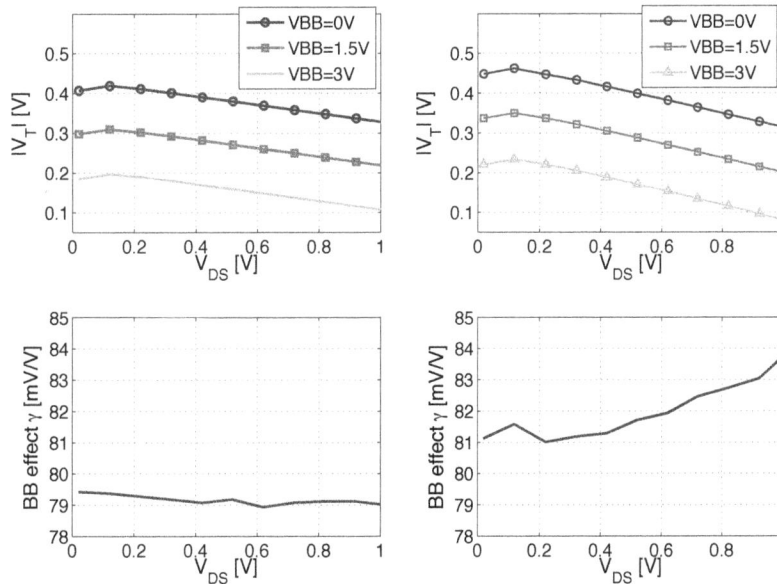

3. Back Biasing Analysis at the Gate Level

3.1. Scaling and Back Biasing Impact on Frequency and Energy Efficiency

To study the evolution of relevant figures-of-merit (FoMs) for circuit design with scaling, we simulated in SPICE an eight-bit benchmark multiplier at 0.35 V from 0.13 μm to 28 nm. We only considered LVT MOSFETs in 28 nm FDSOI for speed concerns and considered different FBB voltages applied to FDSOI (overdrive FBB voltages up to 2 V). In the 65/45-nm bulk, we used upsized gate length (GL) to improve the energy efficiency and functional robustness [6,9].

Figures 5 and 6 shows SPICE simulations of an eight-bit benchmark multiplier at 0.35 V in 130-nm, 90-nm, 65-nm and 45-nm bulk with a GP CMOS process (SVT MOSFETs) and in a 28-nm node with an LP CMOS process, as offered by the foundry, both in bulk and FDSOI technologies, following the framework used in [6,9]. Upsized-GL devices were upsized by 20 nm in 65 nm and by 10 nm in 45 nm. Signa to Noise Margins (SNMs) were simulated as proposed in [7] based on 80 k Monte-Carlo simulations of cross-coupled NAND2-NOR2 gates, and the V_{LIMIT} is defined as the minimum V_{DD}, allowing a 0.1% negative noise margin.

Figure 5. The evolution of the (a) maximum frequency of operation and (b) static power consumption with scaling in the ULV domain.

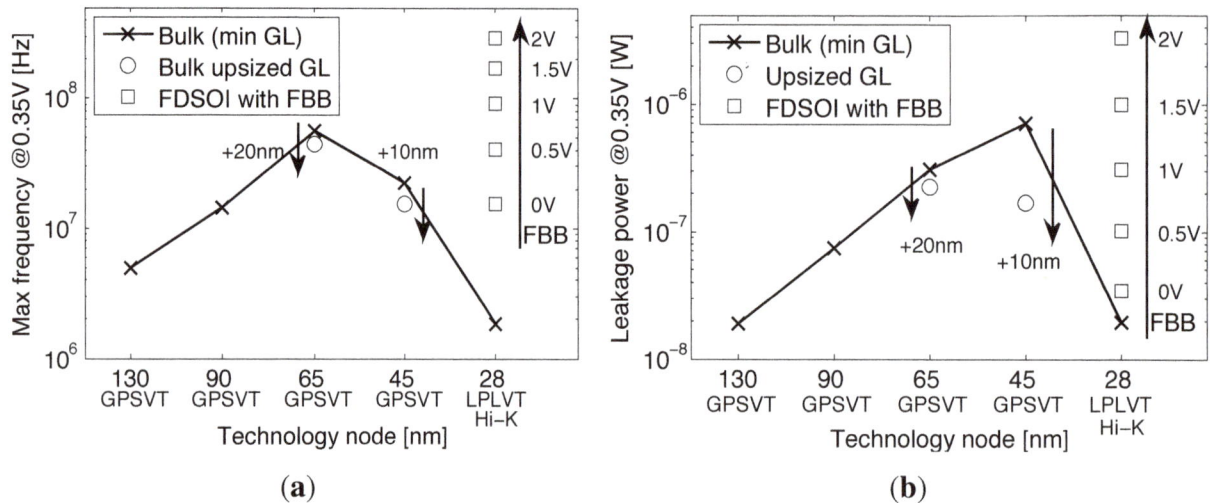

(a)   (b)

Figure 6. The evolution of (a) energy per operation and (b) robustness with scaling in the ULV domain.

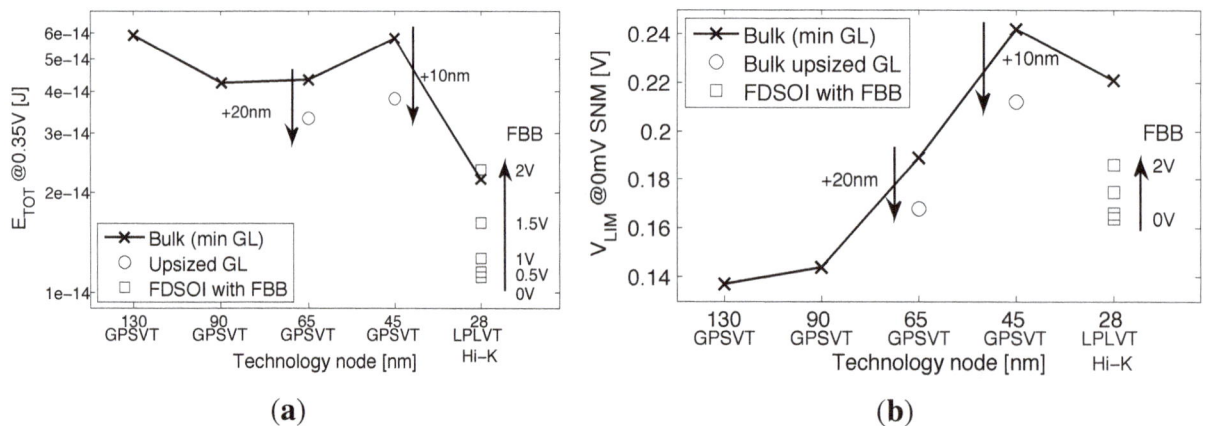

(a)   (b)

As shown on Figure 5a, the maximum frequency increases between 0.13 μm and 65 nm as a result of the reduction in node capacitance and the V_T value. As predicted in [19], the 45-nm node is slower than the 65-nm one. There are three factors in play in this speed reduction: a saturation of the typical subthreshold I_{ON} current [19], a significant increase of the ion variability [19] and an NMOS/PMOS systematic mismatch. These three factors cause a 3σ worst case delay increase, observed in Figure 5a.

The 28-nm bulk implementation is slower than any other bulk nodes, due to the LP CMOS process. Using FBB to reduce the threshold voltage can provide a speed up as high as $17\times$ between a 0-V bias and a 2-V bias. Figure 5b shows the same trend for the leakage power (from both drain and gate leakages). The leakage of the 28-nm bulk is reduced by a factor $35\times$ compared to the 45-nm one, thanks to the LP CMOS process and the high-κ/metal gate. The FBB on FDSOI greatly increases the leakage through the V_T reduction. The total E_{CYCLE} shown in Figure 6a is degraded from 90 to 45 nm, illustrating the energy overhead of short channel effects, variability and gate leakage [9]. Upsized-GL at 65/45 nm improves the total E_{CYCLE} at the expense of a 25/45% speed penalty. The 28-nm technology shows outstanding results in E_{CYCLE}, especially in FDSOI, thanks to the high-κ/metal gate and to the lower short channel effects and, thus, reduced DIBL. Figure 6b further illustrates the degradation of functional robustness at ULV through the evolution of V_{LIMIT} with scaling. As predicted, the low variability due to the undoped channel and the low short-channel effects of FDSOI restores functional robustness at 28 nm by improving V_{LIMIT}.

Based on these results, we can discard the 28-nm bulk technology, as it will be too slow to implement ULV circuits in the MHz range and as it features a robustness limit close to the targeted supply voltage. Contrary to 28-nm bulk, FDSOI technology allows high-speed operation and high robustness at 0.35 V.

3.2. Delay Equalization and Back Biasing Compensation Schemes

Body biasing has already been proposed for bulk technology to mitigate random V_T mismatch under process variation (PV) in [14]. In [15], the authors proposed an analytical framework to analyze the PMOS/NMOS ratio variation with supply voltage and an adaptive scheme to optimize this ratio while compensating for PV. In this work, we use BB to control systematic PMOS/NMOS mismatch over the V_{DD} range without resorting to sizing modifications, allowing the use of standard cell libraries sized at nominal voltage. We show that such adaptive BB can save 18% of the energy per cycle at MEP and improve the gate count for a 95% functional die yield by a factor of six.

To examine the impact of the PMOS/NMOS ratio variation on the relevant FoMs, we model the subthreshold delay and leakage power. In the subthreshold regime, the drain current can be expressed as in [18]:

$$I_{D,sub} = I_0 \times 10^{\frac{V_{GS}-\eta V_{DS}}{S}} \times \left(1 - e^{\frac{-V_{DS}}{U_{th}}}\right) \tag{2}$$

with U_{th}, the thermal voltage, S, the subthreshold swing, and η, the modeling DIBL effect. I_0 regroups the threshold voltage variation due to back gate voltage V_{BS} and bias-independent factors, particularly the body factor n and the zero-bias threshold voltage V_{T0}:

$$I_0 = \mu_0 C_{ox} \frac{1}{L_{eff}} (n-1) U_{th}^2 10^{\frac{-V_{T0}+\gamma V_{BS}}{S}} \tag{3}$$

The NMOS and PMOS I_{ON} are matched at nominal V_{DD} by upsizing the PMOS, but a mismatch appears in the subthreshold domain and for varying BB as the drive current becomes exponentially dependent on the threshold voltage. Figure 7 shows a divergence between NMOS and PMOS I_{ON} below 0.4 V.

There are several ways to modelize the gate delay based on IdVd simulations. We compared four modelizations expressed in Equations (4)–(7).

$$T_{del,inv} \propto \frac{CV_{DD}}{mean(I_{ON,N}, I_{ON,P})} \tag{4}$$

$$T_{del,inv} \propto \frac{CV_{DD}}{mean(I_{eff,N}, I_{eff,P})} \tag{5}$$

$$T_{del,inv} \propto \frac{CV_{DD}^2}{mean(\int_0^{V_{DD}} I_{D,N@V_{GS}-V_{DD}} dV_{DS}, \int_0^{V_{DD}} I_{D,P@V_{GS}=V_{DD}} dV_{DS})} \tag{6}$$

$$T_{del,inv} \quad \propto \quad \frac{CV_{DD}}{mean(\alpha_N, \alpha_P)} \tag{7}$$

$$\text{with} \quad \alpha = \frac{1}{2}\int_0^{V_{DD}} I_{D@V_{GS}=V_{DD}}dV_{DS} + \frac{1}{2}\int_0^{V_{DD}} I_{D@V_{GS}=V_{DD}/2}dV_{DS}$$

Figure 7. NMOS and PMOS I_{ON} evolution with V_{DD} for LVT MOSFET with FBB = 2 V (W_p = 160 nm, W_n = 80 nm, $L_p = L_n$ = 30 nm and $V_{DS} = V_{GS} = V_{DD}$, $V_{BS} = GND$). The PMOS BB is connected to GND . Simulation results are obtained with the ELDO simulator and 28-nm FDSOI CMOS models from STMicroelectronics.

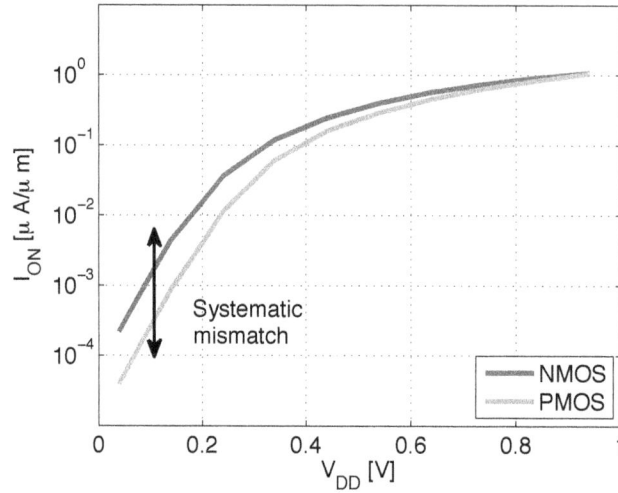

Equation (4) is based on the ON current and Equation (5) is based on the effective current first defined in [20] as:

$$I_{eff} = \frac{1}{2}\{I_D(V_G = V_{DD}/2; V_D = V_{DD}) + I_D(V_G = V_{DD}; V_D = V_{DD}/2)\} \tag{8}$$

Equations (6) and (7) are based on the integration of the IdVd curve at $V_G = V_{DD}$ or at $V_G = V_{DD}/2$. Figure 8 shows the modelization error compared to the simulated delay of an FO1 inverter chain depending on V_{DD}. The modelization based on the effective current produces good matching at ULV and, thus, is chosen to study the impact of systematic mismatch on energy efficiency.

The subthreshold leakage power can be expressed as:

$$P_{leak} = I_0 10^{\frac{\eta V_{DD}}{S}} \times V_{DD} \tag{9}$$

Based on the models in the subthreshold regime, we can study the impact of effective current mismatch on leakage energy. For an N-inverter chain, the total delay can be predicted as the product between the load capacitance and the mean effective current:

$$T_{del} \propto \frac{2 \times N \times C_{load} \times V_{DD}}{(I_{eff,sub,NMOS} + I_{eff,sub,PMOS})} \tag{10}$$

The total leakage current can be written as the arithmetic mean between the two devices' leakage current:

$$P_{leak} = \frac{N}{2}(I_{0,N} 10^{\frac{\eta_N V_{DD}}{S_N}} + I_{0,P} 10^{\frac{\eta_P V_{DD}}{S_P}}) \times V_{DD} \tag{11}$$

Figure 8. Comparison between delay modelization from Equations (4)–(7) and the simulated delay of an FO1 inverter chain.

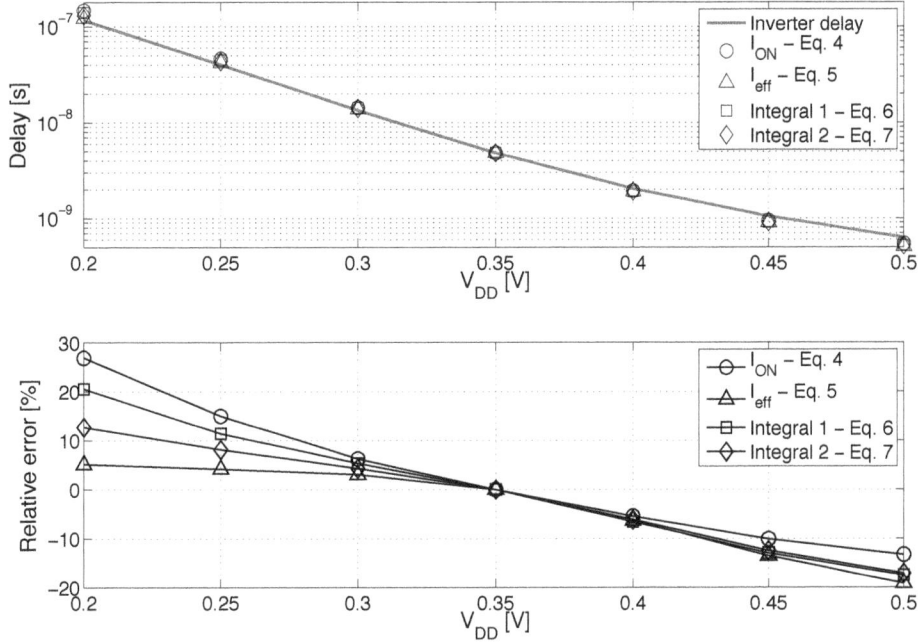

If the systematic mismatch ratio between the NMOS and PMOS effective current $\beta_{I_{eff}} = I_{eff,N}/I_{eff,P}$ is large, the delay can be, at best, divided by two, compared to a unitary ratio situation. The leakage current of the faster device will be increased, while the slower device leakage will be kept constant. The total leakage current will then increase by a factor $N/2 \times \beta_{I_{leak}}$; with $\beta_{I_{leak}} = I_{leak,N}/I_{leak,P}$, the leakage current ratio between NMOS and PMOS. The leakage energy:

$$E_{leak} = T_{del} \times I_{leak} \tag{12}$$

will be increased by a factor $N/4 \times \beta_{I_{leak}}$ that represents the E_{leak} loss caused by device mismatch as $\beta_{I_{leak}}$ is greater than one. A similar conclusion was drawn by Ono *et al.* [21] for PMOS/NMOS mismatch, due to local fluctuations.

PMOS and NMOS devices feature different I_0, subthreshold swings and DIBL factors, which lead to larger than one $\beta_{I_{eff}} = I_{eff,N}/I_{eff,P}$ that is commonly corrected through upsizing PMOS devices targeting equal rising and falling inverter delays. Implementing subthreshold circuits with standard library cells implies the use of cells balanced at nominal V_{DD}. Unfortunately, as the subthreshold I_{ON} divergence suggests, the $\beta_{I_{eff}}$ ratio will diverge from its value at nominal V_{DD} when the supply voltage is shrunk, leading to performance degradation. Figure 9 illustrates the validation of this theoretical trend by simulations and shows the $\beta_{I_{eff}}$ spreading in the subthreshold domain that will cause an E_{leak} increase.

Based on these conclusions, we developed rise/fall delay mismatch compensation technique by using the wide range of BB as a new degree of freedom offered by FDSOI technologies. By varying the relative biasing between the NMOS and PMOS devices, we can use the back gate to modify the $I_{0,NMOS}/I_{0,PMOS}$ ratio through the parameter γ in Equation (3) and control the effective current mismatch.

Figure 9. NMOS-PMOS I_{eff} ratio evolution with V_{DD} for RVT MOSFET and LVT MOSFET with different FBB voltages ($W_p = W_n = 80$ nm, $L_p = L_n = 30$ nm). The PMOS BB is connected to GND . Simulation results are obtained with ELDO simulator and 28-nm FDSOI CMOS models from STMicroelectronics.

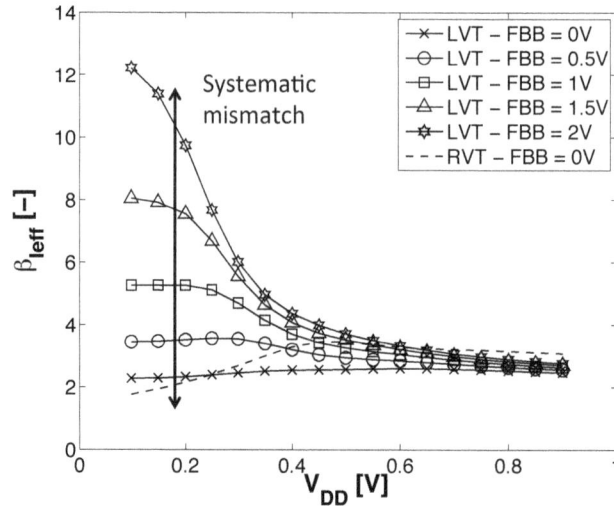

In order to study the back biasing potential in the subthreshold domain, we selected five levels of forward back biasing to be applied on the LVT device: from 0 V up to 2 V. In addition to the FBB, three adaptive asymmetric delta forward back biasing (DFBB) schemes, illustrated in Figure 10, were investigated:

- DFBBVDD: The PMOS back gate is connected to GND instead of V_{DD}, as in bulk technology or for RVT devices. The DFBB is equal to V_{DD} (DFBBVDD). In this technology, the PMOS is weaker than the NMOS, and a straightforward way to roughly compensate for the systematic mismatch is to apply a differential FBB equal to the nominal $V_{DD} \cong 0.9 - 1$ V on the PMOS. This scheme is a simple extrapolation of this rule, where the supply voltage of a super-threshold standard cell-based design is scaled down and no other modifications are made;

- ADFBB: As shown in Figure 9, the PMOS I_{eff} boost with DFBBVDD compensation is not strong enough at low V_{DD}, and $\beta_{I_{eff}}$ diverges from its value in the super-threshold domain, where standard cells are designed. In this scheme, the optimal adaptive DFBB (ADFBB) at each V_{DD} is applied to the PMOS device to equalize rising and falling delays for an inverter;

- IADFBB: The goal of this scheme is to achieve the same delay equalization as with ADFBB by reducing the NMOS I_0 with a negative BB instead of boosting the PMOS I_0.

Figure 11 presents the rise-fall delay ratio $\beta_{del} = T_{fall}/T_{rise}$ for each scheme. We can see that the two adaptive schemes lead to a very stable rise-fall delay ratio over V_{DD}. At high FBB, the DFBBVDD scheme cannot compensate for the I_{eff} drift between NMOS and PMOS. We then expect the most spectacular energy savings between (I)ADFBB and DFBBVDD to be seen for high speed devices.

For the ADFBB scheme, we can see that optimal DFBB is equal to V_{DD}, plus a constant over every supply voltage. The same conclusion can be drawn for the IADFBB scheme keeping in mind that Figure 11 only features the DFBB applied to the NMOS, which is the constant part, and an FBB equal to V_{DD} is simultaneously applied on the PMOS. This suggests that the systematic mismatch on I_0, S and η between NMOS and PMOS is roughly constant over V_{DD}.

Figure 10. Schematic view of LVT device in 28-nm UTBB FDSOI technology from STMicroelectronics with a description of the symmetric and asymmetric back biasing schemes.

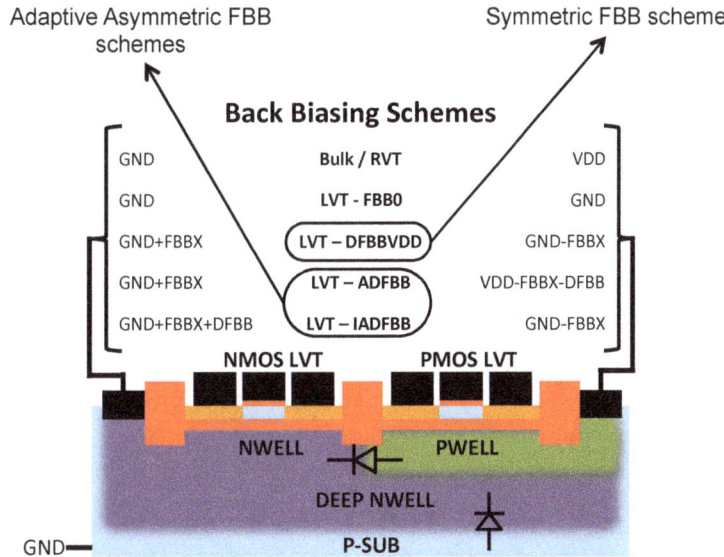

Figure 11. The impact of ADFBB and IADFBB schemes as defined in Figure 10 on the inverter rise-fall delay ratio. From left to right: IADFBB optimal DFBB applied on NMOS, IADFBB rise-fall delay ratio, DFBB V_{DD} (DFBBVDD) rise-fall delay ratio, ADFBB rise-fall delay ratio, ADFBB optimal DFBB applied on PMOS (W_p = 140 nm, W_n = 80 nm, $L_p = L_n = 30$ nm).

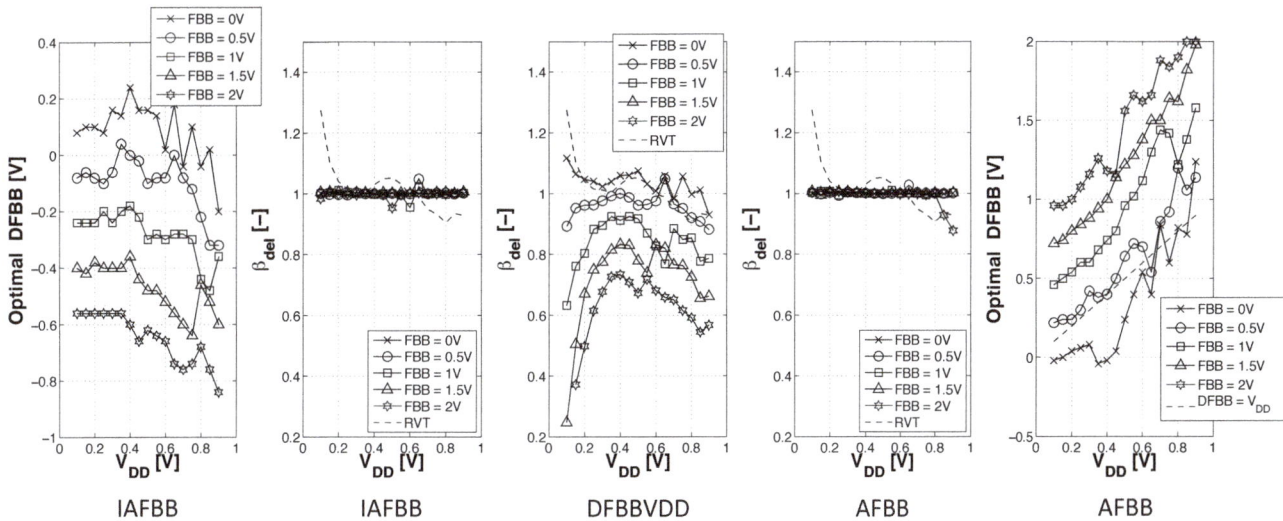

The constant contribution of the optimal DFBB increasing with FBB and the $\beta_{I_{eff}}$ augmentation at high FBB suggest that systematic mismatch is stronger for lower threshold voltages.

The granularity in the optimal DFBB for $V_{DD} > 0.5$ V can be explained by the limited sensitivity of the inverter delay with respect to V_T. When the inverter is not in the subthreshold region, this low derivative causes the optimal DFBB extraction method to reach the precision limit of the simulation.

3.3. Impact of Back Biasing Compensation Schemes on the Minimum Energy Point

To study the evolution of the MEP with FBB and DFBB schemes, we implemented an eight-bit multiplier with RVT devices and LVT-FBB = 0 V devices as baselines. We then applied, on the LVT implementation, an increasing FBB from 0 V to 2 V with the three DFBB schemes described in Section 3. Multiplier delay was extracted by fitting a lognormal distribution over the Monte Carlo simulations and taking the mean value plus 3σ as a worst case value. Multiplier leakage power was estimated by the arithmetic mean over Monte Carlo simulations. All simulations were completed using the ELDO simulator from Mentor with 28-nm FDSOI CMOS models from STMicroelectronics, and only self loading was considered.

Figure 12 shows the E_{MEP} and the V_{DD} at which this energy is attained V_{MEP}. For the (I)ADFBB schemes, the optimal DFBB at $V_{DD} = V_{MEP}$ from Figure 11 was used. We can see that E_{MEP} is increasing as the FBB voltage is increased. As V_T is reduced, the MOSFETs enter the near-threshold regime, which results in higher E_{leak} [18].

Figure 12. (**Left**) minimal E_{Cycle} for five FBB and for the three DFBB schemes; (**right**) V_{DD} at the minimum energy point (MEP) for five FBB and for the three DFBB schemes.

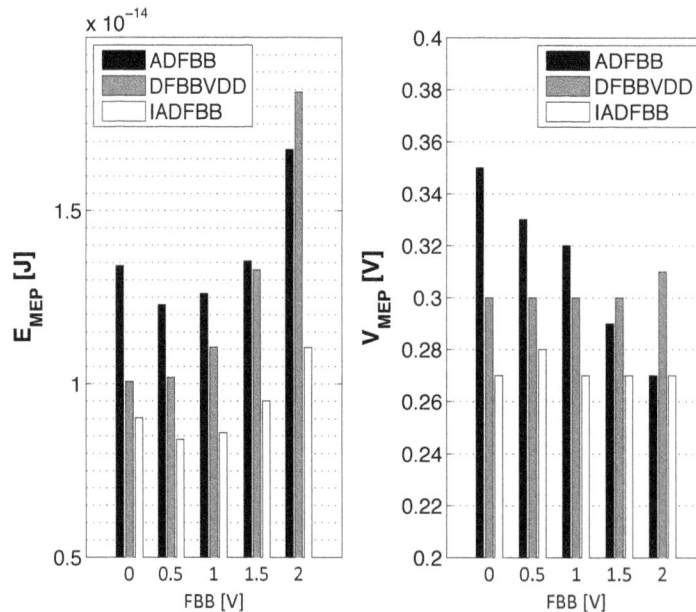

Different results were obtained by [9] on the same simulation framework for 45-nm bulk technology. In bulk nodes, E_{MEP} improvement for low-V_T devices is linked to different phenomena compared to FDSOI. The V_T reduction in bulk technologies is done by lowering channel doping, which narrows the channel depletion region, improves the subthreshold slope, reduces the gate leakage contribution, but increases the DIBL effect. Therefore, in the bulk process, an optimal V_T exists, resulting from the tradeoff between DIBL mitigation, gate leakage reduction and subthreshold current improvement [9].

In the FDSOI process, V_T variation is obtained by variating the back plane (BP) doping below the buried oxide [22], as shown in Figure 10, leaving the channel undoped, and V_T variations will not modify the depletion of the channel. As shown by measures realized on an analog FDSOI process in [23], both S and η are increased with FBB. Increasing the V_T on FDSOI technologies will not lead to an E_{MEP} reduction.

Figure 13 shows the evolution of E_{MEP} and f_{MEP} with FBB. Having access to a continuous V_T design space gives the possibility to shift MEP depending on the computational workload. Traditional subthreshold circuits try to approach MEP by scaling down V_{DD}, but are forced to move away from it if a higher throughput is needed. Dynamically adaptive FBB could help keep the circuit near the MEP under a varying target frequency.

Figure 13. The MEP position in the target frequency design space for five FBB and for the three DFBB schemes.

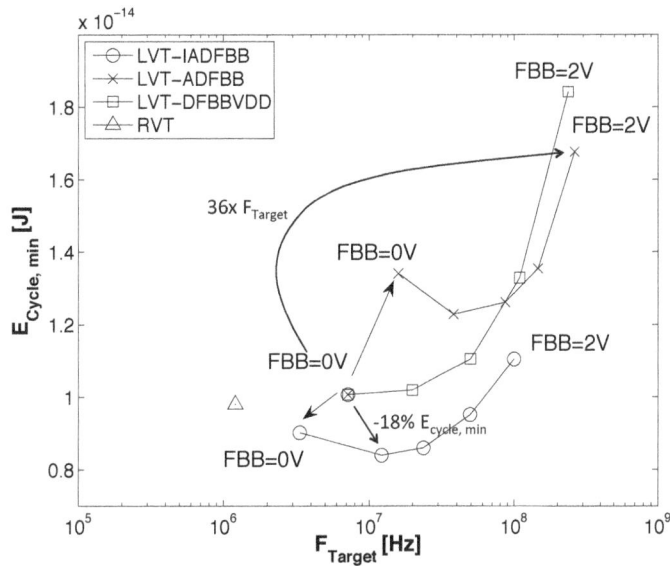

For small FBB, using ADFBB leads to degraded E_{MEP} levels, but at high FBB, which will allow higher f_{MEP}, E_{MEP} with ADFBB is below E_{MEP} with DFBBVDD. This result is coherent with observations made from Figures 9 and 11 that the DFBBVDD scheme is not adequate at high FBB. As ADFBB consists of boosting PMOS devices, the f_{MEP} increases. On the opposite, IADFBB slows down NMOS devices to match the PMOS delay, leading to a reduced f_{MEP}. As shown in Figure 12, IADFBB can offer an E_{MEP} level reduction of 11% at FBB = 0 and up to 18% at FBB = 0.5 V.

To objectively compare the interest of ADFBB and IADFBB schemes, we represented in Figure 13 the position of the MEP as a function of the frequency at which this energy optimum is attained. Results indicate that IADFBB always leads to a lower E_{MEP} value, but at the cost of a lower f_{MEP}. IADFBB can reduce by 18% the MEP level compare to DFBBVDD. On the contrary, ADFBB allows higher f_{MEP} at the cost of E_{MEP} degradation at low FBB, but with E_{MEP} improvement at high FBB. Figure 13 also shows that FBB = 2 V with ADFBB can improve the MEP frequency by 36× compared to FBB = 0 V with DFBBVDD.

The Pwell/Nwell and Nwell/Psub diodes were not simulated. For a positive value of FBB and with the DFBB values shown in Figure 9, these diodes are reverse-biased. To ensure that their leakage current is negligible, we estimated the diodes' area and perimeter for the eight-bit multiplier implemented with standard cells. In the worst case, for high a FBB value and for the three DFBB schemes, the total leakage power is below 30 fW at 25 °C and below 70 pW at 125 °C. This confirms that the leakage power of well diodes is negligible compared to the power of the multiplier.

3.4. Impact of Back Biasing Compensation Schemes on Robustness

Along with its impact on minimum energy, systematic mismatch between NMOS and PMOS currents leads to the reduced robustness of logic levels against crosstalk, radiations or random variability. In the subthreshold domain, where the I_{ON}/I_{OFF} ratio is small, these phenomena, including within die V_T variability, can lead to insufficient noise margins (NM_L and NM_H) and to functional failure, as defined in [7]. At first order and without the DIBL effect, we can reduce Equation (2) and express ON-state and OFF-state currents as:

$$I_{ON} = I_0 \times 10^{\frac{V_{DD}}{S}} \ , \quad I_{OFF} = I_0 \tag{13}$$

We consider $I_{ON,N}/I_{OFF,P}$ and $I_{ON,P}/I_{OFF,N}$ ratios as figures of the merit of robustness, which is a variation of the equivalent resistance model developed in [24]. We then discuss the theoretical impact of I_0, S and η mismatch on these ratios:

I_0: As the two ratios depend linearly on $I_{0,N}/I_{0,P}$ and $I_{0,P}/I_{0,N}$, having $I_{0,N} \neq I_{0,P}$ will favor one noise margin over the other and lead to a decreased global noise margin. Indeed, the gate noise margin is defined as $min[NM_L, NM_H]$;

S: If $S_N > S_P$, the $I_{ON,N}/I_{OFF,P}$ ratio will be reduced, as it can be written as $10^{(-V_{T0}+\gamma V_{BS})/S_N} \times 10^{V_{DD}/S_N} \times 10^{(V_{T0}-\gamma V_{BS})/S_P}$. A subthreshold swing mismatch will then lead to favoring one noise margin over the other;

η: In the subthreshold domain, the DIBL effect increases both NMOS and PMOS currents in the same proportions. If the DIBL effect between the two devices is different, one drive current will be increased, but when the high DIBL device does not drive, the high leakage and low drive device will lead to reduced robustness.

In each case, maximizing the $I_{ON,N}/I_{OFF,P}$ and $I_{ON,P}/I_{OFF,N}$ ratios is done with balanced devices.

In order to evaluate the impact of ADFBB and IADFBB on yield, we estimate the noise margins as proposed in [7] based on 80 k Monte Carlo simulations of a NAND2 gate cross-coupled with a NOR2 gate. As the NOR2 gate features a stacked pull-up network, it has the most stringent constraint on low input noise margin V_{IL}, and similarly, the NAND2 gate has the most stringent noise margin constraint on its low output level V_{OL}. Without precise information about the circuit paths, the considered benchmark can give an interesting figure of merit of the functional failure probability at the gate level.

To obtain an image of functional die yield based on this gate SNM extraction technique, we used a framework developed in [6] where the authors extrapolate die yield η_{die} as follows:

$$\eta_{die} = \eta_{gate}^{N_{gates}/2} \tag{14}$$

where η_{gate} is the gate yield computed with the Monte Carlo simulations.

To consider the impact of noise and crosstalk, gate functional failure was defined as NOR2-NAND2 SNM below two thresholds: 30 mV and 40 mV. Using this model, we can compare the die yield depending on the gate count using or not using the (I)ADFBB compensation as plotted in Figure 14.

Results indicate that having a 95% die yield with RVT devices limits the gate count to 80 k gates for 40 mV minimum SNM, while using LVT devices can bring this limit to 320 k gates. As predicted, balancing the I_0 between NMOS and PMOS has a positive impact on die yield. Figure 14 shows that, at a 30-mV minimum SNM, we can improve by a factor of six the number of gate at a 95% die yield. At 40-mV minimum SNM, the improvement factor between LVT and LVT with (I)ADFBB is higher than $4\times$.

The inset in Figure 14 indicates that a similar maximum number of gates at a 95% die yield can be achieved with IADFBB and with ADFBB, which is coherent with the idea that these two techniques are two equivalent ways to balance $I_{eff,N}$ and $I_{eff,P}$. The inset in Figure 14 also shows the optimal value of DFBB to be applied to maximize robustness. The two maxima appears at different DFBB values, because they are applied differently in IADFBB and in ADFBB, as shown on Figure 10. We can also see that the optimum DFBB values in each case for robustness is close to the optimum DFBB values found in Section 3.3 for energy minimization.

Figure 14.   Functional yield computed with the model described by Equation (14) for an eight-bit multiplier @V_{DD} = 0.3 V implemented with RVT and LVT devices (FBB = 0 V). The gate noise margins were considered with two thresholds: 30 mV and 40 mV. Inset: the maximum number of gates to maintain a 95% die yield with the ADFBB and IADFBB schemes.

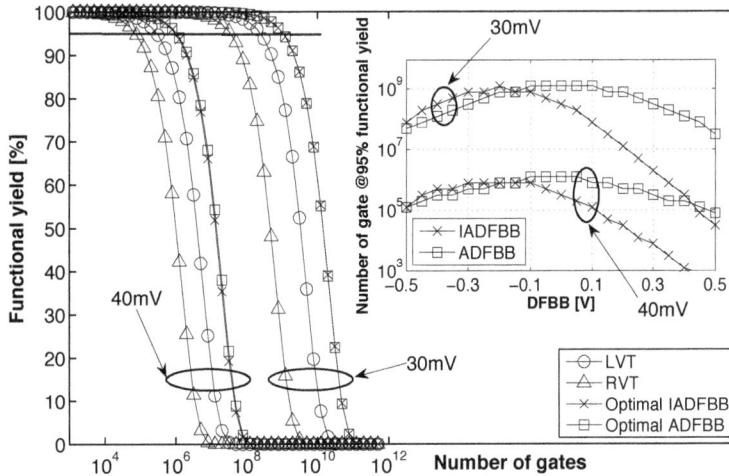

4. Back Biasing Analysis at the Standard Cell Library Level

For 28-nm FDSOI, we recharacterized the LVT standard-cell libraries from STMicroelectronics at 0.35 V using the Liberate tool from Cadence and the ELDO simulator from Mentor. Three versions of the library were computed at FBB = 0, 1 and 2 V. In order to characterize the impact of FBB on the lib, we examine the 20 most used cells in the two microcontroller cores synthesized and presented in Section 5. Figure 15a shows a mean rise-fall transition time drop of 20% to 40% from FBB'= 0 V to FBB = 1 V and a 30% to 50% delay reduction between FBB = 0 V and FBB = 2 V, depending on the cells.

Figure 15c shows that the mean leakage between FBB = 0 V and FBB = 1 V is around $10\times$ and as high as $100\times$ between FBB = 0 V and FBB = 2 V, due to the I_{OFF} exponential dependance on V_T.

Figure 15. Timing, switching power and leakage power factors of merit for the three recharacterized libraries for the 20 most used cells in the synthesized microcontroller core from Section 5. (**a**) Mean of rise-fall timing entries for 20 cells of the recharacterized libraries; (**b**) mean of rise-fall power entries for 20 cells of the recharacterized libraries; (**c**) mean of the leakage entries for 20 cells of the recharacterized libraries.

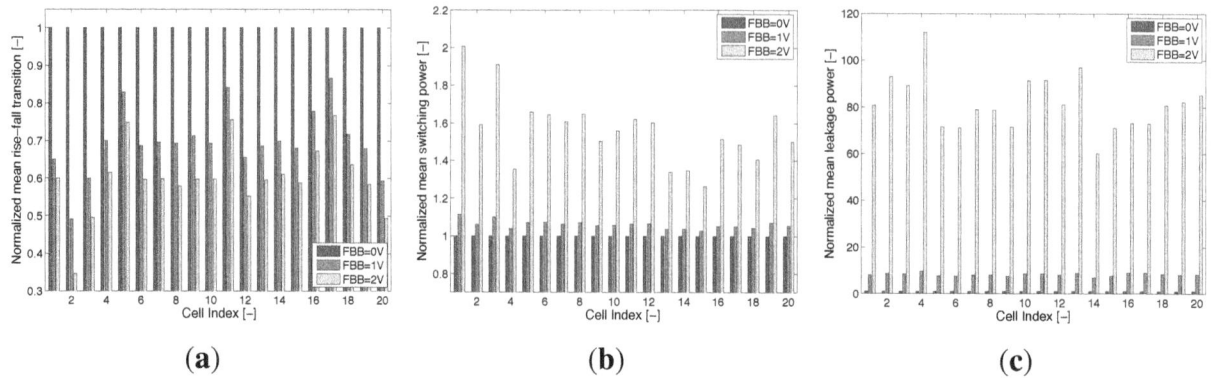

(**a**)   (**b**)   (**c**)

5. Back Biasing Analysis at the IP Level

To asses these FoMs at the block level for ULV SoCs with high computing capabilities, let us consider two benchmark circuits: a 16-bit MSP430-compatible core from [25] as modified at UCL in the SleepWalker SoC [4] (including its instruction cache memory $I\$$ and some peripherals) and the commercially-available ARM Cortex-M0 DesignStart 32-bit core. We ran a synthesis of these two cores, both in 65-nm bulk CMOS and in 28-nm FDSOI at 0.35 V with the recharacterized libraries presented in Section 4. For 65 nm, standard-cell libraries with UGL from [4] were considered. Synthesis was considered for each of these versions separately and with a mix of the three versions.

Figures 16a,b shows that synthesis with high FBB can achieve the timing closure at high frequencies. The M0 can be clocked at 440 MHz in 28-nm FDSOI with FBB = 2 V achieving a speed up of $8.8\times$ compared to 65 nm. However, as the core leaves the subthreshold region (V_T becomes lower than V_{DD}, resulting in a higher leakage energy), the E_{CYCLE} is higher compared to the energy achieved at a lower FBB. At a low frequency, high FBB implementations suffer from energy overhead, due to the integration of the high leakage power over a long cycle time. By allowing the synthesis tool to use a mix between the 3FBB library versions, we manage to keep the energy per cycle below 3 μW/MHz through almost all of the frequency design space. The evolution of the gate count breakdown between the three FBB library versions with frequency shows that faster gates are introduced in the design when the timing closure cannot be met with the previous FBB. Two key results are identified: the strict minimum energy point that is reached in the frequency range using mainly gates with 0- and 1-V FBB and the high-speed minimum energy point using also 2-V FBB gates. Figure 17a illustrates the speed-up and energy savings between 65-nm bulk and 28-nm FDSOI implementations. We achieve on the M0 a 42% E_{CYCLE} reduction and a $7\times$ speed up if we target high performance or a 64% E_{CYCLE} reduction and a $2.4\times$ speed up if we target energy minimization, compared to the 65-nm baseline. The difference between the E_{CYCLE} reductions of the two cores comes from the different contributions of leakage and switching energy to E_{CYCLE}.

Figure 16. Energy per cycle *vs.* the synthesis frequency for the Cortex M0 and for the swMSP430 with each FBB library version and with the mix of all three versions. (**a**) ARM Cortex M0; (**b**) swMSP430.

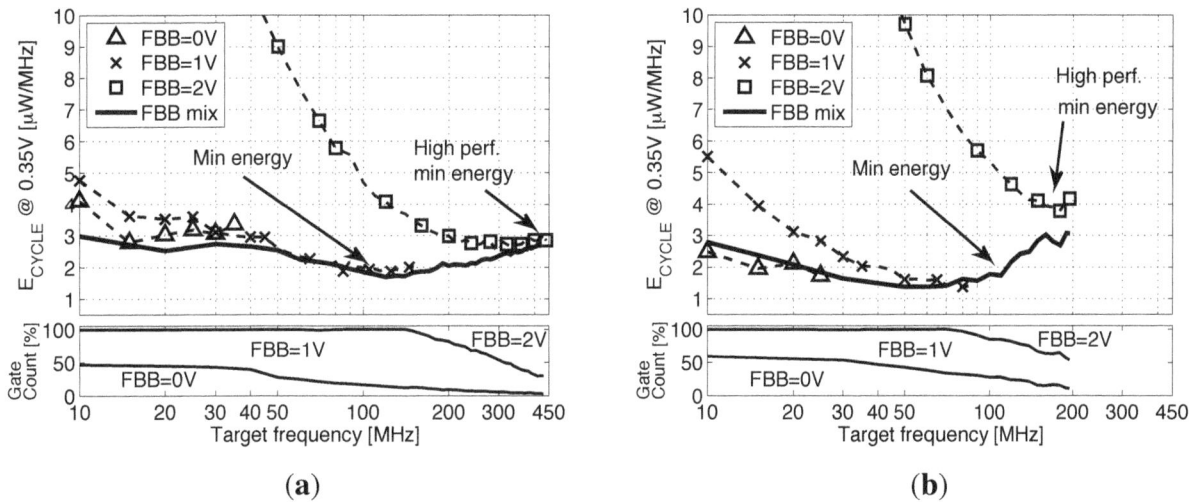

(**a**)   (**b**)

Figure 17. Energy per cycle of synthesized cores depending on the frequency or depending on the FBB at synthesis. (**a**) Evolution of the two energy minima at 0.35 V for swMSP430 and Cortex M0 between the 65-nm baseline and the 28-nm FDSOI optimized implementation; (**b**) energy per cycle for the Cortex M0 synthesized using FBB = 0 V, 1 V or 2 V at 30 MHz and running at different speeds, depending on the applied FBB.

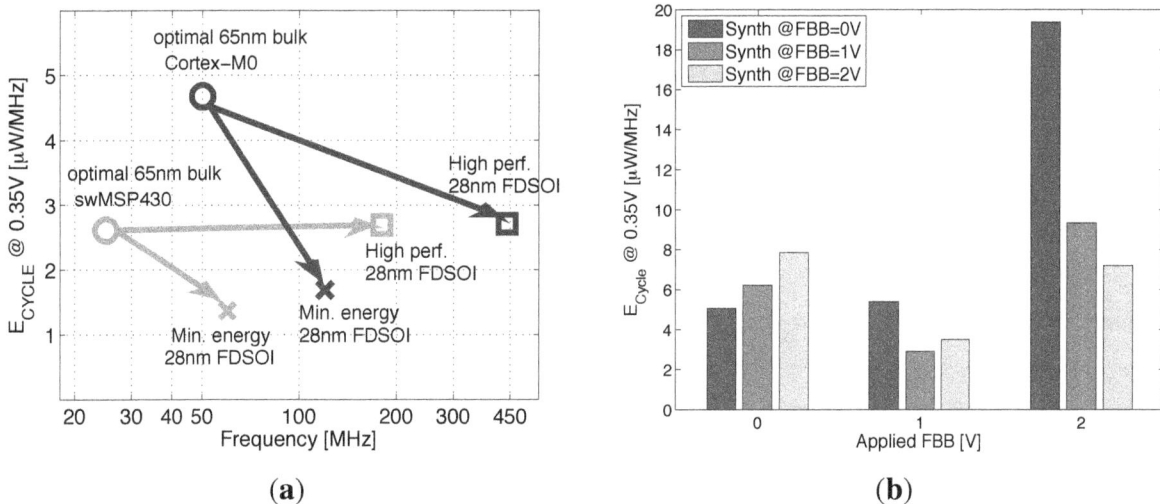

(**a**)   (**b**)

In practical applications, it would be interesting to know how the FBB used during the synthesis influences the energy efficiency in the operation. For a given target frequency, the designer has the choice of synthesizing the IP with a low or high FBB library. A slow synthesized core can then be speeded up by applying a higher FBB during operation. Conversely, the core synthesized with a high speed library can be slowed down with lower FBB in operation. Figure 17b shows the energy per cycle for the ARM Cortex M0 synthesized and operated with the three FBB libraries. The core synthesized at FBB = 0 V is energy inefficient for high FBB, high speed operations, and the core synthesized at

FBB = 2 V maintain good energy efficiency over the entire range of FBB. In conclusion, in order to take advantage of the electrical control over the frequency of operation offered by the forward back biasing and still maintain good energy efficiency, the synthesis operation should be conducted with a library characterized at high FBB.

6. Conclusions

The development of ultra-low-power embedded devices, such as wireless sensor nodes, calls for the Internet-of-Things for both very low energy per operation and high optimal frequency. Increasing the impact of short channel effects and variability on advanced bulk nodes degrades the minimum energy level at ULV in the sub-90-nm CMOS node. At the gate level, we showed that 28-nm UTBB FDSOI with a smaller impact of the short channel effects and a reduced variability can provide the MEP level reduction by a factor 5.5 compared to the 45-nm bulk node. We proposed to use the FBB to shift the MEP toward higher operating frequencies with a limited energy penalty. Dynamic FBB modifications can thus be used to dynamically adapt the clock frequency to the fluctuating computational workload. We also provide a new method for systematic mismatch cancellation between NMOS and PMOS subthreshold currents for different FBB voltages. Adaptive and inverse adaptive FBB further extend the range of frequencies, which leads to an energy minimum. Such adaptive schemes that are used to modify the speed/energy trade-off can reduce the MEP value by 18%, increasing the speed $36\times$.

With the development of complex subthreshold circuits, keeping a reasonable functional yield will became more challenging. The ADFBB and IADFBB schemes can provide a gate count improvement for a 95% yield up to a factor of $6\times$.

At the IP level, we showed that UTBB FDSOI can provide either leakage reduction at low-speed or high-speed operations, while maintaining E_{CYCLE} below the E_{CYCLE} achieved in 65 nm and improving robustness in ULV. Compared to 65-nm bulk, synthesis results show that the speed of ULV microcontrollers in 28-nm FDSOI can be boosted by a factor of $7\times$ in 28-nm FDSOI, and the E_{CYCLE} can be reduced by 64% with a mix of overdrive FBB voltages. We also showed that mixing FBB during synthesis can lead to an energy-efficient implementation for a wide range of frequencies of operation and that a microcontroller core synthesized at high speed can be kept energy efficient, even at low speed, when a lower FBB is applied in operation.

An interesting research direction for future work would be to first investigate the practical implementation of the FBB and DFBB schemes at the layout level to minimize the die area and energy consumption overhead for FBB generation, routing and optimum tracking.

Acknowledgments

Guerric de Streel is with the Université catholique de Louvain as a research fellow from the National Foundation for Scientific Research (FNRS) of Belgium through a Fonds pour la formation Ãă la Recherche dans l'Industrie et dans l'Agriculture (FRIA) grant.

Conflicts of Interest

The authors declare no conflict of interest.

References

1. Bol, D.; de Vos, J.; Botman, F.; de Streel, G.; Bernard, S.; Flandre, D.; Legat, J.-D. Green SoCs for a Sustainable Internet-of-Things. In Proceedings of the IEEE Faible Tension Faible Consommation (FTFC), Paris, France, 20–21 June 2013 .

2. Wang, A.; Chandrakasan, A.P.; Kosonocky, S.V. Optimal Supply and Threshold Scaling for Subthreshold CMOS Circuits. In Proceedings of the IEEE Computer Society Annual Symposium on VLSI Proceedings, Pittsburgh, PA, USA, 25–26 April 2002; pp. 5–9.

3. Calhoun, B.H.; Wang, A.; Chandrakasan, A. Device Sizing for Minimum Energy Operation for Subthreshold Circuits. In Proceedings of the IEEE Custom Integrated Circuits Conference, City, Country, 3–6 October 2004; pp. 90–95.

4. Bol, D.; de Vos, J.; Hocquet, C.; Botman, F.; Durvaux, F.; Boyd, S.; Flandre, D.; Legat, J. SleepWalker: A 25-MHz 0.4-V Sub-mm^2 7-μW/MHz microcontroller in 65-nm LP/GP CMOS for low-carbon wireless sensor nodes. *IEEE J. Solid-State Circuits* **2013**, *48*, 20–32.

5. Abouzeid, F.; Clerc, S.; Pelloux-Prayer, B.; Argoud, F.; Roche, P. 28 nm CMOS, Energy Efficient and Variability Tolerant, 350 mV-to-1.0V, 10 MHz/700 MHz, 252 Bits Frame Error-Decoder. In Proceedings of the ESSCIRC, Bordeaux, France, 17–21 September 2012; pp. 153–156.

6. Bol, D. Robust and energy-efficient ultra-low-voltage circuit design under timing constraints in 65/45 nm CMOS. *J. Low-Power Electron. Appl.* **2011**, *1*, 1–19.

7. Kwong, J.; Chandrakasan, A. Variation-Driven Device Sizing for Minimum Energy Sub-Threshold Circuits. In Proceedings of the International Symposium on Low Power Electronics and Design (ISLPED'06), Tegernsee, Germany, 4–6 October 2006; pp. 8–13.

8. Alioto, M. Ultra-low power VLSI circuit design demystified and explained: A tutorial. *IEEE Trans. Circuits Syst. I* **2012**, *59*, 3–29.

9. Bol, D.; Flandre, D.; Legat, J.-D. Technology Flavor Selection and Adaptive Techniques for Timing-Constrained 45 nm Subthreshold Circuits. In Proceedings of the 14th ACM/IEEE International Symposium on Low Power Electronics and Design (ISLPED '09), San Fancisco, CA, USA, 19–21 August 2009; pp. 21–26.

10. Flatresse, P. Product Vision on Planar Fully Depleted Technology FD28 nm for Mobile Applications. In Proceedings of the IEEE International SOI Conference, Napa, CA, USA, 1–4 October 2012.

11. Bol, D.; Flandre, D.; Legat, J.-D. Nanometer MOSFET effects on the minimum-energy point of sub-45 nm subthreshold logic—Mitigation at technology and circuit levels. *ACM Trans. Des. Autom. Electron. Syst.* **2010**, *16*, 1–26.

12. Bol, D.; Ambroise, R.; Flandre, D.; Legat, J.-D. Sub-45 nm Fully-Depleted SOI CMOS Subthreshold Logic for Ultra-Low-Power Applications. In Proceedings of the IEEE International SOI Conference (SOI), New Paltz, NY, USA, 6–9 October 2008.

13. Bol, D.; Bernard, S.; Flandre, D. Pre-Silicon 22/20 nm Compact MOSFET Models for Bulk *vs.* FD SOI Low-Power Circuit Benchmarks. In Proceedings of the IEEE International SOI Conference (SOIC), Tempe, AZ, USA, 3–6 October 2011.

14. Pu, Y.; de Gyvez, J.P.; Corporaal, H.; Ha, Y. Vt Balancing and Device Sizing toward High Yield of Sub-Threshold Static Logic Gates. In Proceedings of the 2007 International Symposium on Low Power Electronics and Design (ISLPED), Portland, OR, USA, 27–29 August 2007; pp. 355–358.

15. Hwang, M.E.; Roy, K. ABRM: Adaptive β-ratio modulation for process-tolerant ultradynamic voltage scaling. *IEEE Trans. VLSI Syst.* **2010**, *18*, 281–290.

16. De Streel, G.; Bol, D. Impact of Back Gate Biasing Schemes on Energy and Robustness of ULV Logic in 28 nm UTBB FDSOI Technology. In Proceedings of the IEEE International Symposium on Low Power Electronics and Design (ISLPED), Beijing, China, 4–6 September 2013; pp. 255–260.

17. De Streel, G.; Bol, D. Scaling Perspectives of ULV Microcontroller Cores to 28 nm UTBB FDSOI CMOS. In Proceedings of the IEEE SOI-3D-Subthreshold Microelectronics Technology Unified Conference (S3S), Monterey, CA, USA, 7–10 October 2013; pp. 1–2.

18. Calhoun, B.H.; Wang, A.; Chandrakasan, A. Modeling and sizing for minimum energy operation in subthreshold circuits. *IEEE J. Solid-State Circuits* **2005**, *40*, 1778–1786.

19. Bol, D.; Ambroise, R.; Flandre, D.; Legat, J. Interests and limitations of technology scaling for subthreshold logic. *IEEE Trans. VLSI Syst.* **2009**, *17*, 1508–1519.

20. Na, M.H.; Nowak, E.J.; Haensch, W.; Cai, J. The Effective Drive Current in CMOS Inverters. In Proceedings of the International Electron Devices Meeting (IEDM), San Francisco, CA, USA, 8–11 December 2002; pp. 121–124.

21. Ono, G.; Miyazaki, M. Threshold-voltage balance for minimum supply operation. *IEEE J. Solid-State Circuits* **2003**, *38*, 830–833.

22. Thomas, O.; Noel, J.-P.; Fenouillet-Beranger, C.; Jaud, M.-A.; Dura, J.; Perreau, P.; Boeuf, F.; Andrieu, F.; Delprat, D.; Boedt, F.; *et al.* 32 nm and beyond Multi-VT Ultra-Thin Body and BOX FDSOI: From Device to Circuit. In Proceedings of the IEEE International Symposium on Circuits and Systems (ISCAS), Paris, France, 30 May–2 June 2010; pp. 1703–1706.

23. Kilchytska, V.; Flandre, D.; Andrieu, F. On the UTBB SOI MOSFET Performance Improvement in Quasi-Double-Gate Regime. In Proceedings of the European Solid-State Device Research Conference (ESSDERC), Bordeaux, France, 17–21 September 2012; pp. 246–249.

24. Pu, Y.; de Gyvez, J.P.; Corporaal, H.; Yajun, H. Statistical Noise Margin Estimation for Sub-Threshold Combinational Circuits. In Proceedings of the Asia and South Pacific Design Automation Conference (ASPDAC), Seoul, Korea, 21–24 March 2008; pp. 176–179.

25. Girard, O. OpenMSP430 Project, 2010: Opencore.org.

26. Weber, O.; Faynot, O.; Andrieu, F.; Buj-Dufournet, C.; Allain, F.; Scheiblin, P.; Foucher, J.; Daval, N.; Lafond, D.; Tosti, L.; *et al.* High Immunity to Threshold Voltage Variability in Undoped Ultra-Thin FDSOI MOSFETs and Its Physical Understanding. In Proceedings of the IEEE International Electron Devices Meeting (IEDM), San Francisco, CA, USA, 15–17 December 2008; pp. 1–4.

9

Ultra-Low Voltage Sixth-Order Low Pass Filter for Sensing the T-Wave Signal in ECGs

Panagiotis Bertsias and Costas Psychalinos *

Electronics Laboratory, Physics Department, University of Patras, Rio Patras GR-26504, Greece;
E-Mail: phy5150@upnet.gr

* Author to whom correspondence should be addressed; E-Mail: cpsychal@physics.upatras.

External Editor: Alexander Fish

Abstract: An ultra-low voltage sixth-order low pass filter topology, suitable for sensing the T-wave signal in an electrocardiogram (ECG), is presented in this paper. This is realized using a cascade connection of second-order building blocks constructed from a sinh-domain two-integrator loop. The performance of the filter has been evaluated using the Cadence Analog Design Environment and the design kit provided by the Austria Mikro Systeme (AMS) 0.35-μm CMOS process. The power consumption of filters was 7.21 nW, while a total harmonic distortion (THD) level of 4% was observed for an input signal of 220 pA. The RMS value of the input referred noise was 0.43 pA, and the simulated value of the dynamic range (DR) was 51.1 dB. A comparison with already published counterparts shows that the proposed topology offers the benefits of 0.5-V supply voltage operation and significantly improved power efficiency.

Keywords: CMOS analog integrated circuits; ultra-low voltage circuits; biomedical circuits; sinh-domain filters; ECG signal processing

1. Introduction

Many physical conditions and diseases of the heart can be non-invasively detected through feature extraction from ECG. Significant features of the ECG signal include the P-wave, Q, R, and S waves (QRS) complex and T-wave. The P-wave represents atrial depolarization, the QRS complex left ventricular depolarization and the T-wave left ventricular repolarization. Thus, analysis of the T-wave in the ECG is an essential clinical tool for diagnosis, monitoring and follow-up of patients with heart dysfunction [1,2].

In order to sense the T-wave, a low pass filter with a cutoff frequency of 2.4 Hz should be realized. According to [3], a suitable choice is a sixth-order Bessel filter, and this originates from the fact that this type of filter offers an equal delay of all of the pass band frequencies without affecting the shape of the filtered signal. The topology in [3] operates under a ±1.5-V supply voltage and has been implemented using operational transconductance amplifiers (OTAs) as active elements, where the small signal transconductance parameter (g_m) has been employed for realizing the required time constants. The scheme in [4] has been realized using the concept of sinh-domain filtering, and therefore, the large signal characteristic of transistors has been used for the realization of time constants without the requirement of extra linearization stages. The employed supply voltage for this topology was ±1 V.

A sixth-order low pass filter topology is presented in this paper, realized using the concept of sinh-domain filtering. The reason is that sinh-domain filters have inherent class-AB operation, which allows biasing at current levels significantly lower than that of the maximum signal that can be handled by the system. In addition, the employment of MOS transistors in the subthreshold region offers the advantage of reduced power consumption [5–7]. The topology operates in a 0.5-V supply voltage environment. A value of dynamic range (DR) equal to 51.1 dB has been achieved under this ultra-low voltage environment, and according to the provided comparison results, the proposed topology offers the most power-efficient realization among the filters under consideration. The paper is organized as follows: the basic building blocks are presented in Section 2, while the performance of the filter topology is evaluated in Section 3 through a comparison with the corresponding already published counterparts. Simulations have been performed using the Cadence Virtuoso Analog Design Environment and MOS transistor models provided by the AMS C35 0.35-μm CMOS process.

2. Building Blocks for Sinh-Domain Filtering

Non-linear transconductors are the basic building block for realizing sinh-domain circuits. A typical multiple-output non-linear transconductor cell is depicted in Figure 1a, while the corresponding symbol is given in Figure 1b. Assuming that the MOS transistors operate in the subthreshold region, the expressions of output currents of the transconductor are given by Equations (1) and (2)

$$i_S = 2I_B \cdot \sinh\left(\frac{\hat{\upsilon}_{IN+} - \hat{\upsilon}_{IN-}}{nV_T}\right) \tag{1}$$

$$i_C = 2I_B \cdot \cosh\left(\frac{\hat{\upsilon}_{IN+} - \hat{\upsilon}_{IN-}}{nV_T}\right) \tag{2}$$

where I_B is the bias current of the transconductor, n is the subthreshold slope factor ($1 < n < 2$) of the MOS transistor, V_T is the thermal voltage ($\approx$26 mV at 27 °C) and $\hat{\upsilon}_{IN+}$, $\hat{\upsilon}_{IN-}$ are the voltages at the non-inverting and inverting inputs, respectively [5–7].

Figure 1. Multiple-output non-linear transconductor (S/S'/C cell) (**a**) circuitry; and (**b**) symbols.

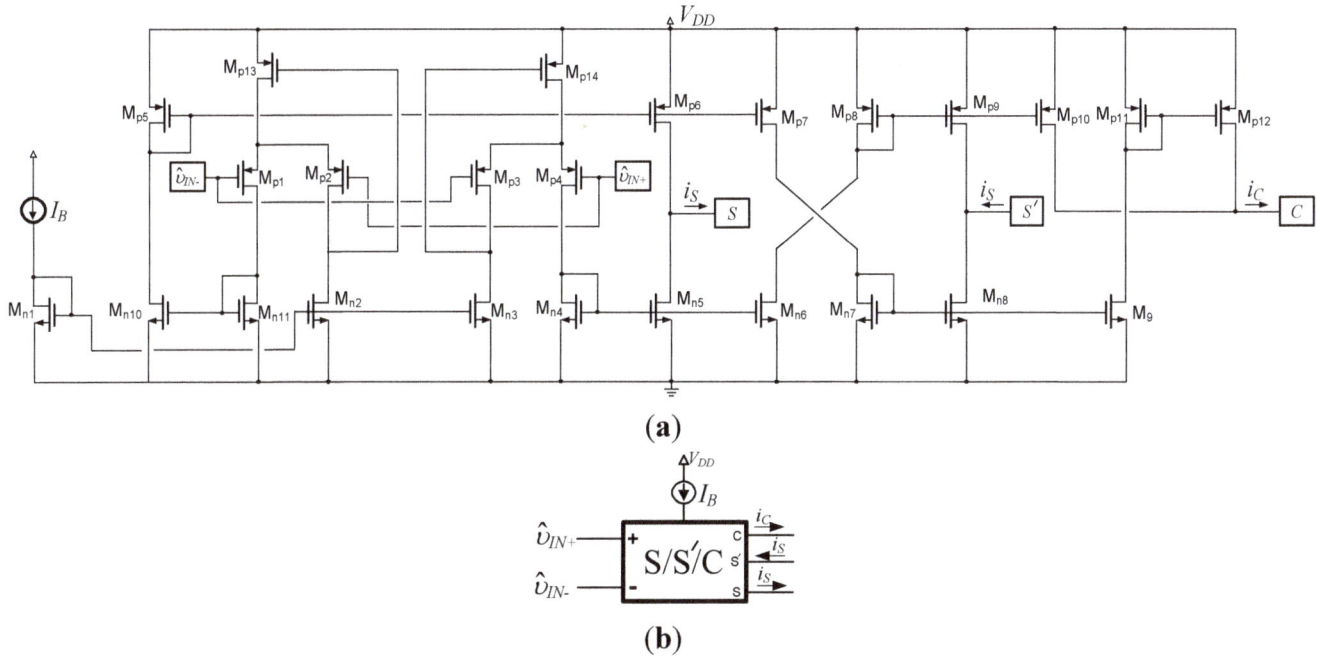

(**a**)

(**b**)

Due to the employment of the current-mirror constructed from transistors, M_{n7}–M_{n8} and M_{p8}–M_{p9}, the current at the S' terminal is an inverted replica of that at the S terminal. Consequently, the expression in Equation (1) is still valid for this current.

The realization of a two-quadrant divider, using single output non-linear transconductor cells, derived from the general topology in Figure 1a by omitting the S' and C outputs, is depicted in Figure 2. The expression for the output current is given by the formula $i_{OUT} = I_{DIV}\cdot(i_1/i_2)$, where I_{DIV} is the bias current of the divider and i_1 and i_2 are two and one-quadrant input currents, respectively [5–7].

Figure 2. Two-quadrant divider (**a**) realization using S cells; and (**b**) symbols.

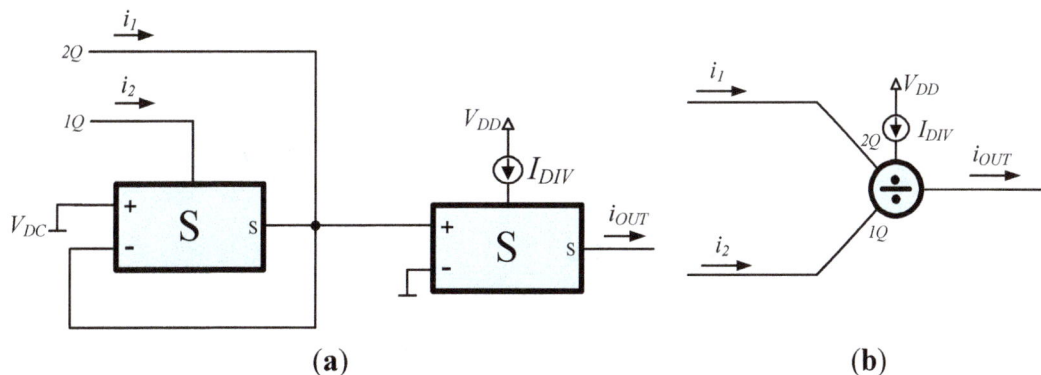

(**a**)   (**b**)

Using the building block in Figure 1, the topology of a lossy (*i.e.*, first-order filter) sinh-domain integrator is demonstrated in Figure 3a, while in Figure 3b, the topology of a lossless integrator is given. The realized transfer functions are given by (3) and (4), respectively, as:

$$H(s) = \frac{1}{\tau \cdot s + 1} \tag{3}$$

$$H(s) = \frac{1}{\tau \cdot s} \tag{4}$$

where the time-constant in both of them is:

$$\tau = \frac{n \cdot C \cdot V_T}{I_{DIV}} \tag{5}$$

An important benefit offered by the topologies in Figure 3 is that the time constant can be electronically adjusted through the DC bias current I_{DIV} without disturbing the DC current I_o used for biasing the multiple-output transconductors. In other words, an orthogonal adjustment between I_{DIV} and I_o is possible, and as a result, relatively large time constants can be realized without affecting the level of currents that can be successfully handled by the system [8,9].The functional block diagram (FBD) of a second-order low pass filter implemented through the utilization of a two-integrator loop is given in Figure 4a, while the corresponding sinh-domain realization derived using the integrators in Figure 3 is demonstrated in Figure 4b.

Figure 3. Sinh-domain integrators (**a**) lossy; and (**b**) lossless.

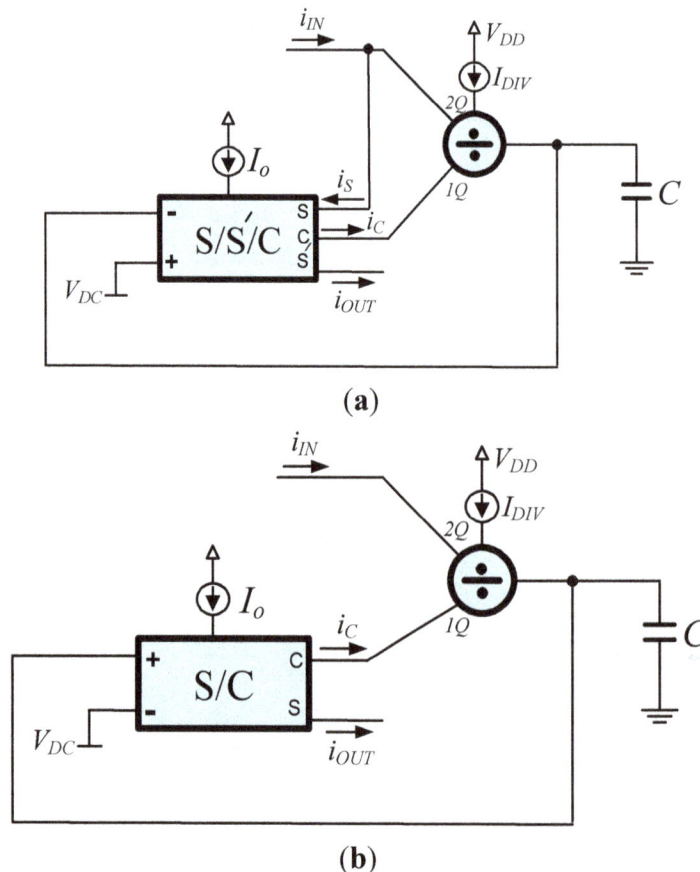

(a)

(b)

Figure 4. Second-order low pass filter (**a**) functional block diagram; and (**b**) realization using the concept of sinh-domain filtering.

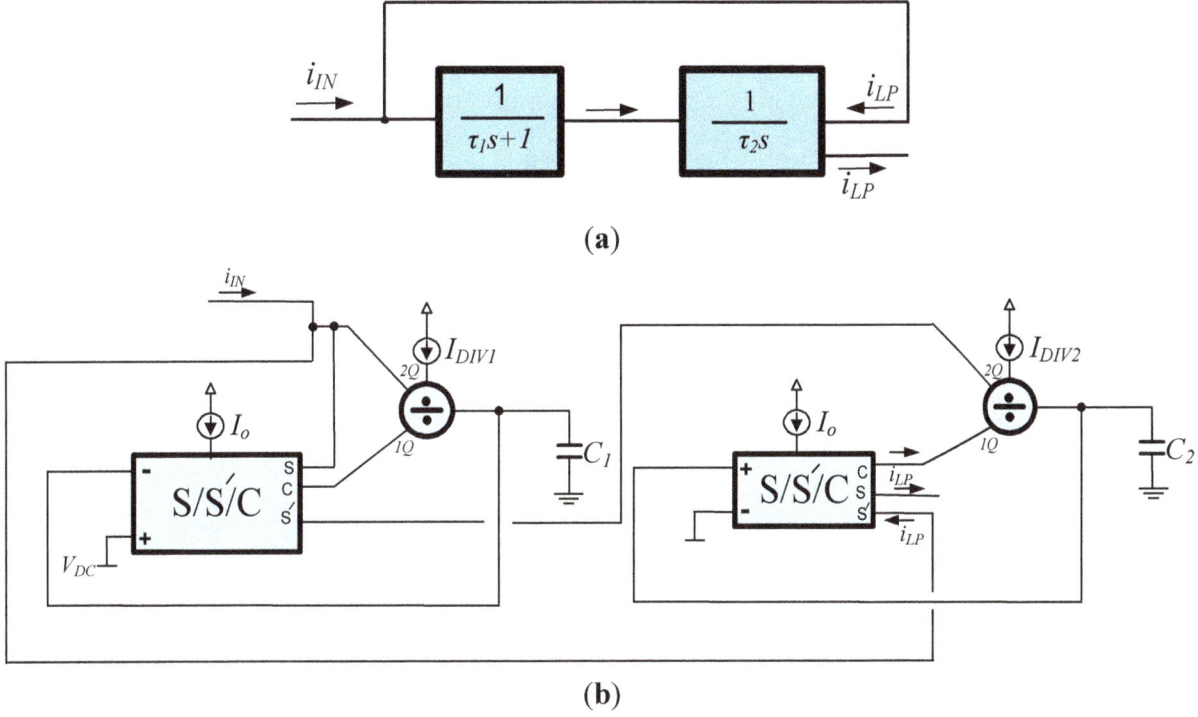

(**a**)

(**b**)

The realized transfer function is:

$$H(s) = \frac{\dfrac{1}{\tau_1 \cdot \tau_2}}{s^2 + \dfrac{1}{\tau_1} \cdot s + \dfrac{1}{\tau_1 \cdot \tau_2}} \tag{6}$$

According to Equation (6), the resonant frequency and the Q factor of the filter are given by Equations (7) and (8), respectively, as:

$$\omega_o = \frac{1}{\sqrt{\tau_1 \tau_2}} \tag{7}$$

$$Q = \sqrt{\frac{\tau_1}{\tau_2}} \tag{8}$$

Using Equation (5), the expressions in Equations (7) and (8) could be written as:

$$\omega_o = \frac{1}{n \cdot V_T} \cdot \sqrt{\frac{I_{DIV1} \cdot I_{DIV2}}{C_1 \cdot C_2}} \tag{9}$$

$$Q = \frac{1}{n \cdot V_T} \cdot \sqrt{\frac{C_1 \cdot I_{DIV2}}{C_2 \cdot I_{DIV1}}} \tag{10}$$

3. Comparison Results and Biomedical Application Example

The sixth-order low pass filter will be realized through a cascade connection of second-order blocks given in Figure 4b. MOS transistor models provided by the AMS 0.35-μm C35 CMOS process, as well as the Cadence Virtuoso Analog Design Environment will be employed. The bias scheme was $V_{DD} = 0.5$ V, $V_{DC} = 100$ mV and $I_o = 100$ pA. Considering that the MOS transistors operate in the subthreshold region, the corresponding aspect ratios for the non-linear transconductor in Figure 1a are summarized in Table 1. The capacitor values, as well as the bias currents of dividers for all stages of the system are given in Table 2.

Table 1. MOS transistor aspect ratios for the non-linear transconductor in Figure 1.

Transistor	W/L (μm/μm)
M_{P1}–M_{P12}	60/1
M_{p13}–M_{p14}	20/0.4
M_{n1}–M_{n11}	8/1

Table 2. Bias scheme and capacitor values for the stages of the sixth-order low pass filter.

Stage	Values
Low Pass filter #1	$V_{DD} = 0.5$ V, $V_{DC} = 0.1$ V, $I_o = 100$ pA $I_{DIV1} = 15.9$ pA, $I_{DIV2} = 8.24$ pA $C_1 = 10$ pF, $C_2 = 20$ pF
Low Pass filter #2	$V_{DD} = 0.5$ V, $V_{DC} = 0.1$ V, $I_o = 100$ pA $I_{DIV1} = 13.9$ pA, $I_{DIV2} = 10.4$ pA $C_1 = 10$ pF, $C_2 = 20$ pF
Low Pass filter #3	$V_{DD} = 0.5$ V, $V_{DC} = 0.1$ V, $I_o = 100$ pA $I_{DIV1} = 9.3$ pA, $I_{DIV2} = 9.8$ pA $C_1 = 10$ pF, $C_2 = 10$ pF

The DC power dissipation of the filter was 7.21 nW. The obtained frequency response is demonstrated in Figure 5, where the achieved cutoff frequency was 2.4 Hz. The group delay as a function of frequency is plotted in Figure 6. The group delay error within the pass band was 0.2%.

Figure 5. Gain *vs.* frequency (freq) of the sixth-order Bessel low pass filter.

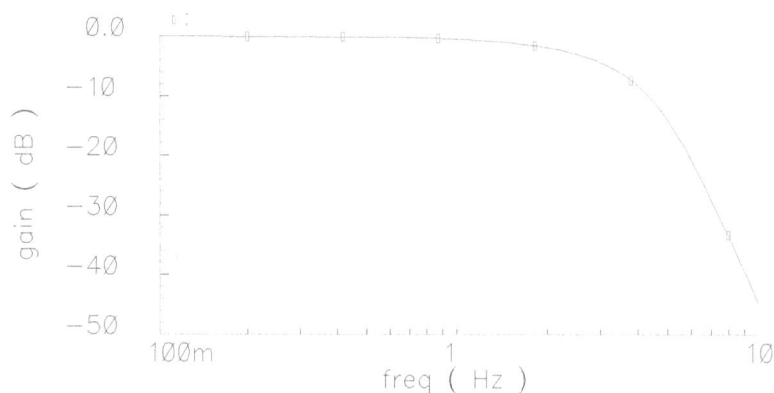

Figure 6. Group delay variation of the sixth-order Bessel low pass filter within the pass band.

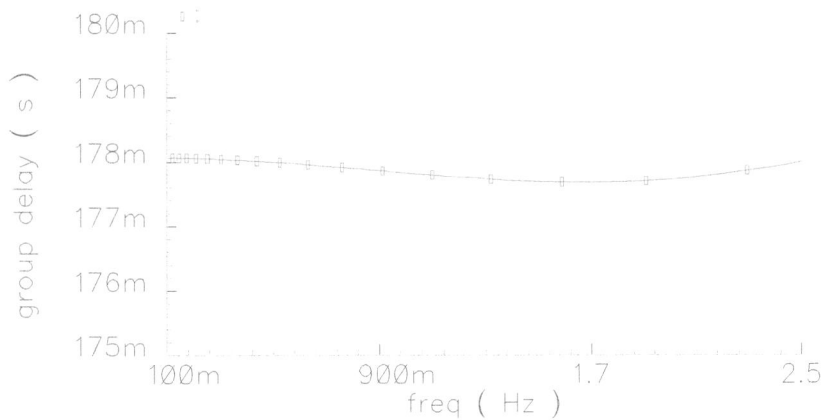

Figure 7. Total harmonic distortion (THD) *vs.* amplitude (ampl) of the input signal.

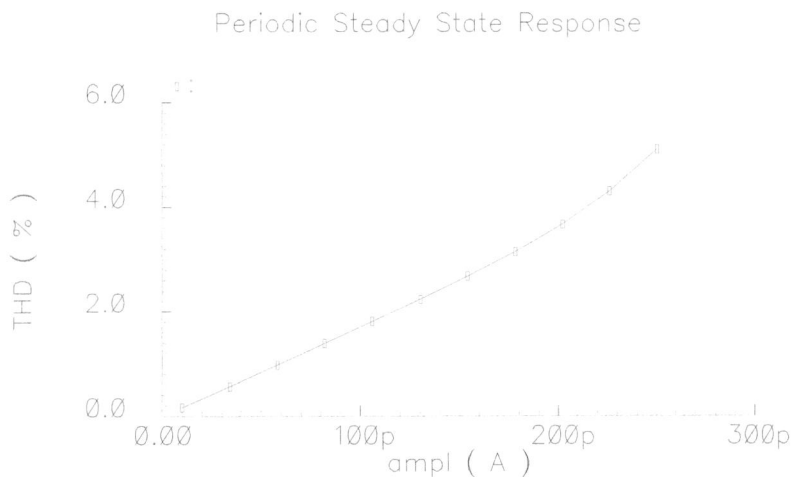

The linear performance of the filter has been evaluated by stimulating its input with a sinusoidal signal with frequency of 0.5 Hz and variable amplitude. The obtained total harmonic distortion (THD) plot is depicted in Figure 7, where THD levels equal to 2% and 4% are achieved for amplitudes of 120 and 220 pA, respectively. The noise has been integrated within the pass band of the filter and the input referenced RMS value of noise was 0.43 pA. For comparison purposes, the input signal that corresponds to a THD level of 4% has been considered, and the predicted value of the DR of the filter is 51.1 dB.

The sensitivity of the filter with regards to the effect of MOS transistors mismatch, as well as process parameter variations has been evaluated using Monte-Carlo analysis offered by the Cadence Analog Design Environment. The derived statistical plots for 100 runs are given in Figure 8. The standard deviation of the low frequency gain was 0.13, whereas that of the bandwidth was 0.4 Hz. Therefore, the proposed filter topology offers reasonable sensitivity characteristics.

Figure 8. Monte-Carlo analysis results for (**a**) low-frequency gain; and (**b**) the bandwidth (bw) of the low pass filter.

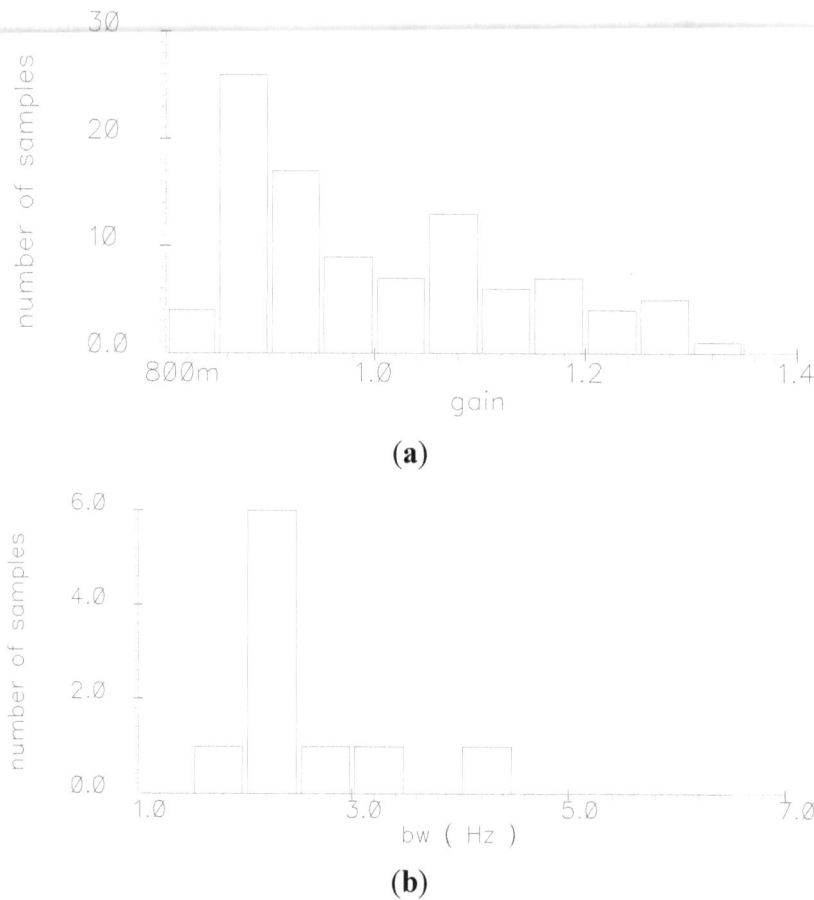

(**a**)

(**b**)

A performance comparison has been done between the proposed filter and those already introduced in [3,4]. The obtained results are summarized in Table 3, where the power efficiency of the topologies under consideration has been evaluated through the employment of the figure of merit (*FoM*) defined by Equation (11) as:

$$FoM = \frac{P}{n \cdot f_o \cdot (DR)} \tag{11}$$

where P is the power dissipation of the filter, n the number of poles, f_o the cutoff frequency and DR the dynamic range. According to the results in Table 3, the proposed filter simultaneously offers the benefits of ultra-low voltage operation and power efficiency.

The behavior of the system has been evaluated using the artificial ECG signal depicted in Figure 9a, obtained from the Massachusetts Institute of Technology/Beth Israel Hospital (MIT/BIH) database [10]. The resulting waveform of the filtered signal is demonstrated in Figure 9b, where it is easily observed that it is very close to that theoretically predicted. In order to facilitate the comparison, both input and output waveforms are simultaneously given in the plot of Figure 9c, where the effectiveness of the proposed system for sensing the T-wave of an ECG is readily verified.

Figure 9. Waveforms of: (**a**) input ECG signal; (**b**) filtered output (red: ideal, blue: real); and (**c**) input-output.

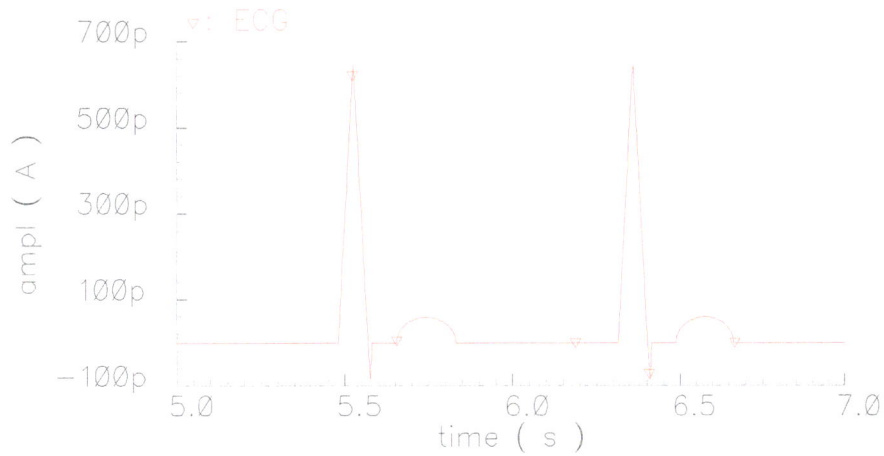

(**a**)

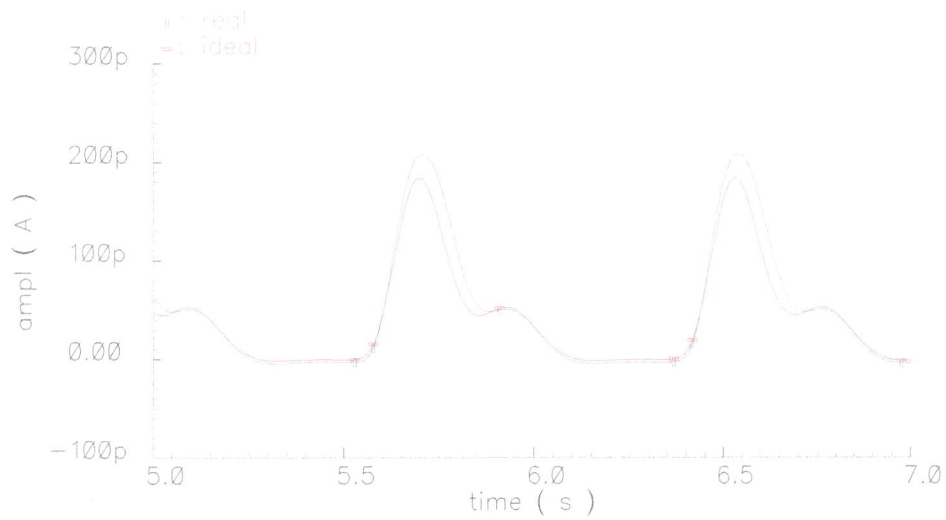

(**b**)

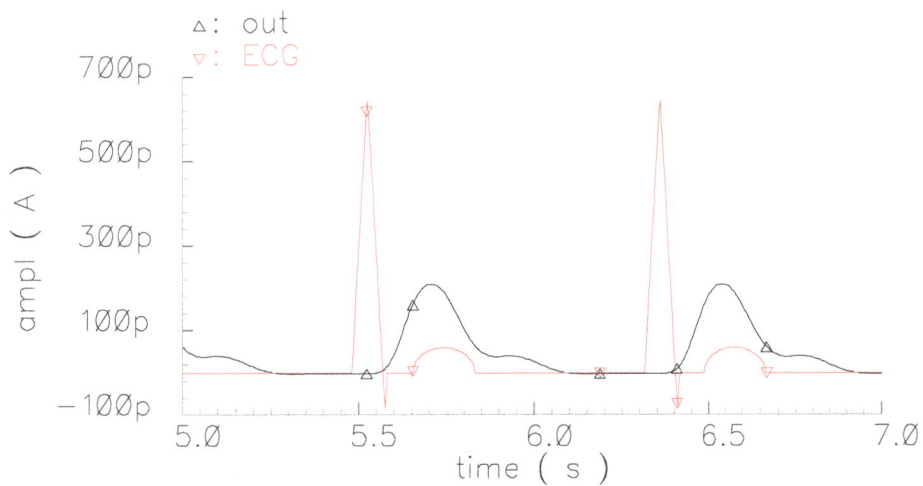

(**c**)

Table 3. MOS transistor aspect ratios for the non-linear transconductor in Figure 1.

Performance characteristics	LP filter in [3]	LP filter in [4]	LP filter in this work
Technology	0.8 μm CMOS	0.35 μm CMOS	0.35 μm CMOS
Power supply voltage	±1.5 V	±1 V	0.5 V
Order of filter	6	6	6
Power dissipation	10 μW	1.5 μW	7.21 nW
Cutoff frequency	2.4 Hz	2.4 Hz	2.4 Hz
Total capacitance	-	120-300 pF	80 pF
Dynamic range	60 dB (0.5% THD)	66 dB (4% THD)	51.1 dB (4% THD)
Group delay error within pass band	-	0.2%	0.2%
Figure of merit	694 pJ	52.2 pJ	1.4 pJ
Standard deviation of gain	-	-	0.13
Standard deviation of bandwidth	-	-	0.4 Hz

In order to verify the robustness of the design, the effect of process variations has been considered through the utilization of Monte-Carlo analysis. The derived output waveforms are demonstrated in Figure 10, where it is evident that the proposed filter preserves its effectiveness for sensing the T-wave of an ECG.

Figure 10. Output waveforms obtained through Monte-Carlo analysis.

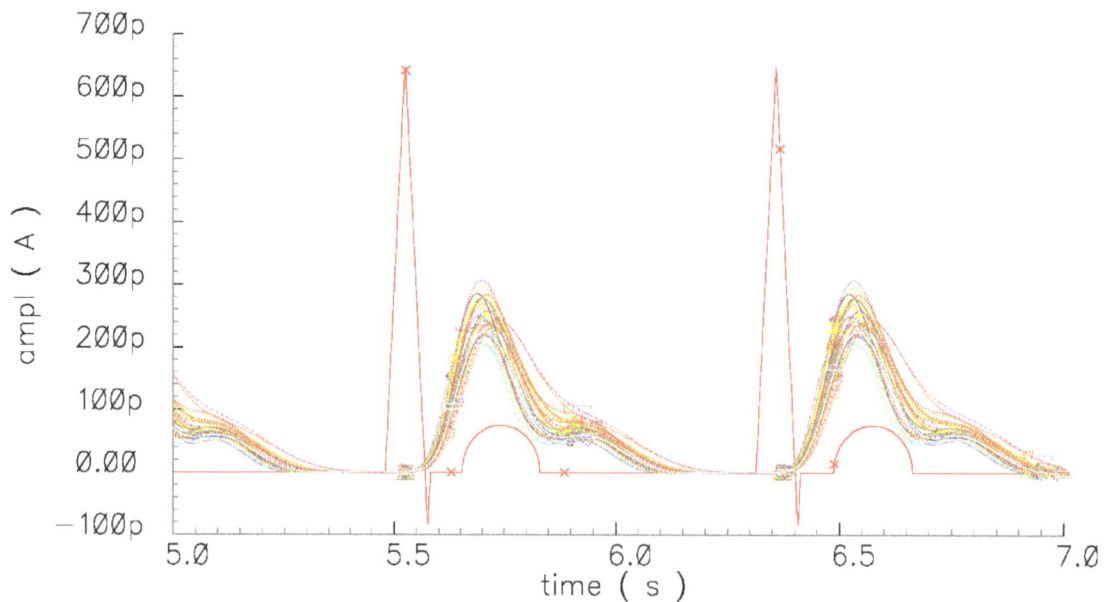

The performance of the proposed filter topology has been also evaluated with regards to the effect of noise. For this purpose, a noisy ECG has been obtained from that in Figure 9a through the addition of Gaussian noise (bandwidth 180 Hz), with a signal-to-noise ratio (SNR) equal to 0 dB, and has been employed as a stimulus for the filter. The derived output waveform is depicted, simultaneously with the input noisy ECG signal, in Figure 11, where it is evident that the filter is capable of detecting the T-wave of a noisy ECG.

Figure 11. Output waveform of the filter stimulated by a noisy ECG.

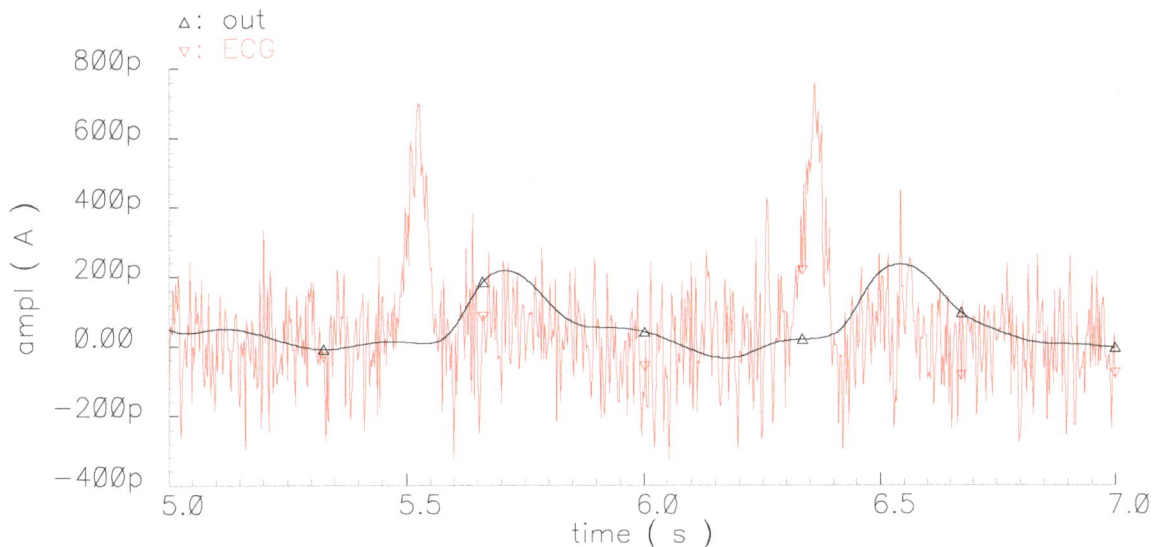

4. Conclusions

The proposed filter topology operates in a 0.5-V supply voltage environment, while the corresponding published topologies operate in ±1.5 V and ±1 V, respectively. In addition, the achieved DR is 51.1 dB under 7.21 nW power dissipation, while the corresponding factors for the schemes in the literature were 60 dB at 10 μW and 60 dB at 1.5 μW. Thus, the proposed topology simultaneously offers the benefits of ultra-low voltage operation and power efficiency. As a result, it can be considered an attractive candidate for realizing modern high-performance biomedical systems with enhanced battery life.

Author Contributions

Panagiotis Bertsias was responsible for the design and testing of the circuits described here. Costas Psychalinos helped to guide this research, to review the proposed circuits and to edit this paper.

Conflicts of Interest

The authors declare no conflict of interest.

References

1. Bailey, J.; Berson, A.; Garson, A.; Horan, L.; Macfarlane, P.; Mortara, D.; Zywietz, C. Recommendations for standardization and specifications in automated electrocardiography: Bandwidth and digital signal processing. *J. Am. Heart Assoc. Circ.* **1990**, *81*, 730–739.
2. Kligfield, P.; Gettes, L.; Bailey, J.; Childers, R.; Deal, B.; Hancock, W.; van Herpen, G.; Kors, J.; Macfarlane, P.; Mirvis, D.; *et al*. Recommendations for the standardization and interpretation of the electrocardiogram part I: The electrocardiogram and its technology. *J. Am. Heart Assoc. Circ.* **2007**, *115*, 1306–1324.

3. Solis-Bustos, S.; Silva-Martinez, J.; Maloberti, F.; Sanchez-Sinencio, E. A 60-dB dynamic-range CMOS sixth-order 2.4-Hz low-pass filter for medical applications. *IEEE Trans. Circuits Syst. II* **2000**, *47*, 1391–1398.

4. Kardoulaki, E.M.; Glaros, K.N.; Katsiamis, A.G.; Drakakis, E.M. An 8 Hz, 0.1 µW, 110+ dBs Sinh CMOS Bessel Filter for ECG signals. In Proceedings of International Conference on Microelectronics (ICM), Marrakech, Morocco, 19–22 December 2009; pp. 14–17.

5. Tsirimokou, G.; Laoudias, C.; Psychalinos, C. Tinnitus detector realization using Sinh-Domain circuits. *J. Low Power Electron.* **2013**, *9*, 458–470.

6. Khanday, F.A.; Pilavaki, E.; Psychalinos, C. Ultra low-voltage ultra low-power Sinh-Domain wavelet filer for electrocardiogram signal analysis. *J. Low Power Electron.* **2013**, *9*, 288–294.

7. Tsirimokou, G.; Psychalinos, C.; Khanday, F.A.; Shah, N.A. 0.5 V Sinh-Domain Differentiator. *Int. J. Electron.* **2014**, doi:10.1080/00207217.2014.901425.

8. Kafe, F.; Psychalinos, C. Realization of companding filters with large time-constants for biomedical applications. *Analog Integr. Circuits Signal Proc.* **2014**, *78*, 217–231.

9. Kafe, F.; Psychalinos, C. A 50 mHz Sinh-Domain high-pass filter for realizing an ECG signal acquisition system. *Circuits Syst. Signal Proc.* **2014**, doi:10.1007/s00034-014-9826-1.

10. ECGSYN: A realistic ECG waveform generator. Available online: http://www.physionet.org/physiotools/ecgsyn/ (accessed on 28 September 2014).

10

A Robust Ultra-Low Voltage CPU Utilizing Timing-Error Prevention [†]

Markus Hiienkari [1],*, Jukka Teittinen [1], Lauri Koskinen [1], Matthew Turnquist [2],
Jani Mäkipää [3], Arto Rantala [3], Matti Sopanen [3] and Mikko Kaltiokallio [4]

[1] Technology Research Center, University of Turku, Joukahaisenkatu 1C, 20520 Turku, Finland;
  E-Mails: jukka.teittinen@utu.fi (J.T.); lauri.koskinen@utu.fi (L.K.)

[2] Department of Micro and Nanosciences, Aalto University, Otakaari 5A, 02150 Espoo, Finland;
  E-Mail: matthew.turnquist@aalto.fi

[3] VTT Technical Research Centre of Finland, Tietotie 3, 02150 Espoo, Finland;
  E-Mails: jani.makipaa@vtt.fi (J.M.); arto.rantala@vtt.fi (A.R.); matti.sopanen@vtt.fi (M.S.)

[4] TDK, Keilaranta 8, 02601 Espoo, Finland; E-Mail: mikko.kaltiokallio@epcos.com

[†] This paper is an extended version of our paper published in IEEE S3S Conference 2014.

* Author to whom correspondence should be addressed; E-Mail: markus.hiienkari@utu.fi

Academic Editor: David Bol & Steven A. Vitale

Abstract: To minimize energy consumption of a digital circuit, logic can be operated at sub- or near-threshold voltage. Operation at this region is challenging due to device and environment variations, and resulting performance may not be adequate to all applications. This article presents two variants of a 32-bit RISC CPU targeted for near-threshold voltage. Both CPUs are placed on the same die and manufactured in 28 nm CMOS process. They employ timing-error prevention with clock stretching to enable operation with minimal safety margins while maximizing performance and energy efficiency at a given operating point. Measurements show minimum energy of 3.15 pJ/cyc at 400 mV, which corresponds to 39% energy saving compared to operation based on static signoff timing.

Keywords: timing-error prevention (TEP); Ultra-Low Power (ULP); digital CMOS; clock stretching; near-threshold; variability; energy-efficiency

1. Introduction

With constantly tightening power budgets and incentives to add new features, energy consumption has become one of the most important aspects of portable electronics. The minimum energy consumption of digital logic is traditionally achieved by utilizing a supply voltage which is below the transistor threshold voltage [1]. Operation in this sub-threshold -region has several practical limitations such as radically reduced performance and exponentially increased variability.

Recent manufacturing processes have pushed the minimum energy point (MEP) towards the threshold voltage, which mitigates performance and variance concerns and makes near-threshold operation much more attractive and practical. Near-threshold systems may still need to satisfy temporary high throughput requirements, which can be satisfied by utilizing dynamic voltage and frequency scaling (DVFS). In energy-constrained systems, supply voltage also varies slowly with time (battery, solar cell, *etc.*), creating additional challenges for ensuring efficient and reliable operation with a dynamic performance target. Despite the challenges, near-threshold computing has recently gained increasing amount of attention due to its suitability to IoT and other extremely energy-limited applications which yet have moderate performance requirements.

The goal of this research is to study applicability of timing-error prevention (TEP) to an ultra-low voltage CPU. We present two variants of a 32-bit RISC microprocessor which can operate reliably at near-threshold voltages with minimal safety margins. The CPU cores are optimized for design compatibility and Ultra-Dynamic Voltage Scaling (UDVS) [2] respectively. They are both placed on same die, which is manufactured in 28 nm CMOS technology.

Section 2 of this article describes more closely the operation at ultra-low voltages with background theory and examples, and presents the concepts of adaptive timing. The design process of the CPU cores is described in Section 3, and Section 4 presents the measurement results and analysis. Finally, Section 5 concludes the article.

2. Ultra-Low Voltage Operation

The goal of this section is to make the reader familiar with the principles and incentives of ultra-low voltage operation. This is followed by a description of the most important challenges, and two possible solutions to overcome them.

2.1. Motivation and Challenges

For digital static CMOS logic, when the performance constraints allow, the straightforward solution to energy minimization is to lower the operating voltage all the way to the MEP. This point has been proven to exist around 0.2–0.4 V depending on various factors [3]. With older process nodes, this is in the subthreshold operation region and for newer process nodes, in the near-threshold region. Figure 1 shows the MEP for 65 nm and 28 nm low-power (LP) bulk processes with different activity factors. On the 28 nm process, MEP resides near the threshold voltage. With lower leakage and higher V_t, the 65 nm process has MEP at approximately 100 mV lower voltage which resides in subthreshold region.

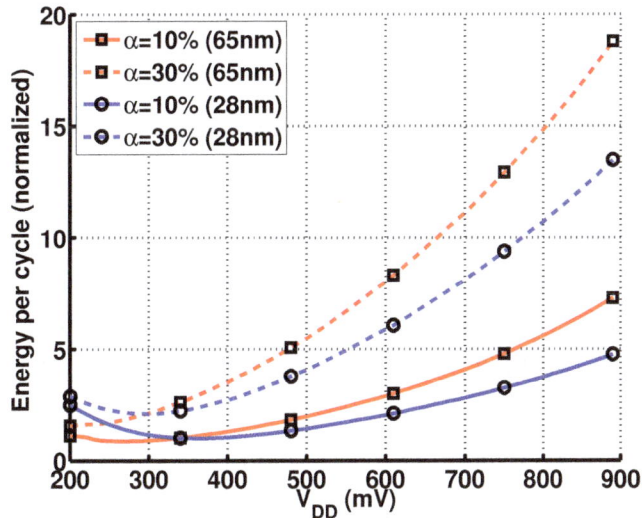

Figure 1. The minimum-energy points for 65 nm and 28 nm low-power (LP) bulk processes using low-V_t (LVT) transistors. α denotes the activity factor. The test circuit here is a 13 stage minimum-size negative-AND -gate (NAND) ring oscillator (based on a circuit presented in [3]).

However, the advantages of operating at the MEP are easily lost due to higher device variability. In subthreshold, process, supply voltage, temperature, and aging (PVTA) induced variance is amplified in the subthreshold region due to exponential dependency of the subthreshold current [3]:

$$I_{ds,sub} = I_0 \exp\left(\frac{V_{gs} - V_t}{nV_{th}}\right)\left(1 - \exp\left(\frac{-V_{ds}}{V_{th}}\right)\right) \tag{1}$$

In Equation (1), I_0 is the drain current when $V_{gs} = V_t$; V_t is the threshold voltage; n is the subthreshold slope factor; and V_{th} is the thermal voltage. While near-threshold operation mitigates the exponential dependency, increased variance still remains when compared to nominal operating voltages.

The theory is verified with a simulation of an example timing path from a test circuit in 28 nm CMOS. The path is simulated with 100 Monte Carlo runs, and Table 1 shows resulting nominal, minimum and maximum delays for near-threshold voltages. The results demonstrate the design challenge of near-threshold logic, as local variation can cause up to 30% delay deviation from nominal. Even without the exponential dependence on transistor parameters of subthreshold, robust operation clearly demands overly large design margins or individual post-fabrication measurements of the components; the former negating the minimum energy operation, and the latter increasing production costs considerably.

Table 1. Delay of an example critical path in 28 nm CMOS with 100 Monte-Carlo iterations.

V_{DD}	T_{avg} (ns)	T_{min} (ns)	T_{max} (ns)	σ (ns)
0.3 V	305	237 (−22.2%)	393 (+28.9%)	32.8 (10.8%)
0.4 V	40.1	31.9 (−20.4%)	52.0 (+29.7%)	3.9 (9.7%)
0.5 V	6.63	5.57 (−16.0%)	8.20 (+23.7%)	0.5 (7.3%)
0.6 V	1.90	1.71 (−10.0%)	2.17 (+14.2%)	0.09 (4.7%)

2.2. Adaptive Timing Methods

Here, we present 2 alternative solutions for solving the issue of large timing margins. Several researches have successfully applied these methods to a number of test chips.

2.2.1. Timing-Error Detection

Timing-Error Detection (TED)-based systems have been shown to be effective in largely removing the variation-incurred timing margins [4–6]. The achieved lower margins have then enabled energy savings (*i.e.*, lower V_{DD} [4]) or a higher yield [5]. The TED methodology is based on having the system operate at a voltage and frequency point in which the timing of critical paths fails intermittently. These failed timing occurrences are detected and handled. The overhead of detection and handling has to be lower than the energy savings resulting from lower V_{DD} or higher frequency. A classic example of a TED system is the instruction replay, where an instruction with a failed timing path is replayed [4].

2.2.2. Timing-Error Prevention

Timing-Error prevention (TEP) is a variant of TED which utilizes time borrowing (TB). When a system is timed for zero TB at normal operation, fractionally late signals can be tolerated without errors occurring in the system. However, TB sets timing requirements on the subsequent pipeline stages; namely the cumulative TB of consecutive stages must never exceed the TB window and therefore careful design time planning is required. Combining TED with TB into TEP conceives a system which can tolerate late coming signals, but which does not require special arrangements with regards to stage lengths. The method works as follows: When a late signal arrives, time borrowing occurs normally. Time borrow events (TBE) are detected with latches or special time borrow flip-flops. A recovery is necessary to prevent borrowed time from accumulating and generating a timing error. This can be done by moving the clock phase [7] or gating the clock on per-stage basis [8].

Since late signals are allowed with both TED and TEP technique, setup-timing margins are eliminated thereby allowing energy reduction. When TEP is integrated into a dual-phase latch pipeline, the resulting system can not only tolerate late signals, but does not require additional hold buffers on fast paths [8]. This is a large advantage compared to a traditional TED system. As shown in Table 1, minimum delay uncertainty is also clearly worsened at near-threshold voltages, which would significantly increase the demand for extra hold buffers on fast paths. A dual-phase latch pipeline can also be driven by 2 non-overlapping clocks, which increases skew tolerance and decreases general requirement for hold buffers.

With a balanced pipeline, a TEP system has margins but only when they might be required (when time borrowing is detected). Importantly, the system is a zero margin device at design time and adaptive margin device at runtime.

3. Design of a 32-bit RISC CPU with TEP

The CPU cores of our system are modified versions of a freely available, open-core LatticeMico32 CPU [9]. LM32 is a configurable medium-scale RISC microprocessor with 6 pipeline stages, full

GCC toolchain support and sufficient performance (1.14 DMIPS/MHz, 1.83 Coremark/MHz) even for demanding sensor network applications.

As described in [10,11], we have enhanced the CPUs with critical path monitoring combined with timing-error prevention. Our TEP system enables time borrowing on all paths. Time borrow events are detected with time borrow detector (TBD) circuits, which are integrated into critical path latches as illustrated in Figure 2a. The circuit indicates when TB occurs, and TBEs are combined and propagated to a clock control circuit, which is responsible for error prevention. The clock control circuit is a state machine consisting of a small number of flip-flops and logic gates (Figure 2b). It is able to shift the global clock phase after a TBE as illustrated in Figure 2c. Therefore, the TEP system prevents the stacking of TB, which could otherwise lead to a timing error. In addition to per-cycle timing error prevention, the real-time feedback from critical paths (combined TBEs) is output from the chip, allowing it to be used in tuning the microprocessor voltage/frequency in order to minimize die characterization effort.

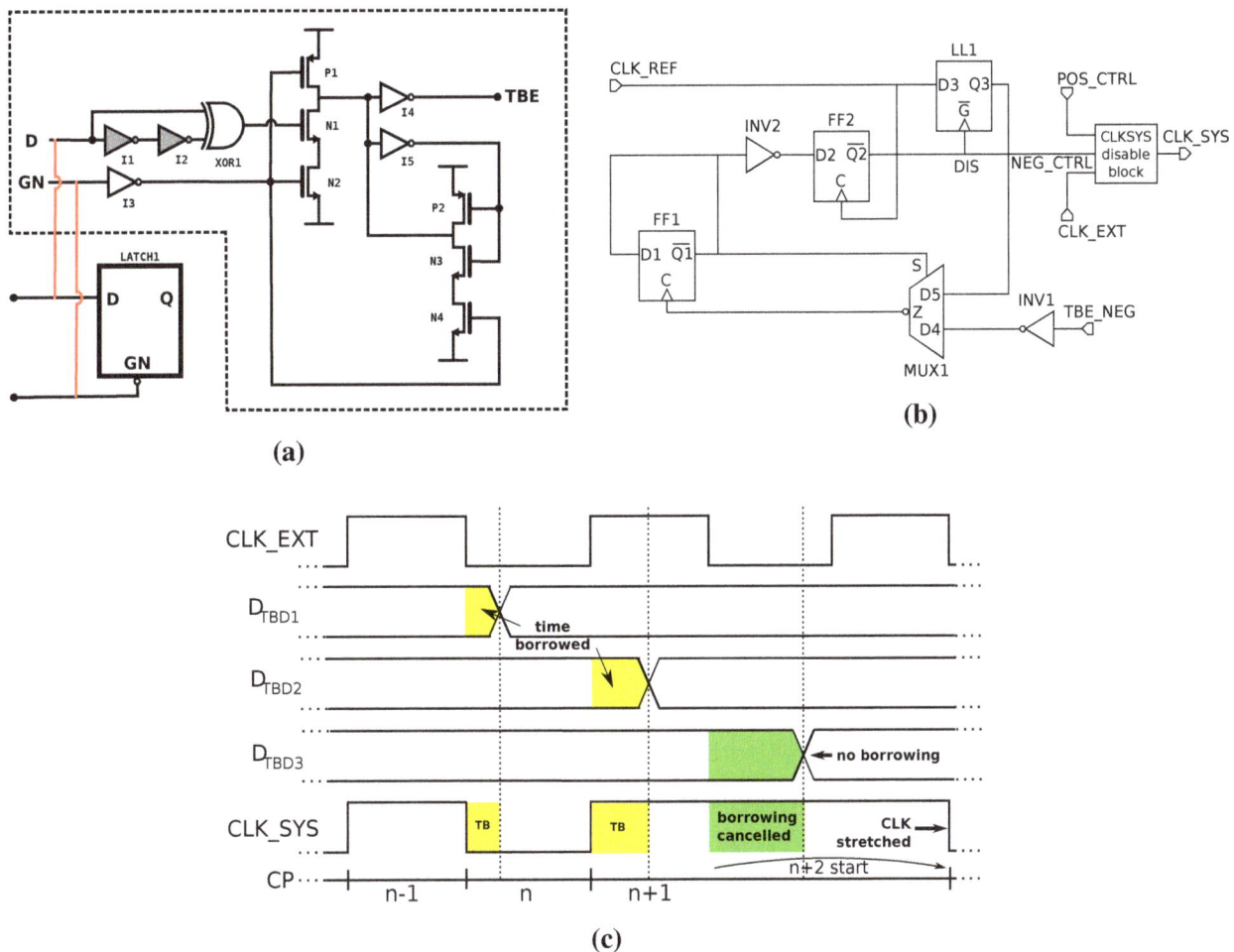

Figure 2. An overview of the timing-error prevention system. Based on [11] by the authors of this article. (a) The time borrow detector (TBD) circuit; (b) the clock control circuit (negative branch in detail); (c) a high-level illustration of adaptive clocking, where clock stretching prevents time borrow accumulation. © 2015 IEEE.

Since the unmodified LM32 is a standard edge-sensitive pipeline, additional design steps are required to make it suitable for TEP. Out of the compatible sequential cells, pulse-latches and TB flip-flops would be the most straightforward to integrate into an edge-sensitive design, but they would add minimum-delay constraints into fast paths and raise robustness concerns at near-threshold voltages. Standard latches are readily available in foundry design kits and have no robustness issues, but an edge-sensitive design needs to be fully transformed into a dual-phase latch pipeline to avoid half-cycle minimum delay requirement for critical paths. We created 2 CPU variants (CPU1 and CPU2) with dual-phase latch pipelines to study the applicability and effectiveness of different implementation strategies. As shown in Table 2, different design techniques were applied to the 2 CPUs. The fully custom TBD cell has 3.2× area of a minimum-sized latch, and the total TBD area overhead specified in the table is the percentage of combined TBD area out of standard cell area. The area overheads of OR-tree and clock control block are insignificant in comparision and are not listed.

Table 2. Design targets and results for the CPUs.

Design attribute	CPU1	CPU2
Pipeline implementation	Latch transformation + retime	RTL rewrite
Memory type	SRAM	Latch array
Supply voltage (mV)	400	400
Frequency target (MHz)	0.89	0.67
Sequential cells	3623 latches	2006 latches
Time-borrow detectors	272	256
TBD area overhead (%)	7.5	10.9
Die area (excl. memory) (mm^2)	0.030	0.022

3.1. CPU1

CPU1 uses automatic latch transformation and EDA-tool backed fixed-phase retiming [12] during synthesis, which is illustrated (for a single flip-flop) in Figure 3. The flip-flops are split into 2 latches, and the whole design is retimed in order to balance the paths. As the method is mostly automatic, it has been demonstrated to be applicable in larger designs [8]. However, there are some problems in practice. The balancing performed by the EDA tools is not guaranteed to be optimal—a carefully optimized flip-flop pipeline may turn into a pipeline where the path between the original master and slave latch becomes consistently shorter than the paths between the latch pairs, reducing overall efficiency. Also, some of the tools are not able to retime designs with multiple voltage domains. The method is also prone to increase clock network complexity, as the number of registers in the design increases.

As the target of CPU1 is to demonstrate applicability of TEP in a typical RISC processor, the core is supplemented with standard on-chip SRAM memory. This creates a new challenge, since it adds a hard edge-sensitive block inside otherwise level-sensitive design. As with [8], our design employs a SRAM wrapper, which moves reads and writes to the falling clock edge to ensure that SRAM input signals are always valid. However, this reduces the allowable time on the paths into subsequent latches. It also prevents usage of TBD latches at the end of these paths, since the data always transitions during their transparent phase, which would generate a possibly false positive. However, the vendor-provided SRAM in our design operates in a separate power domain of higher voltage due to retention requirements, making it significantly faster than the core logic and thereby eliminating the issue of stricter timing requirements. Based on simulations, a core voltage up to 0.5 V was shown to cause no SRAM performance bottleneck in the system.

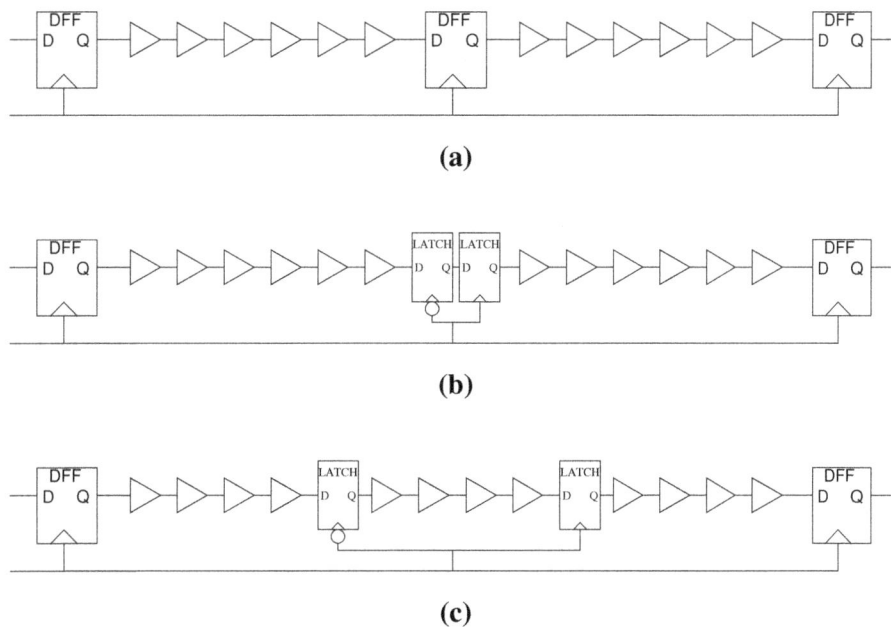

(a)

(b)

(c)

Figure 3. An example of latch transformation and retiming. This method is used in CPU1 implementation. (**a**) A standard pipeline with flip-flops; (**b**) The center flip-flop is split into 2 latches; (**c**) Retiming applied.

3.2. CPU2

A more optimal starting point for a TEP implementation would be a CPU design which has two-phase latch pipeline to start with, and a more flexible memory interface. In order to study the benefits of such base design, we created another variant of the core. CPU2 is manually transformed at register-transfer level (RTL) into dual-phase system before synthesis, and it uses a small latch array memory instead of SRAM. The former removes EDA tool limitations as the registers inside the design are synthesized directly to latches. The latter property removes the need of wrapper logic and memory path length constraining. Moreover, due to the lack of SRAM retention issues, the system is implemented in a single voltage domain. This relaxes voltage scaling limitations, and thus allows operation at an ultra-wide voltage range.

Figure 4 illustrates the RTL rewrite process through a simple 3-stage pipeline example. The edge-sensitive processes are changed into level-sensitive, and consecutive stages are assigned to alternating clock phases. Any paths starting and terminating on a stage with the same polarity—including the case where source and target register is the same (*pc_f* in the example)—must be supplemented by an additional synchronizing latch stage. Despite the additional latches, the total clock network load will be smaller than the original flip-flop design due to latch being smaller and simpler than a flip-flop. Changing a pipeline stage to operate on a single phase halves the achievable clock frequency, but also cuts the pipeline latency into half. It should be noted that conditional constructs in a level-sensitive block will explicitly synthesize into clock-gated latches. This is in contrast to an edge-sensitive block, where a similar construct would synthesize to a circular feedback path by default, unless clock-gating was enabled in the synthesis tool.

```
// These will synthesize into flip-flops
reg [31:0] pc_f;
reg [31:0] inst_f;
reg [9:0] x_result;
reg [9:0] oper_x_0;
reg [9:0] oper_x_1;

// Fetch, decode and execute stages
always @(posedge clk)
begin
    inst_f <= #1 memory[pc_f];
    pc_f <= #1 pc_f + 4;
    oper_x_0 <= #1 inst_f[9:0];
    oper_x_1 <= #1 inst_f[19:10];
    x_result <= #1 oper_x_0 + oper_x_1;
end
```

(a)

```
// These will synthesize into latches
reg [31:0] pc_f;
reg [31:0] pc_f_DEL;
reg [31:0] inst_f;
reg [9:0] x_result;
reg [9:0] oper_x_0;
reg [9:0] oper_x_1;

// Fetch stage (POSITIVE phase)
always @(*)
if (clk == 'TRUE)
begin
    inst_f <= #1 memory[pc_f_DEL];
    pc_f <= #1 pc_f_DEL + 4;
end

// Syncronizing latch stage for pc_f
always @(*)
if (clk == 'FALSE)
    pc_f_DEL <= #1 pc_f;

// Decode stage (NEGATIVE phase)
always @(*)
if (clk == 'FALSE)
begin
    oper_x_0 <= #1 inst_f[9:0];
    oper_x_1 <= #1 inst_f[19:10];
end

// Execute stage (POSITIVE phase)
always @(*)
if (clk == 'TRUE)
    x_result <= #1 oper_x_0 + oper_x_1;
```

(b)

Figure 4. An example of register-transfer level (RTL) rewrite for dual-phase latch pipeline. This method is used in CPU2 implementation. (**a**) Standard edge-sensitive pipeline; (**b**) equivalent level-sensitive pipeline.

4. Measurement Results and Discussion

Based on the simulations and estimated average activity, the CPUs were synthesized and place & routed at the estimated MEP of 400 mV. With the exception of TBD and level shifters (in CPU1), all gates were from vendor standard cell library, which was re-characterized at the target operation point. As shown in Table 3 and explained in Section 3.2, the frequency of CPU2 is lower due to its pipeline structure, but its area is also smaller as shown in the chip microphotograph in Figure 5.

Silicon measurements of 6 test chips verified correct operation for both CPUs at their signoff frequencies, but due to safety margins and inaccuracies in timing libraries, optimal energy or performance was not achieved. This was evidenced by studying timing feedback from critical paths, which showed that TEP was not activated. Further testing verified that the CPUs could be run at significantly higher frequencies until time borrowing started to occur. Since the clock stretching halves the effective frequency when activated, the optimal frequency was set so that time borrow events only occurred rarely.

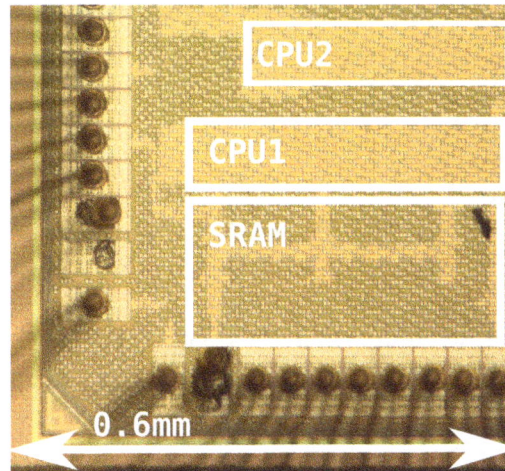

Figure 5. Test chip microphotograph. Based on [10] by the authors of this article. © 2015 IEEE.

Table 3. Comparison with state of the art. Based on [10] by the authors of this article. © 2015 IEEE.

Design metric	MIT, JSSC [13]	Sleepwalker [14]	This Work (CPU2)
Technology	65 nm CMOS	65 nm CMOS LP/GP	28 nm CMOS LP
CPU	16-bit MSP430 compatible	16-bit MSP430 compatible	32-bit LatticeMico32
CPU area (mm^2)	0.14	approx. 0.12	0.022
Ext. supply (V)	1.2	1.0–1.2	1.0 V–1.5 V
Int Vdd (V)	0.3–0.6	0.32–0.48	0.25–0.5
Frequency (MHz)	0.43 (25 C)	25	2.4 (0.4 V/25 C)
CPU energy (pJ/cyc)	6–10	2.6	3.15
DMIPS/MHz	0.45	0.45	1.14
DMIPS/mJ (estim.)	60	173	362

Voltage was scaled next to study energy/performance tradeoff at various operation points (Figure 6a,b). The lower bound (250 mV) was based on the simulated functional failure rate as described in [10], while the upper bound of CPU1 was limited by fixed SRAM performance. Voltage of CPU2 could be scaled up to 750 mV, after which further performance improvements were not possible as the externally generated clock became the limiting factor. If an on-chip PLL was used for clock generation, we estimate that CPU2 would operate approximately at 300 MHz with a nominal voltage of 1.0 V.

Figures 6a–c illustrate power distribution, performance and energy of the CPUs running a stress test at the measured voltage/frequency ranges. The dynamic energy consumed by logic is similar for both CPUs as both are inherently same architecture, but the more complicated clock network of CPU1 results to higher total dynamic energy. CPU1 also has higher leakage power, but when integrated over a clock period, the resulting leakage energy is slightly smaller because of higher clock frequency. Thus, when leakage power is not dominant (from 350 mV upwards), CPU2 is over 30% more energy-efficient than CPU1.

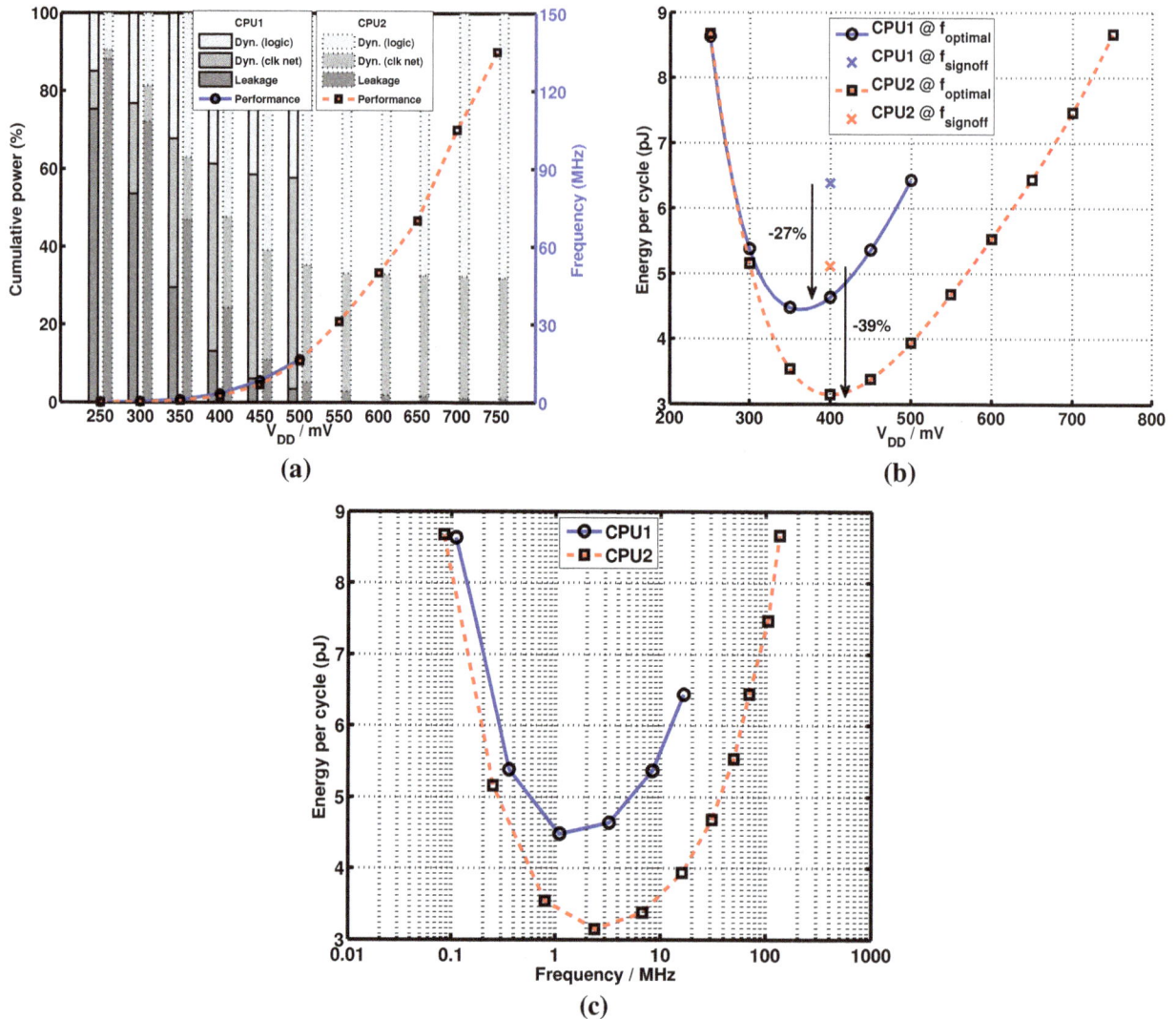

Figure 6. Measurement results for CPU1 and CPU2. (**a**) Power distribution and performance at optimal frequency; (**b**) E_{cyc} at optimal and signoff-estimated frequencies; (**c**) energy versus performance.

The frequency of CPU1 ranges from 110 kHz to 16.5 MHz, and its minimum energy point (4.48 pJ/cyc) is at around 350 mV. The ultra-wide voltage scale of CPU2 allows it to operate up to 135 MHz, making it suitable for a wide number of usage scenarios. Minimum energy consumption (3.15 pJ/cyc) is achieved at a near-threshold voltage of 400 mV. Compared to operation based on static signoff timing, TEP-assisted DVFS allows reducing the energy consumption by 27% and 39% for CPU1 and CPU2 respectively.

The results demonstrate that TEP can utilized successfully in a typical RISC processor with a few automated extra steps during the design flow (CPU1). However, largest benefits are achieved with a processor, which is designed TEP in mind (CPU2). Such processor design would ideally have a dual-phase level-sensitive pipeline, and no hard edge-sensitive blocks. The measured energy consumption is compared with state of the art in Table 3, which shows that CPU2 achieves over $2\times$ lower effective energy than competition with smaller area.

5. Conclusions

Ultra-low voltage CPUs are not adopted widely in commercial systems due to complex design and reliability concerns. Shown here, is a simple adaptive technique, timing-error prevention. This technique helps the two presented CPU variants to operate reliably with minimal safety margins in order to ensure optimal performance and energy consumption. The first presented CPU is converted from a standard RISC design to a TEP-compatible system with minimal amount of manual intervention, demonstrating the technique's applicability to contemporary designs. The second CPU variant is specifically designed with ultra-low voltage operation and TEP in mind, enabling higher energy savings and ultra-wide voltage operation from 250 mV to 750 mV. Compared to operation at static signoff-based frequency, the TEP system reduces energy consumption of the presented CPUs by 27% and 39%.

Acknowledgments

Technology Industries of Finland Centennial Foundation and Academy of Finland #270585.

Author Contributions

Markus Hiienkari designed and implemented majority of the digital parts of the system, and wrote test software and the manuscript. Matthew Turnquist designed analog blocks residing on the same chip and assisted in the optimization of TBD cell. Jani Mäkipää designed the clock control block, and Jukka Teittinen assisted in layout-related tasks. Matti Sopanen characterized the standard cell libraries used in the design, and Arto Rantala designed pad cells for the chip. Mikko Kaltiokallio helped in generating on-chip power supply network, and Lauri Koskinen was the system developer and coordinated the research project.

Conflicts of Interest

The authors declare no conflict of interest.

References

1. Bol, D.; Ambroise, R.; Flandre, D.; Legat, J.D. Interests and Limitations of Technology Scaling for Subthreshold Logic. *IEEE Trans. Very Large Scale Integr. (VLSI) Syst.* **2009**, *17*, 1508–1519.
2. Calhoun, B.; Chandrakasan, A. Ultra-Dynamic Voltage scaling (UDVS) using sub-threshold operation and local Voltage dithering. *IEEE J. Solid-State Circuits* **2006**, *41*, 238–245.
3. Wang, A.; Calhoun, B.H.; Chandrakasan, A.P. *Sub-Threshold Design for Ultra Low-Power Systems (Series on Integrated Circuits and Systems)*; Springer-Verlag New York, Inc.: Secaucus, NJ, USA, 2006.

4. Das, S.; Tokunaga, C.; Pant, S.; Ma, W.H.; Kalaiselvan, S.; Lai, K.; Bull, D.; Blaauw, D. RazorII: *In Situ* Error Detection and Correction for PVT and SER Tolerance. *IEEE J. Solid-State Circuits* **2009**, *44*, 32–48.

5. Bowman, K.; Tschanz, J.; Lu, S.; Aseron, P.; Khellah, M.; Raychowdhury, A.; Geuskens, B.; Tokunaga, C.; Wilkerson, C.; Karnik, T.; *et al.* A 45 nm Resilient Microprocessor Core for Dynamic Variation Tolerance. *IEEE J. Solid-State Circuits* **2011**, *46*, 194–208.

6. Mäkipää, J.; Turnquist, M.J.; Laulainen, E.; Koskinen, L. Timing-Error Detection Design Considerations in Subthreshold: An 8-bit Microprocessor in 65 nm CMOS. *J. Low Power Electron. Appl.* **2012**, *2*, 180–196.

7. Chae, K.; Mukhopadhyay, S. A Dynamic Timing Error Prevention Technique in Pipelines with Time Borrowing and Clock Stretching. *IEEE Trans. Circuits Syst. I* **2014**, *61*, 74–83.

8. Fojtik, M.; Fick, D.; Kim, Y.; Pinckney, N.; Harris, D.; Blaauw, D.; Sylvester, D. Bubble Razor: Eliminating Timing Margins in an ARM Cortex-M3 Processor in 45 nm CMOS Using Architecturally Independent Error Detection and Correction. *IEEE J. Solid-State Circuits* **2013**, *48*, 66–81.

9. LatticeMico32: Open, Free 32-Bit Soft Processor. Available online: http://www.latticesemi.com/ en/Products/DesignSoftwareAndIP/IntellectualProperty/IPCore/IPCores02/LatticeMico32.aspx (accessed on 20 Febuary 2015).

10. Hiienkari, M.; Teittinen, J.; Koskinen, L.; Turnquist, M.; Kaltiokallio, M. A 3.15 pJ/cyc 32-bit RISC CPU with Timing-Error Prevention and Adaptive Clocking in 28 nm CMOS. In Proceedings of the 2014 IEEE Custom Integrated Circuits Conference (CICC), San Jose, CA, USA, 15–17 September 2014; pp. 1–4.

11. Hiienkari, M.; Teittinen, J.; Koskinen, L.; Turnquist, M.; Kaltiokallio, M.; Makipaa, J.; Rantala, A.; Sopanen, M. Ultra-Wide Voltage Range 32-bit RISC CPU with Timing-Error Prevention in 28 nm CMOS. In Proceedings of the 2014 IEEE SOI-3D-Subthreshold Microelectronics Technology Unified Conference (S3S), Millbrae, CA, USA, 6–9 October 2014; pp. 1–2.

12. Yoshikawa, K.; Hagihara, Y.; Kanamaru, K.; Nakamura, Y.; Inui, S.; Yoshimura, T. Timing Optimization by Replacing Flip-Flops to Latches. In Proceedings of the 2004 Asia and South Pacific Design Automation Conference, ASP-DAC '04, Singapore, 27–30 January 2004; IEEE Press: Piscataway, NJ, USA, 2004; pp. 186–191.

13. Kwong, J.; Ramadass, Y.; Verma, N.; Chandrakasan, A. A 65 nm Sub-Vt Microcontroller with Integrated SRAM and Switched Capacitor DC-DC Converter. *IEEE J. Solid-State Circuits* **2009**, *44*, 115–126.

14. Bol, D.; de Vos, J.; Hocquet, C.; Botman, F.; Durvaux, F.; Boyd, S.; Flandre, D.; Legat, J. SleepWalker: A 25-MHz 0.4-V Sub-mm^2 7-μW/MHz Microcontroller in 65-nm LP/GP CMOS for Low-Carbon Wireless Sensor Nodes. *IEEE J. Solid State Circuits* **2013**, *48*, 20–32.

11

Multi-Threshold NULL Convention Logic (MTNCL): An Ultra-Low Power Asynchronous Circuit Design Methodology

Liang Zhou [1], Ravi Parameswaran [2], Farhad A. Parsan [2], Scott C. Smith [3,*] and Jia Di [4]

[1] Advanced Micro Devices, Inc., Sunnyvale, CA 94089, USA; E-Mail: kingdom701@gmail.com

[2] Department of Electrical Engineering, University of Arkansas, Fayetteville, AR 72701, USA;
E-Mails: ravisp.g@gmail.com (R.P.); fparsan@uark.edu (F.A.P.)

[3] Department of Electrical & Computer Engineering, North Dakota State University, Fargo,
ND 58108, USA

[4] Department of Computer Science & Computer Engineering, University of Arkansas, Fayetteville,
AR 72701, USA; E-Mail: jdi@uark.edu

* Author to whom correspondence should be addressed; E-Mail: scott.smith.1@ndsu.edu

Academic Editor: Alexander Fish

Abstract: This paper develops an ultra-low power asynchronous circuit design methodology, called Multi-Threshold NULL Convention Logic (MTNCL), also known as Sleep Convention Logic (SCL), which combines Multi-Threshold CMOS (MTCMOS) with NULL Convention Logic (NCL), to yield significant power reduction without any of the drawbacks of applying MTCMOS to synchronous circuits. In contrast to other power reduction techniques that usually result in large area overhead, MTNCL circuits are actually smaller than their original NCL versions. MTNCL utilizes high-V_t transistors to gate power and ground of a low-V_t logic block to provide for both fast switching and very low leakage power when idle. To demonstrate the advantages of MTNCL, a number of 32-bit IEEE single-precision floating-point co-processors were designed for comparison using the 1.2 V IBM 8RF-LM 130 nm CMOS process: original NCL, MTNCL with just combinational logic (C/L) slept, Bit-Wise MTNCL (BWMTNCL), MTNCL with C/L and completion logic slept, MTNCL with C/L, completion logic, and registers slept, MTNCL with Safe Sleep architecture, and synchronous MTCMOS. These designs are compared in terms of throughput, area, dynamic energy, and idle power, showing the tradeoffs between the various MTNCL architectures, and that the best MTNCL design is much better than the original NCL design in all aspects, and much better than the synchronous MTCMOS design in terms of area, energy per operation, and idle power, although the synchronous design can operate faster.

Keywords: NULL convention logic (NCL); multi-threshold CMOS (MTCMOS); sleep convention logic (SCL)

1. Introduction

With the current trend of semiconductor devices scaling into the deep submicron region, design challenges that were previously minor issues have now become increasingly important. Where in the past, dynamic switching power has been the predominant factor in CMOS digital circuit power dissipation, recently, with the dramatic decrease of supply and threshold voltages, a significant growth in leakage power demands new design methodologies for digital Integrated Circuits (ICs). The main component of leakage power is sub-threshold leakage, caused by current flow through a transistor even if it is supposedly turned off. Sub-threshold leakage increases exponentially with decreasing transistor feature size.

Among the many techniques proposed to control or minimize leakage power in deep submicron technology, Multi-Threshold CMOS (MTCMOS) [1], which reduces leakage power by disconnecting the power supply from the circuit during idle (or sleep) mode while maintaining high performance in active mode, is very promising. MTCMOS incorporates transistors with two or more different threshold voltages (V_t) in a circuit. Low-V_t transistors offer fast speed but have high leakage, whereas high-V_t transistors have reduced speed but far less leakage current. MTCMOS combines these two types of transistors by utilizing low-V_t transistors for circuit switching to preserve performance, and high-V_t transistors to gate the circuit power supply to significantly decrease sub-threshold leakage.

Quasi-delay-insensitive (QDI) NULL Convention Logic (NCL) circuits [2], designed using CMOS, exhibit an inherent idle behavior since they only switch when useful work is being performed; however, there is still significant leakage power during idle mode. This paper combines the MTCMOS technique with NCL to sleep the NCL circuit during idle mode, in lieu of the NULL cycle, to yield a fast ultra-low power asynchronous circuit design methodology, called Multi-Threshold NULL Convention Logic (MTNCL), also referred to as Sleep Convention Logic (SCL), which requires less area than the original NCL circuit.

Section 2 provides an overview of NCL, MTCMOS, and previous MTNCL implementations. Section 3 develops enhancements to the MTNCL technique that allows for both registration and completion logic to be slept along with the combinational logic (C/L). Section 4 compares the various implementations; and Section 5 provides conclusions.

2. Previous Work

2.1. Introduction to NCL

NCL circuits utilize multi-rail logic, such as dual-rail, to achieve delay-insensitivity. A dual-rail signal, D, consists of two wires, D^0 and D^1, which may assume any value from the set {DATA0, DATA1, NULL}. The DATA0 state ($D^0 = 1$, $D^1 = 0$) corresponds to a Boolean logic 0, the DATA1 state ($D^0 = 0$, $D^1 = 1$) corresponds to a Boolean logic 1, and the NULL state ($D^0 = 0$, $D^1 = 0$) corresponds to the empty set, meaning that the value of D is not yet available. The two rails are mutually exclusive, such that both rails can never be asserted simultaneously; this state is defined as an illegal state. Dual-rail logic is a space optimal 1-hot delay-insensitive code, requiring two wires per bit.

NCL circuits are comprised of 27 fundamental gates [3]. These 27 gates constitute the set of all functions consisting of four or fewer variables. Here, a variable refers to one rail of a multi-rail signal; hence, a four variable function is not the same as a function of four literals, which would consist of eight variables, assuming dual-rail logic. The primary type of threshold gate, shown in Figure 1, is the THmn gate, where $1 \leq m \leq n$. THmn gates have n inputs; at least m of the n inputs must be asserted before the output will become asserted; and NCL threshold gates are designed with *hysteresis* state-holding capability, such that all asserted inputs must be de-asserted before the output will be de-asserted. Hysteresis ensures a complete transition of inputs back to NULL before asserting the output associated with the next wavefront of input data. Therefore, a THnn gate is equivalent to an n-input C-element [4] and a TH1n gate is equivalent to an n-input OR gate. Besides the static NCL gate implementation, shown in Figure 2, there are other CMOS implementations, as detailed in [5,6]. In a THmn gate, each of the n inputs is connected to the rounded portion of the gate; the output emanates from the pointed end of the gate; and the gate's threshold value, m, is written inside of the gate. NCL threshold gates may also include a *reset* input to initialize the output. These resettable gates are used in the design of DI registers [7].

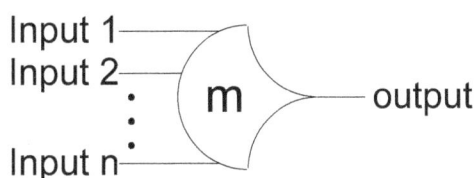

Figure 1. THmn threshold gate.

NCL systems contain at least two DI registers, one at both the input and at the output, and can be finely pipelined by inserting additional registers, as shown in Figure 3. Two adjacent register stages interact through their request and acknowledge signals, K_i and K_o, respectively, to prevent the current DATA wavefront from overwriting the previous DATA wavefront, by ensuring that the two DATA wavefronts are always separated by a NULL wavefront. The acknowledge signals are combined in the Completion Detection circuitry to produce the request signal(s) to the previous register stage, utilizing either the full-word or bit-wise completion strategy [7].

To ensure delay-insensitivity, NCL circuits must adhere to the following criteria: Input-Completeness and Observability [8]. Input-Completeness requires that all outputs of a combinational circuit may not transition from NULL to DATA until all inputs have transitioned from NULL to DATA, and that all outputs of a combinational circuit may not transition from DATA to NULL until all inputs have transitioned from DATA to NULL. In circuits with multiple outputs, it is acceptable according to Seitz's "weak conditions" of DI signaling [9] for some of the outputs to transition without having a complete input set present, as long as all outputs cannot transition before all inputs arrive. Observability requires that no *orphans* may propagate through a gate [10]. An orphan is defined as a wire that transitions during the current DATA wavefront, but is not used in the determination of the output. Orphans are caused by wire forks and can be neglected through the isochronic fork assumption [11,12], as long as they are not allowed to cross a gate boundary. This *observability* condition, also referred to as indicatability or stability, ensures that every gate transition is observable at the output, which means that every gate that transitions is necessary to transition at least one of the outputs.

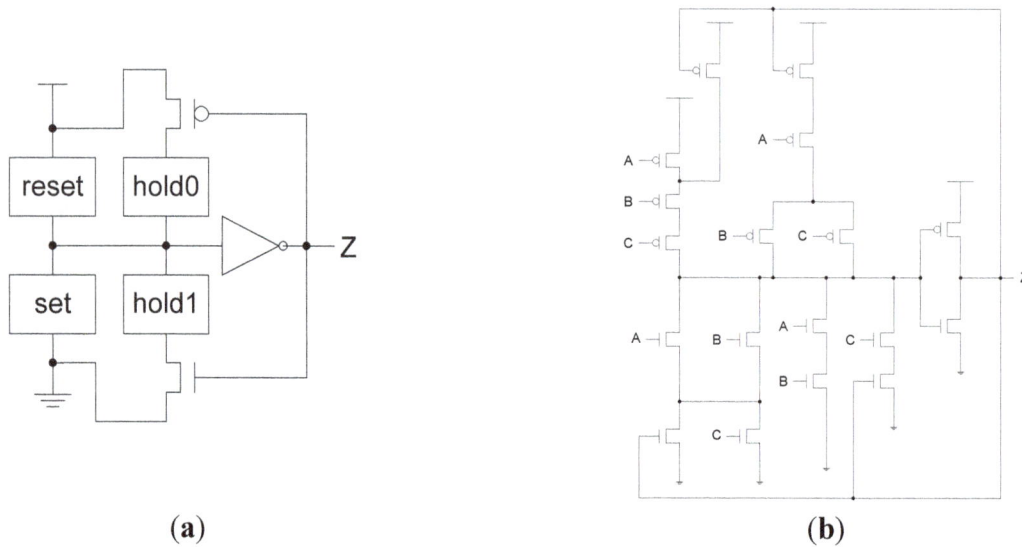

| (a) | (b) |

Figure 2. NCL threshold gate design. (**a**) General static implementation; (**b**) TH23 static implementation.

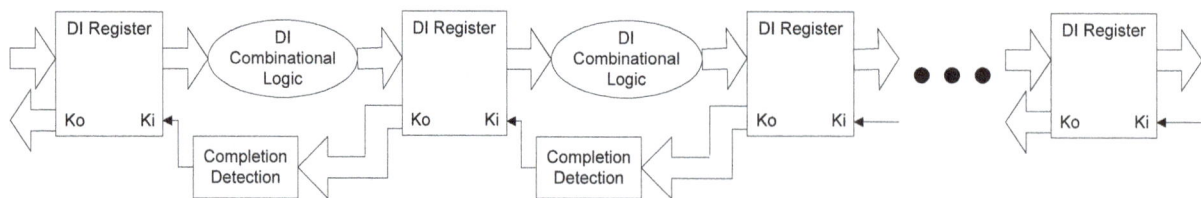

Figure 3. NCL system framework: input wavefronts are controlled by local handshaking and Completion Detection signals.

2.2. Introduction to MTCMOS

MTCMOS reduces leakage power by disconnecting the power supply from the circuit during idle (or sleep) mode while maintaining high performance in active mode, by utilizing different transistor threshold voltages (V_t) [1]. Low-V_t transistors are faster but have high leakage, whereas high-V_t transistors are slower but have far less leakage current. MTCMOS combines these two types of transistors by utilizing low-V_t transistors for circuit switching to preserve performance and high-V_t transistors to gate the circuit power supply to significantly decrease sub-threshold leakage.

One MTCMOS method uses low-V_t transistors for critical paths to maintain high performance, while using slower high-V_t transistors for the non-critical paths to reduce leakage. Besides this path replacement methodology, there are two other architectures for implementing MTCMOS. A course-grained technique investigated in [13] uses low-V_t logic for all circuit functions, and gates the power to entire logic blocks with high-V_t sleep transistors, denoted by a dotted circle, as shown in Figure 4. The sleep transistors are controlled by a *Sleep* signal. During active mode, the *Sleep* signal is de-asserted, causing both high-V_t transistors to turn on and provide a virtual power and ground to the low-V_t logic. When the circuit is idle, the *Sleep* signal is asserted, forcing both high-V_t transistors to turn off and disconnect power from the low-V_t logic, resulting in a very low sub-threshold leakage current. One major drawback of this method is that partitioning the circuit into appropriate logic blocks and sleep transistor sizing is difficult for large circuits. An alternative fine-grained architecture, shown in Figure 5, incorporates the MTCMOS technique within every gate [14], using low-V_t transistors for the Pull-Up Network (PUN) and Pull-Down Network (PDN) and a high-V_t transistor to gate the leakage current between the two networks. Two additional low-V_t transistors are included in parallel with the PUN and PDN to maintain nearly equivalent voltage potential across these networks during sleep mode (*i.e.*, X1 is approximately V_{DD} and X2 is approximately GND). Implementing MTCMOS within each gate solves the problems of logic block partitioning and sleep transistor sizing, since each gate is sized separately comprising the gate library; however, this results in a large area overhead.

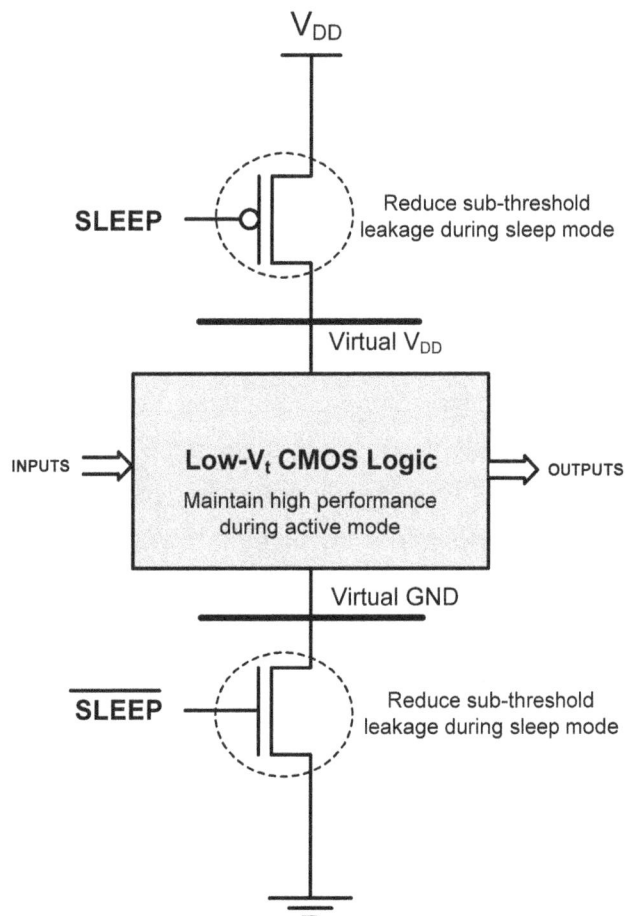

Figure 4. General MTCMOS circuit architecture [13].

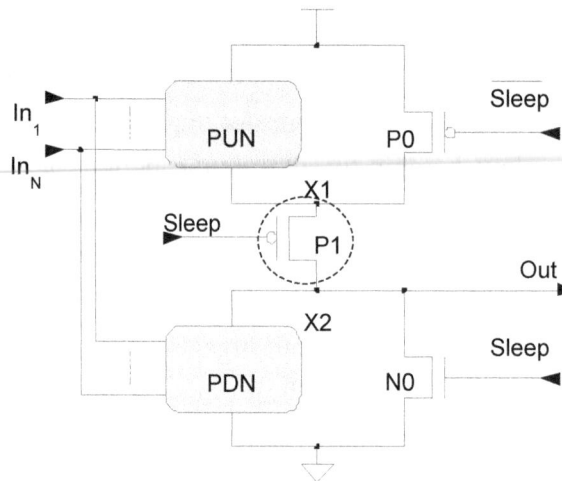

Figure 5. MTCMOS applied to a Boolean gate [14].

In general, three serious drawbacks hinder the widespread usage of MTCMOS in synchronous circuits [13]: (1) the generation of *Sleep* signals is timing critical, often requiring complex logic circuits; (2) synchronous storage elements lose data when the power transistors are turned off during sleep mode, although this can be solved with special register designs; and (3) logic block partitioning and transistor sizing is very difficult for the course-grained approach, which is critical for correct circuit operation, and the fine-grained approach requires a large area overhead. However, all three of these drawbacks are eliminated by utilizing NCL in conjunction with the MTCMOS technique.

2.3. Introduction to MTNCL

MTNCL was originally developed in [15–19], as summarized below, while this paper provides significant enhancements to the original MTNCL concept.

2.3.1. Early-Completion Input-Incomplete (ECII) MTNCL Architecture

NCL threshold gates are larger and implement more complicated functions than basic Boolean gates, such that fewer threshold gates are normally needed to implement an arbitrary function compared to the number of Boolean gates; however, the NCL implementation often requires more transistors. Therefore, incorporating MTCMOS inside each threshold gate facilitates easy sleep transistor sizing without requiring as large of an area overhead. Since floating nodes may result in substantial short circuit power consumption at the following stage, an MTCMOS structure similar to the one shown in Figure 5 is used to pull the output node to ground during sleep mode. When all MTNCL gates in a pipeline stage are in sleep mode, such that all gate outputs are logic 0, this condition is equivalent to the pipeline stage being in the NULL state. Hence, after each DATA cycle, all MTNCL gates in a pipeline stage can be forced to output logic 0 by asserting the sleep control signal instead of propagating a NULL wavefront through the stage, such that data is not lost during sleep mode.

Since the completion detection signal, K_o, indicates whether the corresponding pipeline stage is ready to undergo a DATA or NULL cycle, K_o can be naturally used as the sleep control signal, without requiring any additional hardware, in contrast to the complex *Sleep* signal generation circuitry needed for synchronous MTCMOS circuits. Unfortunately, the direct implementation of this idea using regular

NCL completion compromises delay-insensitivity [18]. To solve this problem, Early Completion [20] can be used in lieu of regular completion, as shown in Figure 6, where each completion signal is used as the sleep signal for all threshold gates in the subsequent pipeline stage. Early Completion utilizes the inputs of register$_{i-1}$ along with the K_i request to register$_{i-1}$, instead of just the outputs of register$_{i-1}$ as in regular completion, to generate the request signal to register $_{i-2}$, Ko_{i-1}. The combinational logic will not be put to sleep until all inputs are NULL and the stage is requesting NULL; therefore, the NULL wavefront is ready to propagate through the stage, so the stage can instead be put to sleep without compromising delay-insensitivity. The stage will then remain in sleep mode until all inputs are DATA and the stage is requesting DATA, and is therefore ready to evaluate. This Early Completion MTNCL architecture, denoted as *ECII*, ensures input-completeness through the sleep mechanism (*i.e.*, the circuit is only put to sleep after all inputs are NULL, when all gates are then simultaneously forced to logic 0, and only makes an evaluation after all inputs are DATA), such that input-incomplete logic functions can be used to design the circuit, which decreases area and power and increases speed. Note that sleeping the C/L in lieu of propagating a NULL wavefront compromises observability, such that all C/L gates whose outputs are not inputs to the subsequent register are unobservable. While this makes the architecture less theoretically delay-insensitive, in practice this additional delay sensitivity will not cause problems, as long as the sleep trees are constructed such that there is not a large delay between sleeping the C/L output gates and gates internal to the C/L. This sleep tree timing requirement is easily achievable.

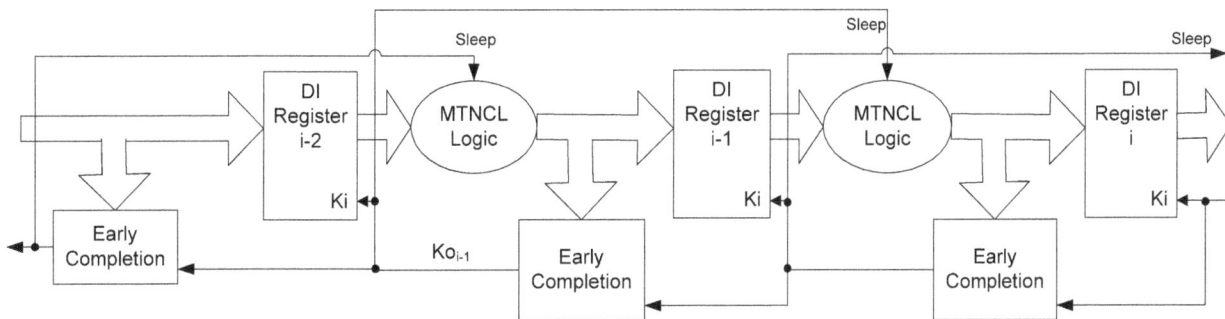

Figure 6. MTNCL pipeline architecture using Early Completion.

2.3.2. MTNCL Threshold Gate Design for ECII Architecture

The MTCMOS structure is incorporated inside each NCL threshold gate, and actually results in a number of the original transistors no longer being needed. As shown in Figure 7a, the *reset* circuitry is no longer needed, since the gate output will now be forced to NULL by the MTCMOS sleep mechanism, instead of by all inputs becoming logic 0. *hold1* is used to ensure that the gate remains asserted, once it has become asserted, until all inputs are de-asserted, in order to guarantee input-completeness with respect to the NULL wavefront; however, since the ECII architecture guarantees input-completeness through the sleep mechanism, as explained in Section 2.3.1, it follows that NCL gate hysteresis is no longer required. Hence, the *hold1* circuitry and corresponding NMOS transistor are removed, and the PMOS transistor is removed to maintain the complementary nature of CMOS logic (*i.e.*, *set* and *hold0* are complements of each other), such that the gate is never floating.

(a)

(b)

(c)

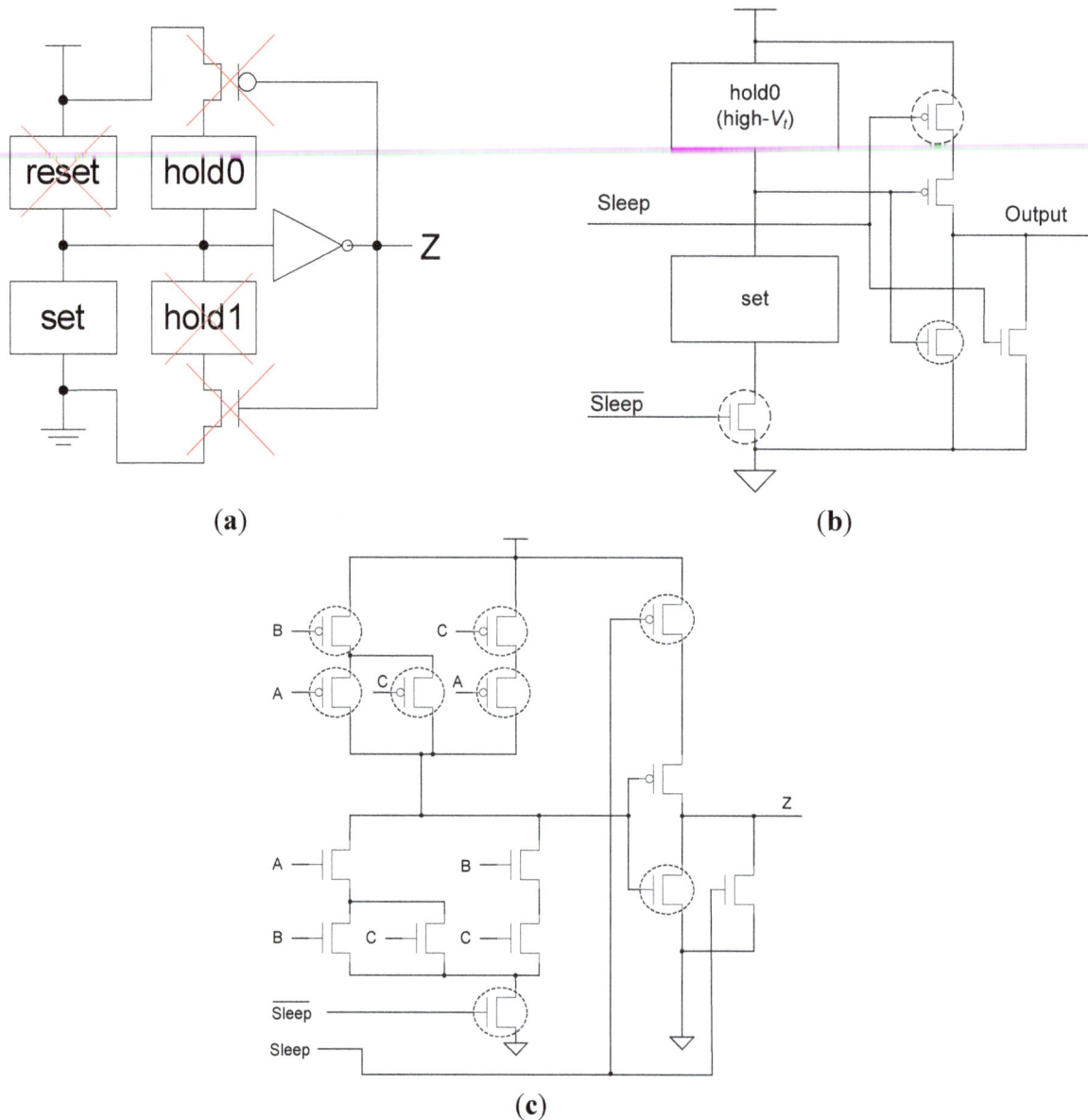

Figure 7. (**a**) Incorporating MTCMOS into NCL threshold gates; (**b**) SMTNCL gate structure; (**c**) TH23 implementation.

Improved from the direct MTCMOS NCL threshold gate implementation [15], similar to the structure shown in Figure 5, a modified Static MTNCL threshold gate structure, referred to as SMTNCL, is shown in Figure 7b. This modification eliminates the output wake-up glitch by moving the power gating high-V_t transistor to the PDN, and removing the two bypass transistors. All PMOS transistors except the output inverter are high-V_t, because they are only turned on when the gate enters sleep mode and the inputs become logic 0, and remain on when the gate exits sleep mode, until the gate's *set* condition becomes true. In both cases, the gate output is already logic 0; therefore, the speed of these PMOS transistors does not affect performance, so high-V_t transistors are used to reduce leakage current. During active mode, the *Sleep* signal is logic 0 and $\overline{Sleep}$ is logic 1, such that the gate functions as normal. During sleep mode, *Sleep* is logic 1 and $\overline{Sleep}$ is logic 0, such that the output low-V_t pull-down transistor is turned on quickly to pull the output to logic 0, while the high-V_t NMOS gating transistor is turned off to reduce leakage. Note that since the internal node, between *set* and *hold0*, is logic 1 during sleep mode and the output is logic 0, the NMOS transistor in the output inverter is no longer on the critical path and therefore can be a high-V_t transistor. As an example, this SMTNCL implementation of the static TH23 gate is shown in Figure 7c.

2.3.3. MTNCL Threshold Gate Design for ECII Architecture

Combining the ECII architecture with the SMTNCL gate structure, results in a delay-sensitivity problem, as shown in Figure 8. After a DATA cycle, if most, but not all, inputs become NULL, this Partial NULL (PN) wavefront can pass through the stage's input register, because the subsequent stage is requesting NULL, and cause all stage outputs to become NULL, before all inputs are NULL and the stage is put to sleep, because the *hold1* logic has been removed from the SMTNCL gates. This violates the input-completeness criteria, discussed in Section 2.1, and can cause the subsequent stage to request the next DATA while the previous stage input is still a partial NULL, such that the preceding wavefront bits that are still DATA will be retained and utilized in the subsequent operation, thereby compromising delay-insensitivity, similar to the problem when using regular completion instead of Early Completion for MTNCL [18].

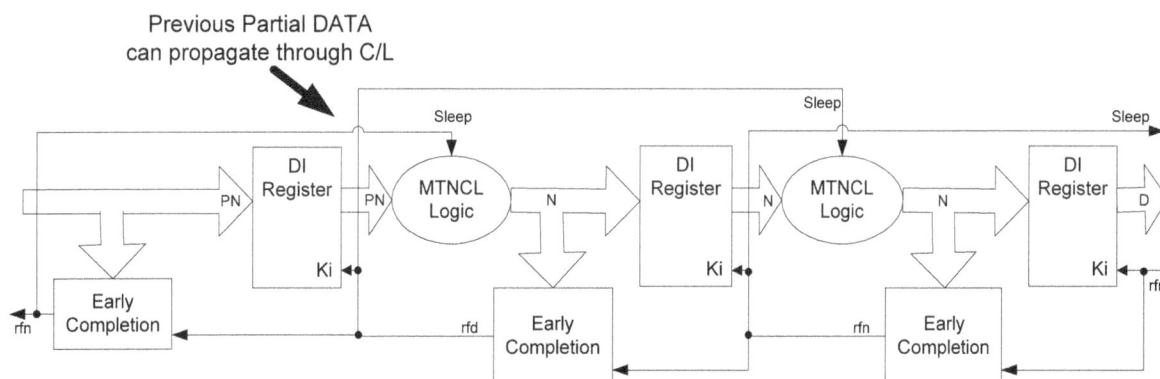

Figure 8. Delay-sensitivity problem combining ECII architecture with SMTNCL gates.

There are two solutions to this problem, one at the architecture level and the other at the gate level. Since the problem is caused by a partial NULL passing through the register, this can be fixed at the architecture-level by ensuring that the NULL wavefront is only allowed to pass through the register after all register inputs are NULL, which is easily achievable by using the stage's inverted sleep signal as its input register's K_i signal. This Fixed Early Completion Input-Incomplete (FECII) architecture is shown in Figure 9. Compared to ECII, FECII is slower because the registers must wait until all inputs become DATA/NULL before they are latched. Note that a partial DATA wavefront passing through the register does not pose a problem, because the stage will remain in sleep mode until all inputs are DATA, thereby ensuring that all stage outputs will remain NULL until all inputs are DATA.

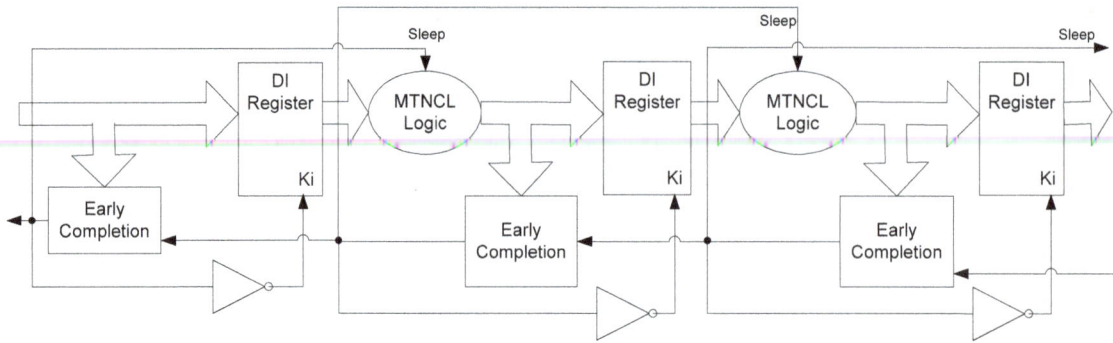

Figure 9. Fixed Early Completion Input-Incomplete (FECII) architecture.

This problem can also be solved at the gate level by adding the *hold1* logic back into each SMTNCL gate, to ensure input-completeness with respect to NULL, such that a partial NULL wavefront cannot cause all outputs to become NULL. Note that this requires the PMOS transistor between *hold0* and V_{DD} to be re-added to prevent a direct path from V_{DD} to ground when both *hold1* and *hold0* are simultaneously asserted. Also note that the *hold1* transistors not shared with the *set* condition can be high-V_t transistors, since they are not on the critical path. This Static MTNCL implementation with *hold1* is shown in Figure 10, and is denoted as SMTNCL1.

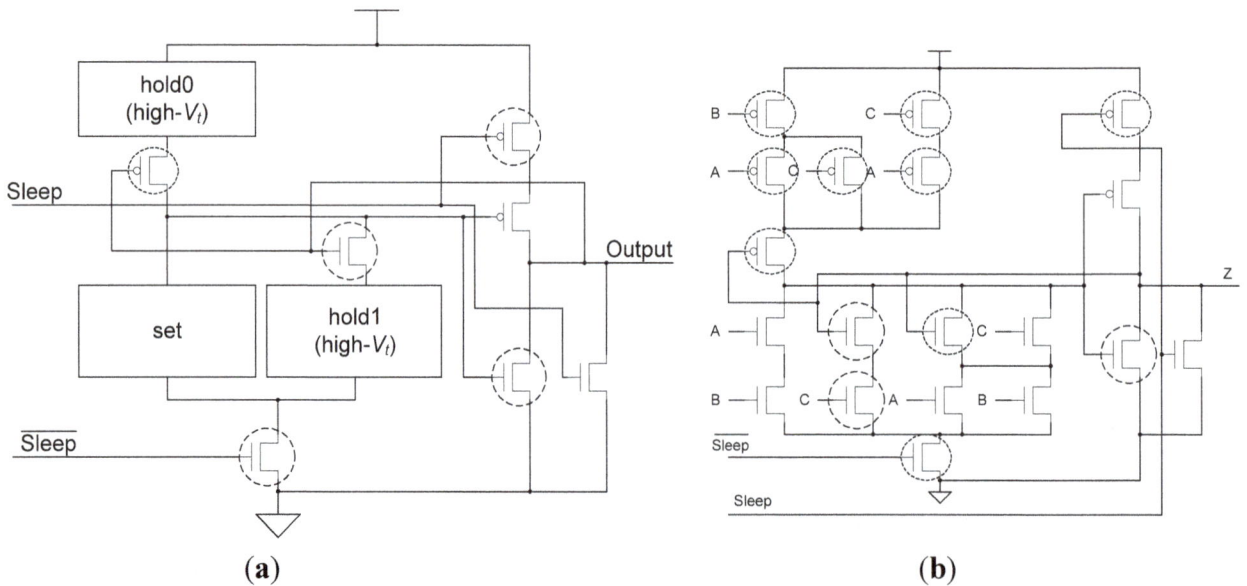

Figure 10. (a) SMTNCL1 gate structure; (b) TH23 implementation.

To summarize, the ECII architecture only works with SMTNCL1 gates, which include the *hold1* function. The FECII architecture works with both SMTNCL and SMTNCL1 gates; however, SMTNCL gates would normally be used with FECII since they require fewer transistors. Additionally, the ECII architecture is faster than FECII, when both use the same MTNCL gates.

2.3.4. Bit-Wise MTNCL

Bit-Wise MTNCL (BWMTNCL) was developed in [19] to yield an ultra-low power methodology for bit-wise pipelined [7] NCL systems. Direct application of the MTNCL concept (*i.e.*, utilizing Early Completion and sleeping gates in lieu of the NULL cycle) to bit-wise pipelined NCL systems resulted in excessive overhead [19]. So, BWMTNCL instead utilizes the regular NCL architecture, shown in Figure 3, along with the regular NCL gate design, shown in Figure 7a, modified to utilize the minimum number of high-V_t transistors such that all paths from V_{DD} to ground contain a high-V_t transistor. As an example, the BWMTNCL TH23 gate is shown in Figure 11. Even though BWMTNCL was originally developed for bit-wise pipelined NCL systems, the BWMTNCL gates can also be used for full-word pipelined NCL systems, as demonstrated in [19] and included for comparison herein.

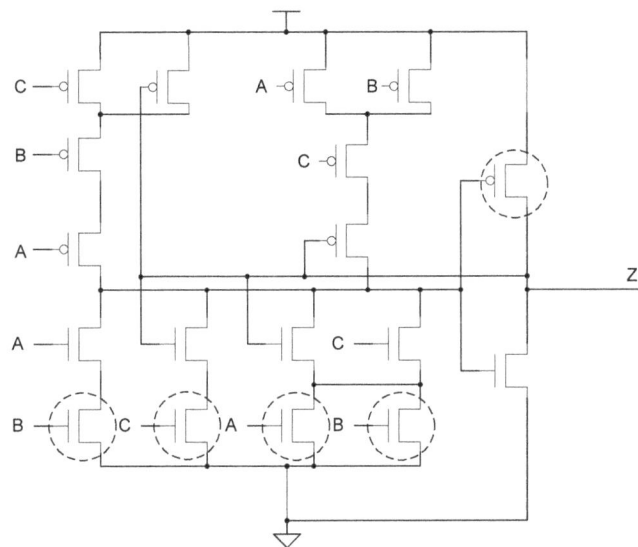

Figure 11. BWMTNCL applied to TH23 gate.

3. MTNCL Enhancements

The previous SMTNCL1 gate, shown in Figure 10, requires a significant number of additional transistors to implement the *hold1* functionality; however, the number of additional transistors can be significantly reduced. Additionally, the previous MTNCL architecture only allows for the combinational logic (C/L) to be slept, whereas this paper develops two modified MTNCL architectures, where (1) the completion logic can also be slept; and (2) both the registration and completion logic can also be slept.

3.1. New SMTNCL1 Gate

Figure 12 shows the new SMTNCL1 gate, which only requires two additional transistors *vs.* the SMTNCL gate. The difference between the new SMTNCL1 gate in Figure 12 and the previous version in Figure 10 is that the *hold1* logic has been removed. The feedback NMOS transistor is sufficient to hold the output at *logic 1*, without the hold1 circuitry, because this ensures that once the gate output has been asserted due to the current DATA wavefront, that it will only be de-asserted when the gate is put to sleep (*i.e.*, when all circuit inputs are NULL), and will not be de-asserted due to a partial NULL wavefront.

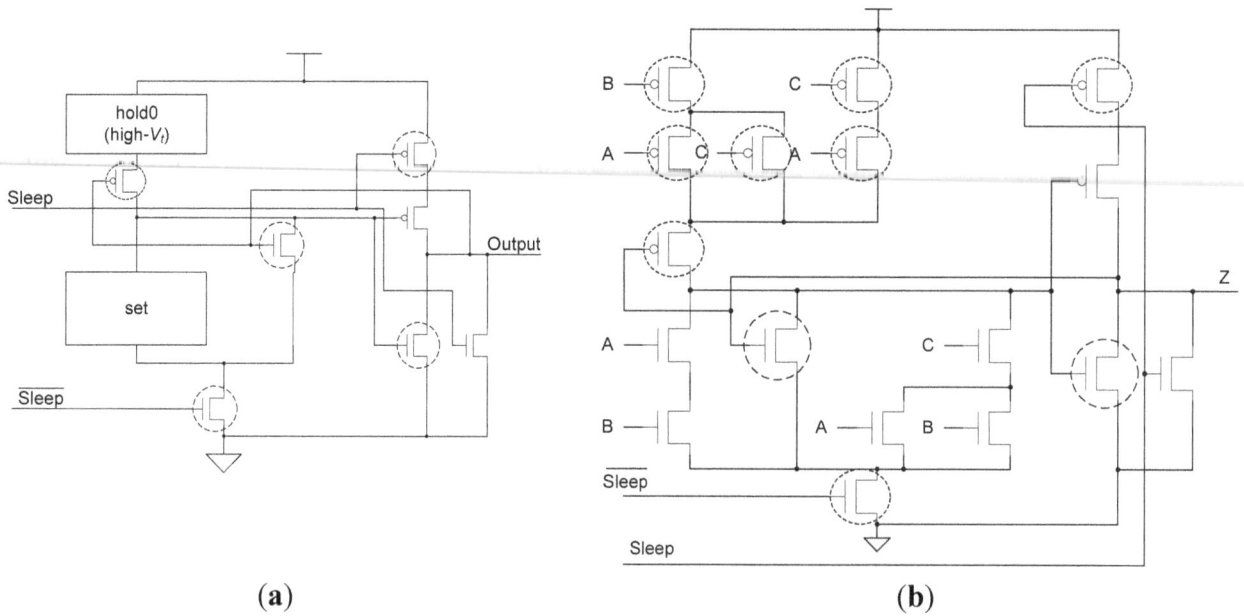

Figure 12. (a) New SMTNCL1 gate structure; (b) TH23 implementation.

3.2. Sleep Completion and Registration Logic

Section 2.3 described the MTNCL architecture where an NCL circuit's C/L was slept in lieu of the NULL cycle to significantly reduce leakage power. However, during sleep mode the circuit's completion and registration logic remains active, which for a fine grain pipelined circuit may be a significant portion of the logic. Therefore, it would be very beneficial to be able to sleep the completion and registration logic in addition to the C/L. The completion logic can be slept by modifying the ECII architecture, shown in Figure 6, to include a sleep input to the completion logic and use SMTNCL1 gates to implement the completion logic, as shown in Figures 13 and 14, respectively. Note that the final inverting TH22 gate is a regular NCL gate, which is not slept. This is consistent with the NULL cycle, where the internal completion component gates are all logic 0, except for the final inverting TH22 gate.

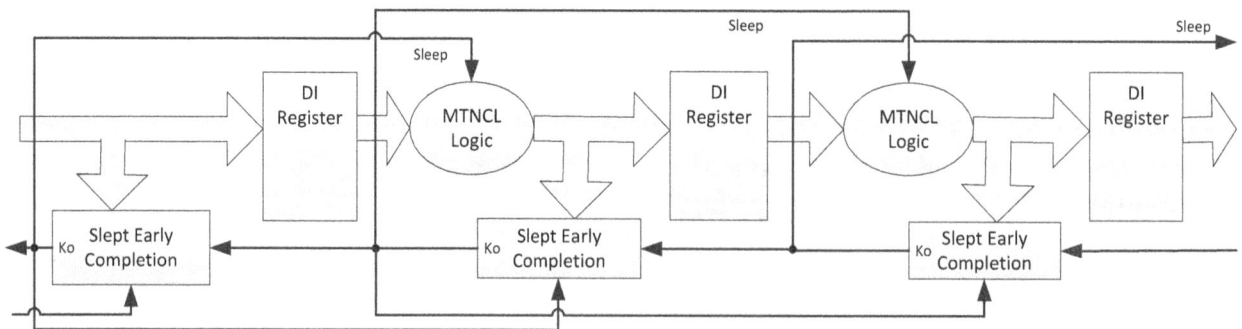

Figure 13. SECII architecture with Completion Logic slept.

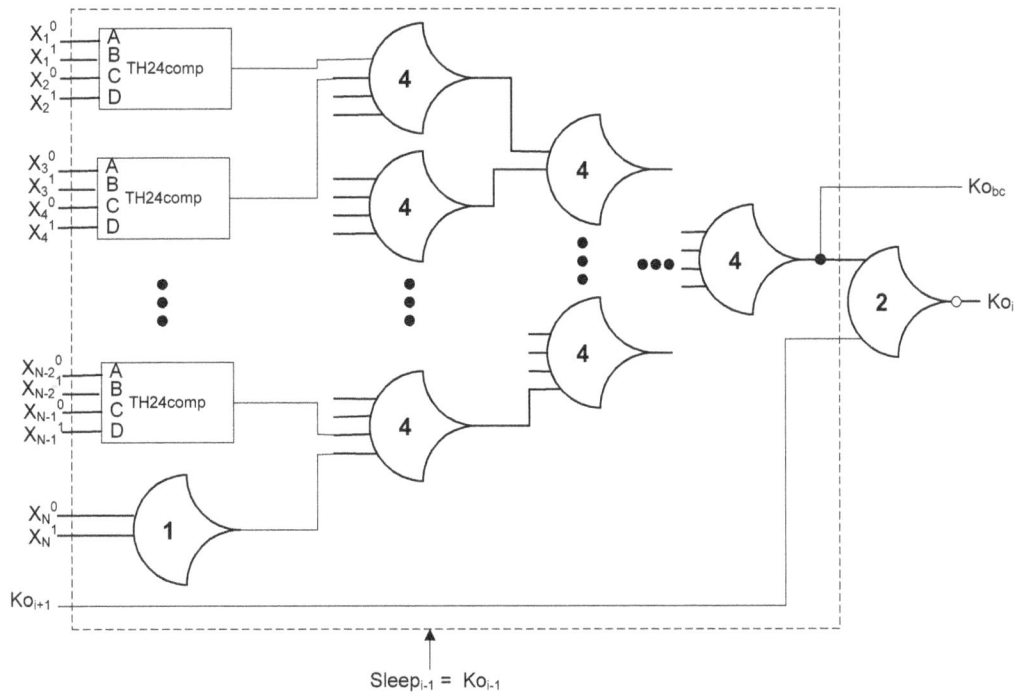

Figure 14. Early completion component with *Sleep* input.

During a NULL cycle, the register output is also NULL, so it too can be slept, as shown in Figure 15. Instead of using two SMTNCL1 TH22 gates to implement the register, the sleep transistors for each rail can be combined, such that a dual-rail register is implemented as a single component in order to reduce the area, as shown in Figure 16a. Note that this architecture is similar to the FECII architecture shown in Figure 9, which does not allow a partial NULL to propagate through the register, such that the C/L can be implemented with the smaller SMTNCL gates instead of SMTNCL1 gates.

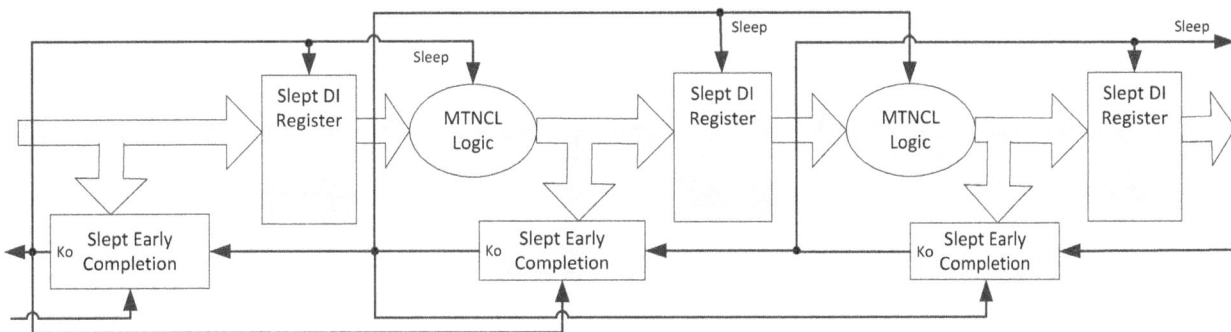

Figure 15. SECRII architecture with Completion Logic and Registration slept.

3.3. Combine SECRII with BWMTNCL

The SMTNCL gates utilized in the SECRII architecture require both a *Sleep* and $\overline{Sleep}$ input, each of which necessitates a large buffer tree. Hence, eliminating one of these inputs would decrease area and energy. The $\overline{Sleep}$ (*nsleep*), input can be eliminated from the SMTNCL gate by combining the SMTNCL architecture in Figure 7 with the BWMTNCL architecture in Figure 11, as shown in Figures 16b and 17. Instead of utilizing a high-V_t transistor to gate the *set* logic from ground, the *set* logic is implemented in BWMTNCL fashion utilizing the minimum number of high-V_t transistors such that all paths through the *set* function to ground contain a high-V_t transistor.

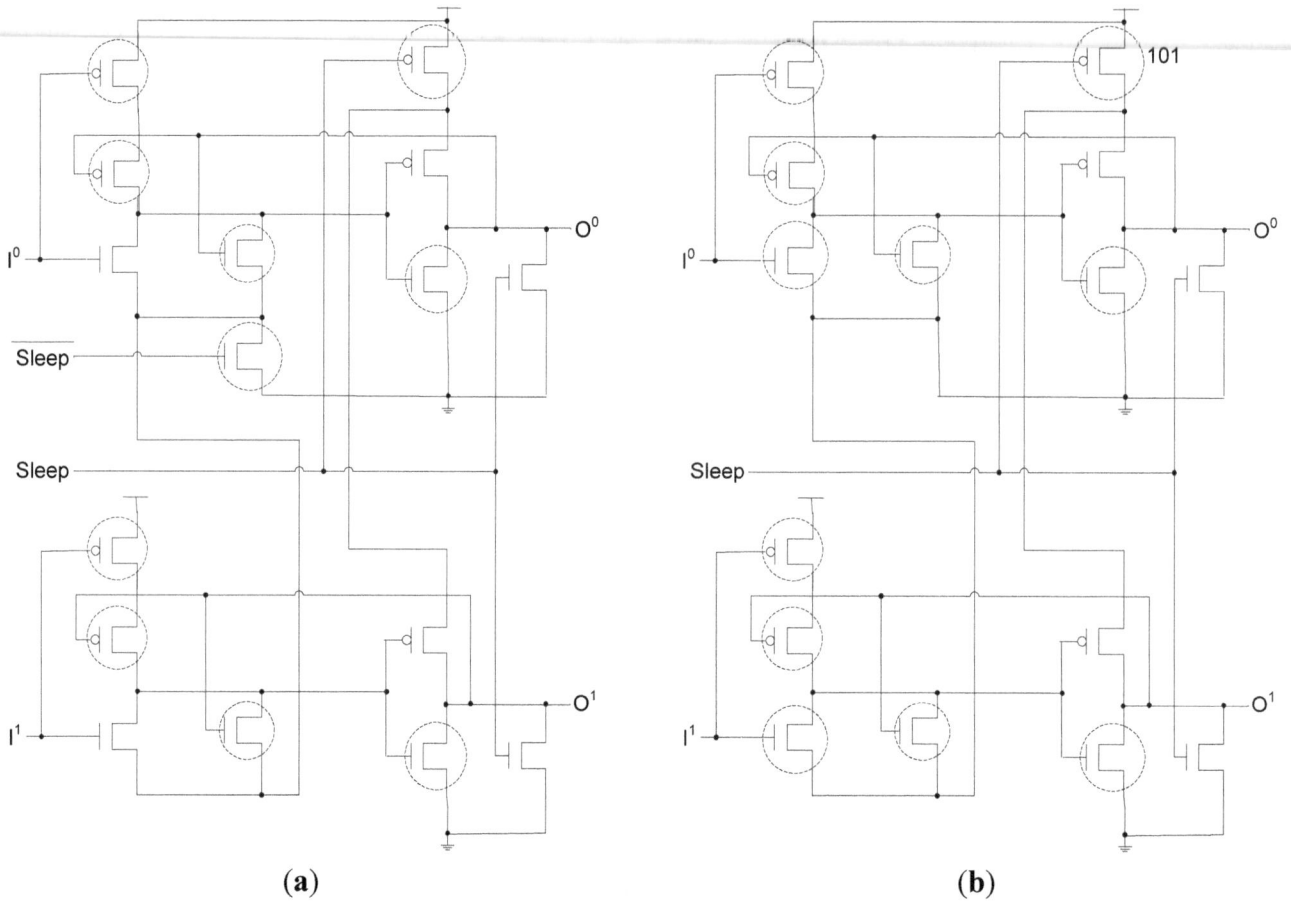

Figure 16. Slept DI register: (**a**) with both *Sleep* and *nsleep* inputs; (**b**) w/o *nsleep* input.

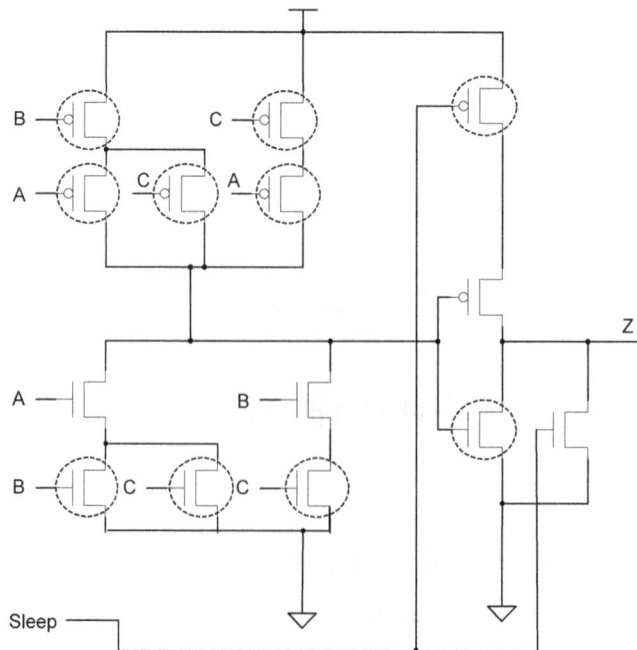

Figure 17. SMTNCL w/o *nsleep* applied to TH23 gate.

3.4. Safe SECRII Architecture

The SECRII w/o nsleep architecture pushes the benefits of MTNCL to its limits, providing for the smallest, fastest, and lowest energy consuming MTNCL architecture, as discussed in Section 4. The SECRII architecture, however, comes with more timing sensitivity due to a potential race condition between sleeping a stage and latching DATA for the next stage. For example, when register$_i$'s input is DATA, and DATA is being requested by the next stage, register$_i$'s Slept Early Completion component output will change from logic 1 to logic 0, which will wake up register$_i$ to latch the DATA. If the input to the previous register, register$_{i-1}$, is already NULL, the 1 to 0 transition of register$_i$'s Slept Early Completion component will cause register$_{i-1}$'s Slept Early Completion component output to change from logic 0 to logic 1, which will sleep the C/L preceding register$_i$. Hence, a race condition exists between register$_i$ latching the DATA before it becomes NULL by sleeping the preceding C/L that generated the DATA, such that the time to pass DATA through register$_i$ once its *sleep* input transitions from 1 to 0 must be less than the time to transition register$_{i-1}$'s Slept Early Completion component inverting TH22 output gate from 0 to 1 followed by the time to sleep the C/L preceding register$_i$ to NULL. The first path is a single MTNCL gate (the register); whereas the second path consists of an NCL TH22 gate, followed by an inverter, followed by an MTNCL gate; therefore, this potential race condition is easily mitigated. Note that in practice, both of these paths also contain a buffer tree for the sleep signal; hence, a simple analysis of each sleep network is needed to ensure that its timing is not generating a problematic race condition.

This potentially problematic race condition can be avoided by using the safe SECRII architecture, as described in [21] and shown in Figure 18. This safe architecture requires the DATA wavefront to completely propagate through the next stage before sleeping the current stage, hence, avoiding the race condition. This can be easily achieved by using the signal at the output of the completion tree (before the final inverting TH22 gate, referred to as Ko$_{bc}$ in Figure 14) of the next stage as the K_o input to the Early Completion/Sleep Generation circuit for the current stage. This signal is shown as an additional output for each Early Completion component in Figure 18, and must be inverted before connecting to the K_o input of the previous Early Completion component. The main drawback of this safe architecture is that it significantly decreases performance.

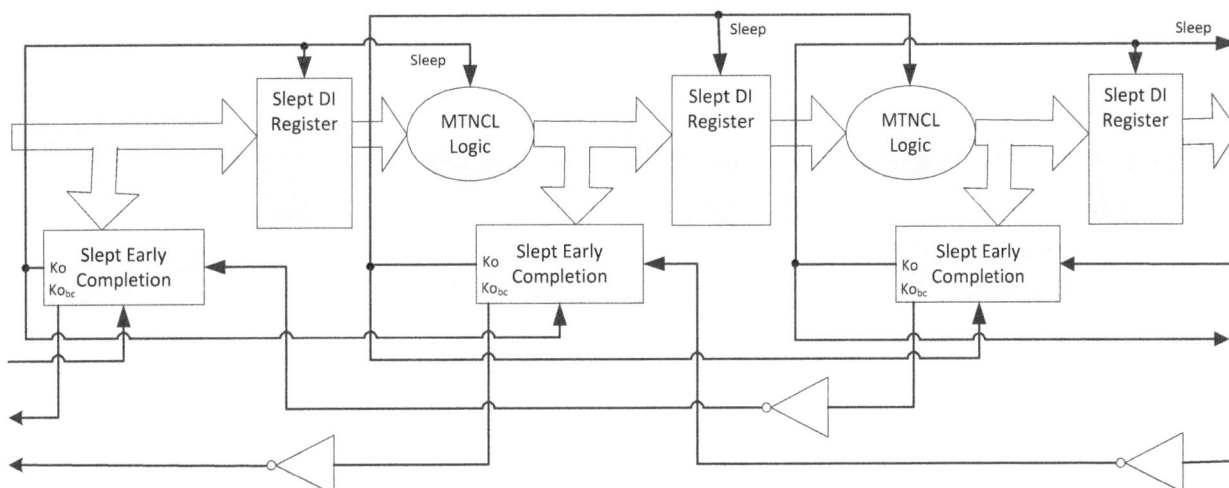

Figure 18. Safe SECRII architecture.

4. Simulation Results

To compare the various MTNCL architectures, a number of four-stage pipelined 32-bit IEEE single-precision floating-point co-processors, which perform addition, subtraction, and multiplication [22], as shown in Figure 19, were designed using the 1.2V IBM 8RF-LM 130 nm CMOS process, and were simulated at the transistor level, after inserting buffers, using Cadence's UltraSim simulator running a VerilogA controller in mixed-signal mode that utilized 25 sets of randomly selected floating-point numbers for each add/sub and multiply operation, whose results were averaged to generate the data shown in Table 1. Note that all transistors for all designs, both asynchronous and synchronous, are minimum sized except for the buffers. Table 1 lists the MTNCL results and also compares to the regular NCL implementation using all low-V_t transistors and all high-V_t transistors, and the synchronous MTCMOS design. Note that the synchronous MTCMOS design only sleeps after a preset number of identical input datasets are received, which requires a small Power Management Unit (PMU) to compare adjacent datasets and count identical consecutive datasets [23]. The floating-point co-processor has two distinct datapaths, the add/subtract unit and the multiplier, which have different throughput, so the data for each is presented separately, and can be averaged to yield the combined results. T_{DD} is the average DATA plus NULL processing time, which is comparable to the synchronous clock period. To compare to the MTCMOS synchronous design, the clock was set to match the speed of the fastest MTNCL design, even though the synchronous design can operate faster (less than 2X). T_{DD} and Energy/Operation are calculated while the circuit is operating at its maximum speed, while Idle Power is calculated using DC analysis after the pipeline is flushed with all NULL inputs for the asynchronous circuits, and after the PMU sleeps the synchronous circuit. Note that the PMU must always stay active and can never sleep because it needs to continuously monitor the incoming data to know when it changes in order to wake up the rest of the circuitry to process the new data. Hence, for the asynchronous circuits, idle power is equivalent to leakage power, and does not depend on the previous type of operations (*i.e.*, either add/sub or mult), since the following sleep state is the same (*i.e.*, both pipelines are all NULL). However, idle power is slightly different for add/sub *vs.* mult for the synchronous design because the clock period is different.

Comparing the various MTNCL designs shows that the new MTNCL gate with *hold1* (SMTNCL1) requires less area, energy, and power than the previous version in [18], and is slightly faster. Sleeping the completion logic along with the C/L slightly reduces area, energy, and leakage power, and significantly increases speed, while sleeping the C/L, completion logic, and registers significantly decreases area, energy, and leakage power, and slightly increases speed. The SMTNCL with SECRII without nsleep design that combines the SMTNCL with SECRII and BWMTNCL architectures further reduces area and energy while increasing speed, at the cost of a slight increase in leakage power. Note that the FECII circuit is faster than the ECII circuit because the FECII design utilizes the faster SMTNCL gates.

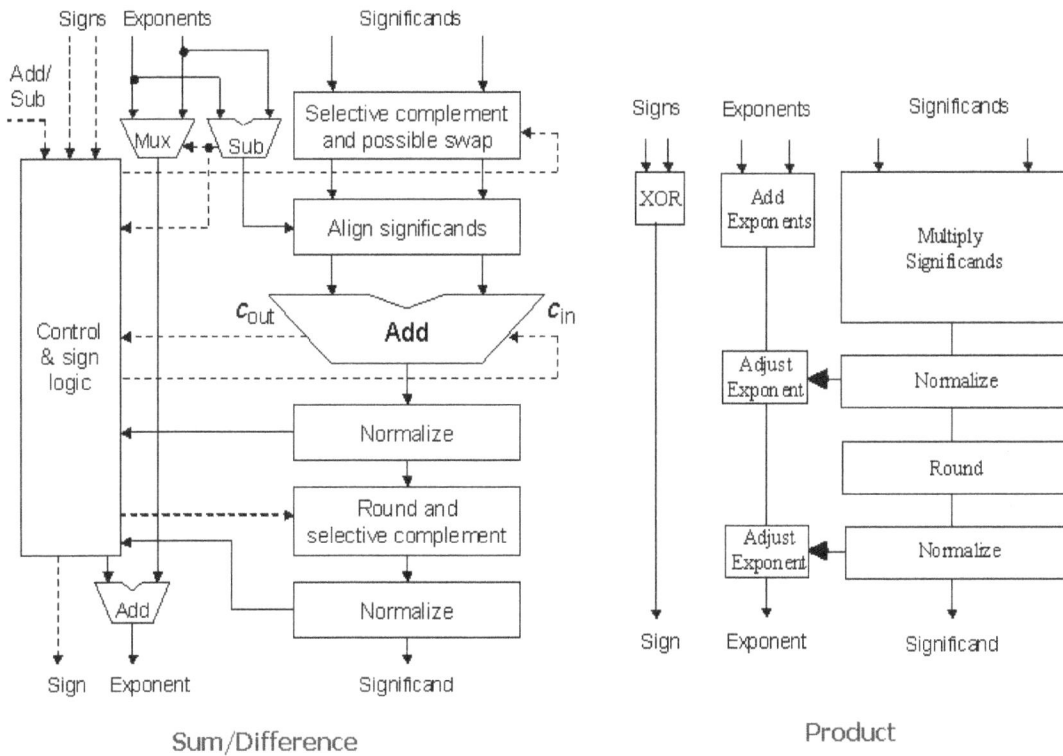

Figure 19. Thirty-two-bit IEEE single-precision floating-point co-processor architecture.

Table 1. MTNCL comparisons.

Circuit Type	# Transistors	T_{DD} (ns)		Energy/Operation (pJ)		Idle Power (nW)	
		add/sub.	Mult.	add/sub.	Mult.	add/sub.	Mult.
NCL Low-V_t	158059	14.1	14.4	27.4	23.7	12,300	12,300
NCL High-V_t	158059	32.7	33.4	28.5	25.1	208.0	208.0
BWMTNCL	158059	17.9	16.2	27.1	23.7	190.7	190.7
SMTCNL with FECII	111506	11.6	15.3	14.9	27.5	115.9	115.9
Original SMTNCL1 with ECII	130476	12.5	16.7	16.0	27.8	140.8	140.8
New SMTNCL1 with ECII	119706	12.1	15.7	14.7	26.1	121.9	121.9
SMTNCL1 with SECII	119244	10.7	15.4	14.6	26.0	121.1	121.1
SMTNCL with SECRII	96640	11.1	14.8	13.5	25.3	111.2	111.2
SMTNCL with SECRII w/o nsleep	90041	10.0	13.9	12.1	21.8	112.1	112.1
SMTNCL with SECRII w/o nsleep (safe architecture)	90049	13.4	16.6	12.3	22.3	113	113
MTCMOS Synchronous	104571	10.0	13.9	124.3	124.7	156,000	132,000

The best MTNCL design, *SMTNCL with SECRII without nsleep*, requires 43% less area, 34% less energy, two orders of magnitude less leakage power, and is 19% faster than the regular low-V_t NCL design, has 46% less leakage power than the regular high-V_t NCL design, and requires 14% less area, 86% less energy, and three orders of magnitude less idle power than the synchronous MTCMOS design while operating at the same speed, although the MTCMOS synchronous design can operate faster (less than 2X). Note that the synchronous circuit's idle power is mostly dynamic power from its PMU, which continues to run when the co-processor is idle to determine when to wake up the co-processor, as mentioned above, which is why it is so much more than the asynchronous designs, which due to their handshaking, do not switch at all when idle, without requiring any additional circuitry. Hence, if a user-controlled sleep input was added, the synchronous circuit would not require a PMU, such that its idle power would be solely comprised of its leakage power, which would be the same order of magnitude as the MTNCL asynchronous designs, since leakage power is proportional to number of gates for MTCMOS based circuits.

Hence, the *SMTNCL with SECRII without nsleep* architecture presented herein vastly outperforms traditional NCL in all aspects, area, speed, energy, and leakage power, and significantly outperforms the MTCMOS synchronous architecture in terms of area, energy, and idle power. Finally, to increase timing robustness, one can optionally use the safe SMTNCL with SECRII without nsleep architecture, which preserves all the area and power advantages of the normal SMTNCL with SECRII without nsleep architecture but reduces performance.

5. Conclusions

This paper describes enhancements to the ultra-low power MTNCL methodology developed in [15–19]. Specifically, a new MTNCL gate with *hold1* capability was designed, which is smaller, faster, and has lower power than the previous version in [18]. Additionally, two new MTNCL architectures were developed that allow for the completion logic to be slept in addition to the C/L and for the C/L, completion logic, and registration to all be slept; and this new architecture was combined with the BWMTNCL architecture in [19] to further reduce area and energy and increase speed by removing the $\overline{Sleep}$ buffer tree. Transistor-level simulation of a four-stage pipelined 32-bit IEEE single-precision floating-point co-processor using the 1.2V IBM 8RF-LM 130 nm CMOS process shows that the new architecture that sleeps the C/L, completion logic, and registration without the $\overline{Sleep}$ buffer tree is superior to the previous MTNCL architectures in all categories (*i.e.*, area, speed, energy, and leakage power), vastly outperforms traditional NCL in all aspects, and significantly outperforms the MTCMOS synchronous architecture in terms of area, energy, and idle power, although the MTCMOS synchronous design can operate faster.

Acknowledgments

The authors gratefully acknowledge the support from DARPA under grant: W15P7T-08-C-V404, and from the National Science Foundation under grant: CCF-1116405.

Author Contributions

Liang Zhou, Scott Smith, and Jia Di developed the MTNCL methodology presented herein; Ravi Parameswaran developed the 32-bit single precision floating-point unit architecture utilized as the test bed herein to compare the various circuits; and Farhad Parsan implemented and simulated the safe SECRII architecture, and wrote the corresponding section.

Conflicts of Interest

The authors declare no conflict of interest.

References

1. Mutoh, S.; Douseki, T.; Matsuya, Y.; Aoki, T.; Shigematsu, S.; Yamada, J. 1-V Power Supply High-Speed Digital Circuit Technology with Multithreshold-Voltage CMOS. *IEEE J. Solid-State Circuits* **1995**, *30*, 847–854.

2. Fant, K.M.; Brandt, S.A. NULL Convention Logic: A Complete and Consistent Logic for Asynchronous Digital Circuit Synthesis. In Proceedings of the International Conference on Application Specific Systems, Architectures and Processors, Chicago, IL, USA, 19–21 August 1996; pp. 261–273.

3. Sobelman, G.E.; Fant, K.M. CMOS Circuit Design of Threshold Gates with Hysteresis. In Proceedings of the 1998 IEEE International Symposium on Circuits and Systems (II), Monterey, CA, USA, 31 May–3 June 1998; pp. 61–65.

4. Muller, D.E. Asynchronous Logics and Application to Information Processing. In *Switching Theory in Space Technology*; Stanford University Press: Redwood City, CA, USA, 1963; pp. 289–297.

5. Parsan, F.A.; Smith, S.C. CMOS Implementation of Static Threshold Gates with Hysteresis: A New Approach. In Proceedings of the IFIP/IEEE International Conference on VLSI-SoC, Santa Cruz, CA, USA, 7–10 October 2012; pp. 41–45.

6. Parsan, F.A.; Smith, S.C. CMOS Implementation Comparison of NCL Gates. In Proceedings of the IEEE International Midwest Symposium on Circuits and Systems, Boise, ID, USA, 5–8 August 2012; pp. 394–397.

7. Smith, S.C.; DeMara, R.F.; Yuan, J.S.; Hagedorn, M.; Ferguson, D. Delay-Insensitive Gate-Level Pipelining. *Elsevier's Integ. VLSI J.* **2001**, *30*, 103–131.

8. Smith, S.C.; DeMara, R.F.; Yuan, J.S.; Ferguson, D.; Lamb, D. Optimization of NULL Convention Self-Timed Circuits. *Integr. VLSI J.* **2004**, *37*, 135–165.

9. Seitz, C.L. System Timing. In *Introduction to VLSI Systems*; Addison-Wesley: Boston, MA, USA, 1980; pp. 218–262.

10. Kondratyev, A.; Neukom, L.; Roig, O.; Taubin, A.; Fant, K. Checking Delay-Insensitivity: 10^4 Gates and Beyond. In Proceedings of the Eighth International Symposium on Asynchronous Circuits and Systems, Manchester, UK, 8–11 April 2002; pp. 149–157.

11. Martin, A.J. Programming in VLSI: From Communicating Processes to Delay-Insensitive Circuits. In *Developments in Concurrency and Communication*; UT Year of Programming Institute on Concurrent Programming, Addison-Wesley: Boston, MA, USA, 1990; pp. 1–64.

12. Van Berkel, K. Beware the Isochronic Fork. *Integr. VLSI J.* **1992**, *13*, 103–128.

13. Kao, J.T.; Chandrakasan, A.P. Dual-Threshold Voltage Techniques for Low-Power Digital Circuits. *IEEE J. Solid-State Circuits* **2000**, *35*, 1009–1018.

14. Lakshmikanthan, P.; Sahni, K.; Nunez, A. Design of Ultra-Low Power Combinational Standard Library Cells Using a Novel Leakage Reduction Methodology. In Proceedings of the IEEE International SoC Conference, Taipei, Taiwan, 24–27 September 2006; pp. 93–94.

15. Bailey, A.D.; Di, J.; Smith, S.C.; Mantooth, H.A. Ultra-Low Power Delay-Insensitive Circuit Design. In Proceedings of IEEE Midwest Symposium on Circuits and Systems, Knoxville, TN, USA, 10–13 August 2008; pp. 503–506.

16. Bailey, A.D.; Al Zahrani, A.; Fu, G.; Di, J.; Smith, S.C. Multi-Threshold Asynchronous Circuit Design for Ultra-Low Power. *J. Low Power Electron.* **2008**, *4*, 337–348.

17. Alzahrani, A.; Bailey, A.D.; Fu, G.; Di, J. Glitch-Free Design for Multi-Threshold CMOS NCL Circuits. In Proceedings of the ACM 2009 Great Lakes Symposium on VLSI, Boston, MA, USA, 10–12 May 2009.

18. Smith, S.C.; Di, J. *Designing Asynchronous Circuits using NULL Convention Logic (NCL)*; Synthesis Lectures on Digital Circuits and Systems; Morgan & Claypool Publishers: San Rafael, CA, USA, 2009; Volume 4, doi:10.2200/S00202ED1V01Y200907DCS023.

19. Zhou, L.; Smith, S.C.; Di, J. Bit-Wise MTNCL: An Ultra-Low Power Bit-Wise Pipelined Asynchronous Circuit Design Methodology. In Proceedings of the IEEE Midwest Symposium on Circuits and Systems, Seattle, WA, USA, 1–4 August 2010; pp. 217–220.

20. Smith, S.C. Speedup of Self-Timed Digital Systems Using Early Completion. In Proceedings of the IEEE Computer Society Annual Symposium on VLSI, Pittsburgh, PA, USA, 25–26 April 2002; pp. 107–113.

21. UNCLE User Manual. Available online: http://www.ece.msstate.edu/~reese/uncle/UNCLE.pdf (accessed on April 2015).

22. Parhami, B. *Computer Arithmetic Algorithms and Hardware Designs*; Oxford University Press: New York, NY, USA, 2000.

23. Thian, R. Multi-Threshold CMOS Circuit Design Methodology from 2D to 3D. Master's Thesis, Computer Science and Computer Engineering Department, University of Arkansas, Fayetteville, AR, USA, December 2010.

12

Impacts of Work Function Variation and Line-Edge Roughness on TFET and FinFET Devices and 32-Bit CLA Circuits [†]

Yin-Nien Chen *, Chien-Ju Chen, Ming-Long Fan, Vita Pi-Ho Hu, Pin Su and Ching-Te Chuang *

Department of Electronics Engineering and Institute of Electronics, National Chiao-Tung University, 1001 University Road, Hsinchu 300, Taiwan; E-Mails: zuzu322.ep97@g2.nctu.edu.tw (C.-J.C.); mlfan.ee95@gmail.com (M.-L.F.); vitabee@gmail.com (V.P.-H.H.); pinsu@faculty.nctu.edu.tw (P.S.)

[†] The original of this paper had been presented in IEEE S3S Conference 2014.

* Authors to whom correspondence should be addressed;
E-Mails: snoopyfairy@gmail.com (Y.-N.C.); chingte.chuang@gmail.com (C.-T.C.)

Academic Editors: David Bol and Steven A. Vitale

Abstract: In this paper, we analyze the variability of III-V homojunction tunnel FET (TFET) and FinFET devices and 32-bit carry-lookahead adder (CLA) circuit operating in near-threshold region. The impacts of the most severe intrinsic device variations including work function variation (WFV) and fin line-edge roughness (fin LER) on TFET and FinFET device I_{on}, I_{off}, C_g, 32-bit CLA delay and power-delay product (PDP) are investigated and compared using 3D atomistic TCAD mixed-mode Monte-Carlo simulations and HSPICE simulations with look-up table based Verilog-A models calibrated with TCAD simulation results. The results indicate that WFV and fin LER have different impacts on device I_{on} and I_{off}. Besides, at low operating voltage (<0.3 V), the CLA circuit delay and power-delay product (PDP) of TFET are significantly better than FinFET due to its better I_{on} and $C_{g,ave}$ and their smaller variability. However, the leakage power of TFET CLA is larger than FinFET CLA due to the worse I_{off} variability of TFET devices.

Keywords: tunnel FET (TFET); FinFET; work function variation (WFV); line-edge-roughness (LER); carry-lookahead adder (CLA)

1. Introduction

Steep subthreshold slope TFET, which utilizes the band-to-band tunneling as the conduction mechanism, is one of the most promising candidates for ultra-low voltage/power applications [1]. Recent research works on TFET-based circuits have shown significant performance improvement and power reduction at low operating voltage [2–4]. With device scaling, the impacts of random variations become more severe. Several studies on the TFET device level variability have been reported [5–8], while other works on TFET circuits employed simple parameter sensitivity methods that neglect physical non-uniformities [2,9,10], and a physics-based TFET performance and variability assessment for large logic circuits is lacking. Among all variation sources, the work function variation (WFV) caused by the granularity of different grain orientations and sizes of the metal gate material and fin Line-Edge-Roughness (LER) due to the resolution limit of the lithography and etching processes have the most significant impacts on TFET and FinFET devices. In this work, we provide an in-depth physics-based assessment on the impacts of WFV and fin LER on TFET and FinFET devices including the detailed comparative analyses on I_{on}, I_{off}, and C_g using three-dimensional atomistic TCAD simulations. To assess the variability on large logic circuits, we build look-up table based Verilog-A models, and examine the variability of TFET- and FinFET-based 32-bit CLA circuits using HSPICE simulations with Verilog-A model calibrated with TCAD simulation results. Our work provides in-depth physics-based understanding on the variability of 32-bit CLA circuits and fundamental guidelines on the implementation of TFET-based large logic circuits considering variability.

2. Device Structures, Characteristics and Simulation Methodology

2.1. Device Structures and Characteristics

The basic TFET structure under study comprises a gated p-i-n tunnel diode under reverse bias with asymmetrical source/drain doping. For N-TFET, the source is p+ region with dominant electron conduction, the channel is gated intrinsic region, and the drain is n+ region. When N-TFET is "OFF" ($V_{GS} = 0$), the valence band edge of the source is below the conduction band edge of the channel, and the band-to-band tunneling probability is low due to lack of available states in the channel region and wide barrier at source-channel junction. When N-TFET is "ON" ($V_{GS} > 0$), the conduction band edge of the channel is pulled down below the valence band edge of the source, and carriers can tunnel into available empty states of the channel region. For P-TFET, the source is n+ region with dominant hole conduction, applying $V_{GS} < 0$ turns P-TFET "ON". The band diagrams of TFET in ON/OFF states are shown in Figure 1.

In this work, we consider the $In_{0.53}Ga_{0.47}As$ homojunction N-TFET and $Ge_{0.925}Sn_{0.075}$ homojunction P-TFET due to their high I_{on} and compatible I_{DS}-V_{GS} characteristic [12,13]. $In_{0.53}Ga_{0.47}As$ N-FinFET and Ge P-FinFET with high mobility are considered for comparison. Figure 2 shows the 3D TFET and

FinFET device structures constructed for atomistic TCAD simulations. The device parameters and doping are shown in Table 1. We use the non-local band-to-band tunneling model which is applicable to arbitrary tunneling barrier with non-uniform electric field for TFET simulations [11], and the parameters used in the model are calibrated with [12,13]. Figure 3a shows the I_{DS}-V_{GS} characteristics of TFETs and FinFETs at $V_{DS} = 0.3$ V and $V_{DS} = 0.03$ V. The DIBL (drain-induced barrier lowering) and DIBT (drain-induced barrier thinning) values *versus* drain current for N-TFET and N-FinFET are shown in Figure 3b. DIBL for the conventional MOSFET device is estimated using the following formula in weak inversion region (subthreshold region):

$$\text{DIBL} = \frac{\Delta V_{TH}}{\Delta V_{DS}} \text{ (mV/V)} \tag{1}$$

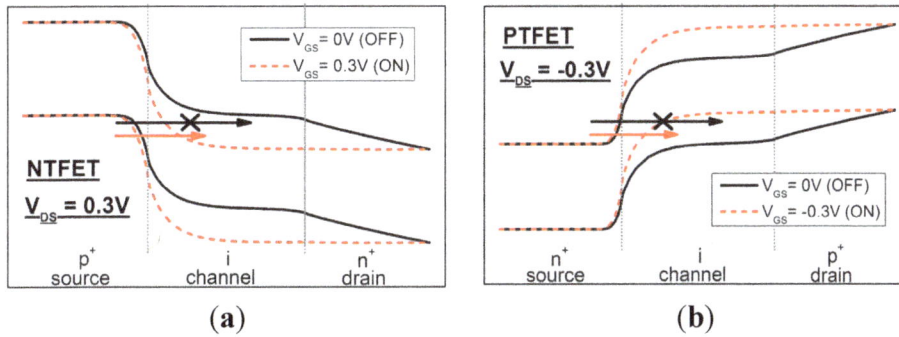

Figure 1. Energy band diagrams of (**a**) n-type and (**b**) p-type TFET in ON/OFF state.

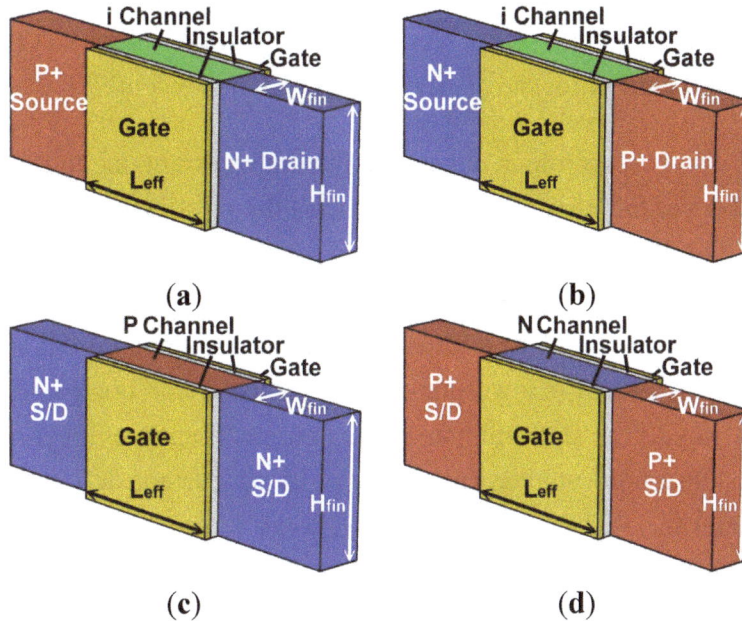

Figure 2. Physical structures of (**a**) In$_{0.53}$Ga$_{0.47}$As homojunction N-TFET; (**b**) Ge$_{0.925}$Sn$_{0.075}$ homojunction P-TFET; (**c**) In$_{0.53}$Ga$_{0.47}$As N-FinFET and (**d**) Ge P-FinFET.

Table 1. Parameters of TFET and FinFET devices.

Devices	TFET		FinFET
L_{eff} = 25 nm	W_{fin} = 7 nm	H_{fin} = 20 nm	EOT = 0.65 nm
	nTFET	*pTFET*	*FinFET*
Material	$In_{0.53}Ga_{0.47}As$	$Ge_{0.925}Sn_{0.075}$	$In_{0.53}Ga_{0.47}As$
Nch (cm^{-3})	undoped	undoped	1×10^{17}
Ns (cm^{-3})	4.5×10^{19} (p-type)	2×10^{19} (n-type)	1×10^{20}
Nd (cm^{-3})	2×10^{17} (n-type)	2×10^{17} (p-type)	1×10^{20}

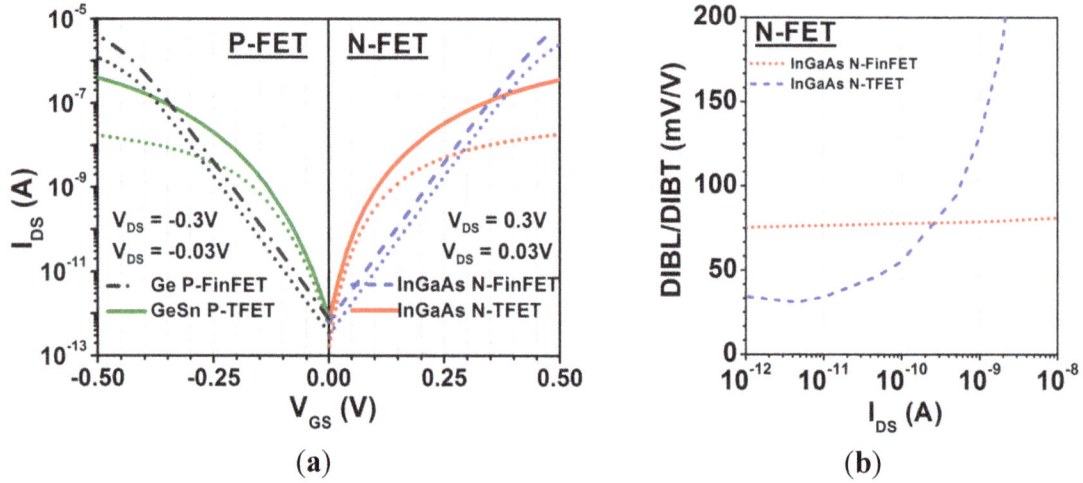

Figure 3. (a) I_{DS}-V_{GS} characteristics at V_{DS} = 0.3 V and V_{DS} = 0.03 V of $In_{0.53}Ga_{0.47}As$ N-TFET, $Ge_{0.925}Sn_{0.075}$ P-TFET, $In_{0.53}Ga_{0.47}As$ N-FinFET and Ge P-FinFET; **(b)** DIBL and DIBT value *versus* drain current for $In_{0.53}Ga_{0.47}As$ N-TFET and N-FinFET.

In TFET, the drain bias also plays a role in enhancing the drain current due to the drain bias induced source-channel tunneling barrier thinning effect. However, as the physics-based method for extracting the threshold voltage of TFET is still under investigation, there is no clear definition for DIBT extraction analogous to DIBL in FiFET device. Hence, for first-order approximation for estimating DIBT in TFET device, we draw the DIBT as a function of drain to source current shown in Figure 3b. As can be seen, the DIBT for TFET shows non-monotonic behavior compared with the FinFET counterpart and increases rapidly as the drain to source current increases beyond 0.2 nA. This is because TFET has smaller threshold voltage (using the constant current defined V_{th}) and enters the saturation region earlier than the FinFET which is in the weak inversion region with DIBL roughly around 80 mV/V.

Figure 4 shows the output characteristics for TFET and FinFET devices. As shown, TFET device shows larger V_{DSAT} [14] as indicated in rhombus symbol due to the fact that TFET can be regarded as a source-channel tunneling junction in series with a resistor (*i.e.*, channel resistance), hence exhibiting an upward-concaved shape in the triode-like region (analogous to FinFET). At moderate and high V_{DS}, TFET provides a better (flatter) saturation characteristic due to reduced carriers in the channel region, and the electric field from the drain side cannot penetrate into the source-channel tunnel junction, so the current increases slowly. For FinFET device, no obvious saturation is observed due to more severe short-channel effect.

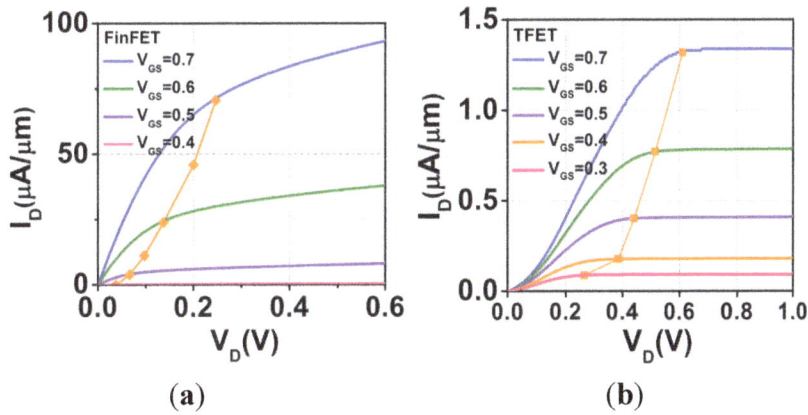

Figure 4. I_{DS}-V_{DS} characteristics at various V_{GS} bias for (**a**) FinFET and (**b**) TFET device with the rhombus symbol showing the extrated V_{DSAT}.

2.2. Simulation Methodology

To assess WFV, we use the Vonoroi grain pattern [15] for TiN gate material, which has two different grain orientations <200> and <111> with the probability of 60% and 40%, respectively, as shown in Figure 5a by the yellow and orange regions, and the relevant parameters are shown in Table 2. To assess fin LER, the rough line edge patterns are generated by Fourier synthesis approach [16] with correlation length (Λ) = 20 nm and root-mean-square amplitude (Δ) = 1.5 nm as shown in Figure 5b. We analyze the impacts of WFV and fin LER on devices using 3D atomistic TCAD mixed-mode Monte-Carlo simulations with 100 samples, respectively.

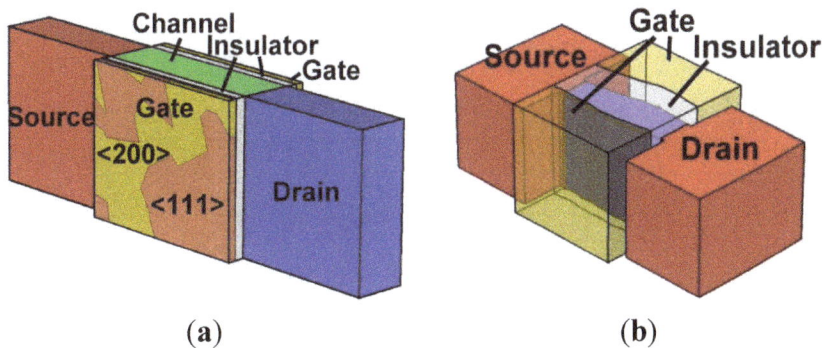

Figure 5. Examples of structures with (**a**) WFV and (**b**) fin LER.

Table 2. Parameters for WFV simulations.

Gate Material = TiN		Grain Size = 5 nm	
Work function (eV)	**Nominal**	**<200> (60%)**	**<111> (40%)**
InGaAs N-TFET	4.53	4.61	4.41
GeSn P-TFET	4.82	4.9	4.7
InGaAs N-FinFET	4.88	4.96	4.76
Ge P-FinFET	4.27	4.35	4.15

TCAD mixed-mode simulations for complex circuits with large transistor counts face the challenges of computation resources, prohibitively long simulation times and convergence problems. To overcome these obstacles, look-up table based Verilog-A model has been employed for TFET circuit simulations in some studies [2,4]. However, these works on TFET circuits employed simple parameter sensitivity methods [2,9], and these sensitivity-based Verilog-A models cannot accurately describe the physical non-uniformities and variability. In this work, we adopt physics-based assessment to account for variability at device and circuit level. The flow chart for physics-based small signal Verilog-A model generation is shown in Figure 6. The transfer characteristics of TFET and FinFET devices and their variability with WFV and fin LER are extracted from atomistic 3D TCAD device simulations with I_{DS} (V_{GS}, V_{DS}), C_{gs} (V_{GS}, V_{DS}) and C_{gd} (V_{GS}, V_{DS}) characteristics across voltage range of interest to build two-dimensional Verilog-A look-up tables. The Verilog-A models of devices with random variations are then employed in HSPICE circuit simulations. The calibrations of Verilog-A models with TCAD results on I-V, C-V characteristics of the nominal cases for TFET and FinFET devices are shown in Figure 7. The almost exact agreements can be clearly seen.

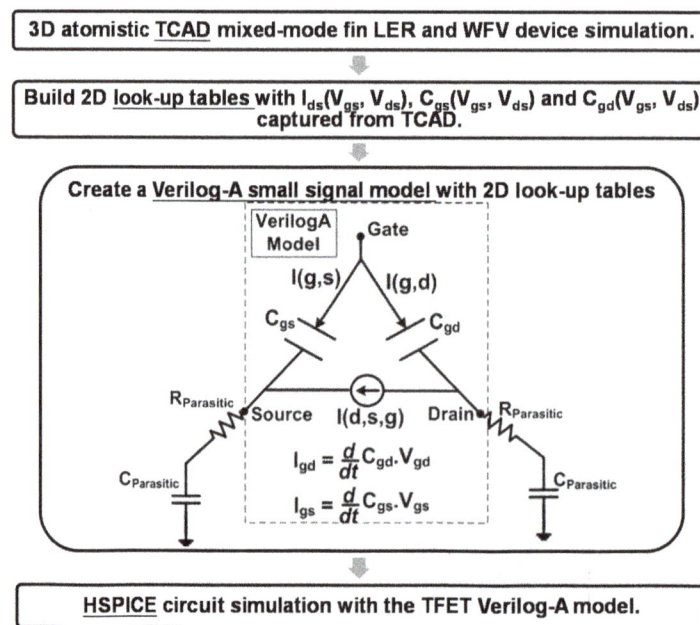

Figure 6. Flowchart for HSPICE look-up table based Verilog-A model generation from atomistic 3D TCAD simulations [2,4].

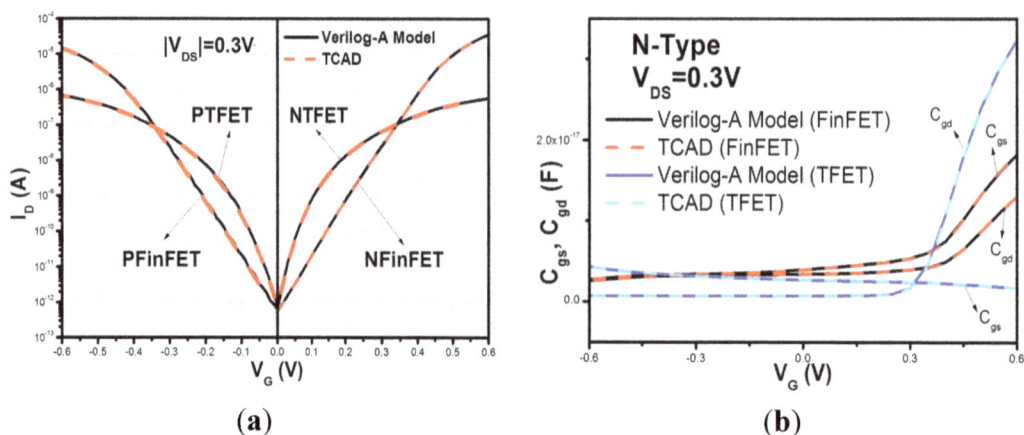

(a)   **(b)**

Figure 7. Calibrations of Verilog-A models with TCAD results on (**a**) I-V and (**b**) C-V charcteristics of the nominal cases for TFET and FinFET deivces at $V_{DS} = 0.3$ V.

3. Device Variability Due to WFV and Fin LER

3.1. I_{off} and I_{on} Variability

Figure 8 shows the impacts of WFV and fin LER on I_{DS}-V_{GS} dispersions of TFET and FinFET devices at V_{DS} = 0.3 V. Figure 9 illustrates the probability distributions of I_{on} (I_{DS} at V_{DS} = V_{GS} = 0.3 V) and I_{off} (I_{DS} at V_{DS} = 0.3 V and V_{GS} = 0 V). Note that, for TFET variability, the different structure constructs used for WFV and fin LER lead to slightly different nominal I_{DS}-V_{GS} curves. Therefore, the corresponding probability distributions show two nominal values. The mean values (μ), standard deviations (σ) and the ratio of the mean-to-standard deviation (μ/σ) are listed in the table with the figures.

For FinFETs, the V_t is a linear function of gate WF, WFV causes a V_t shift of I_{DS}-V_{GS} curves in subthreshold region with almost equal subthreshold swing (S.S.), therefore the I_{on} and I_{off} probability distributions are similar. On the other hand, fin LER influences the effective fin width and electrostatic integrity, thus impacting both V_t and S.S., so the I_{on} and I_{off} probability distributions are quite different. As can be seen, both the μ/σ of I_{on} and I_{off} are worse with fin LER than WFV, especially for I_{off}.

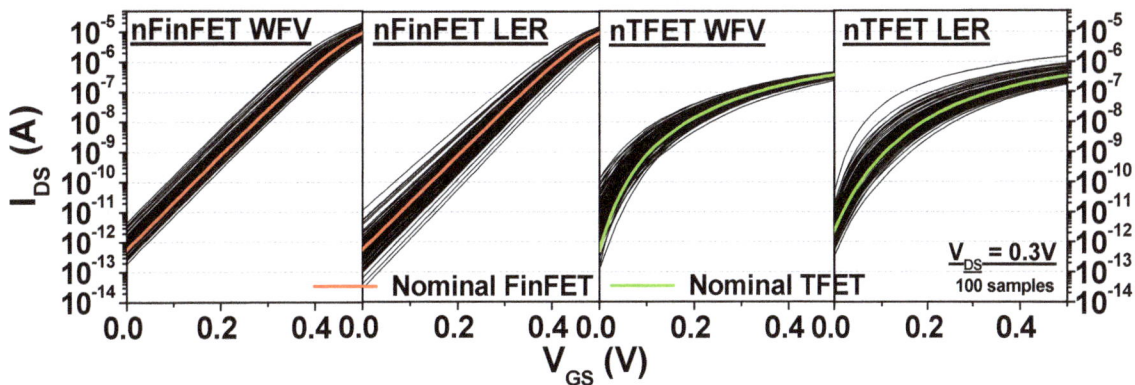

Figure 8. Simulated I_{DS}-V_{GS} characteristics at V_{DS} = 0.3 V for TFET and FinFET with WFV and fin LER.

	μIoff (pA)	σIoff (pA)	μ/σ
N-FinFET WFV	0.968	0.749	1.29
N-FinFET LER	0.856	1.466	0.58

	μIon (nA)	σIon (nA)	μ/σ
N-FinFET WFV	42.508	31.998	1.37
N-FinFET LER	28.13	22.583	1.25

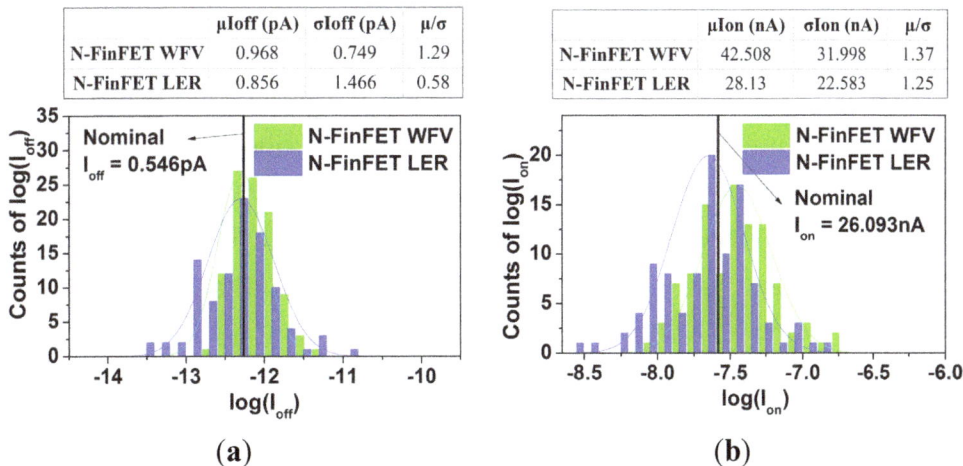

(a)   (b)

Figure 9. *Cont.*

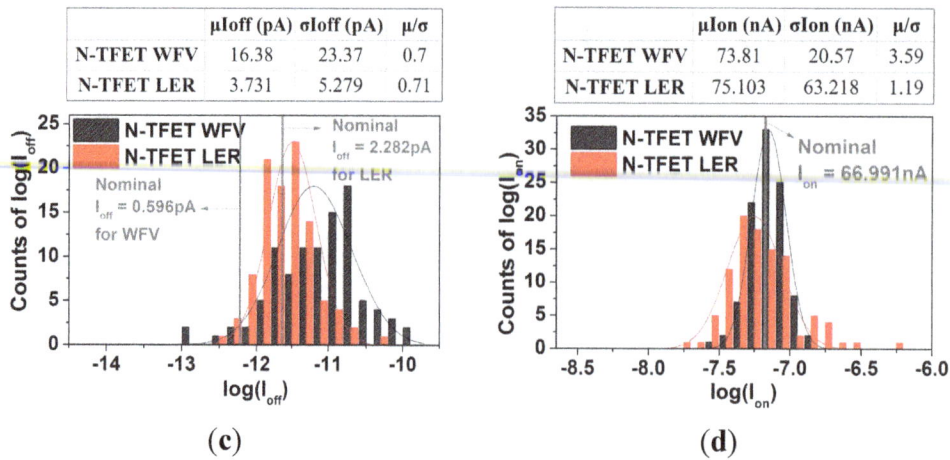

	μIoff (pA)	σIoff (pA)	μ/σ
N-TFET WFV	16.38	23.37	0.7
N-TFET LER	3.731	5.279	0.71

	μIon (nA)	σIon (nA)	μ/σ
N-TFET WFV	73.81	20.57	3.59
N-TFET LER	75.103	63.218	1.19

Figure 9. Probability distribution of (**a**) log(I_{off}); (**b**) log(I_{on}) for FinFET and (**c**) log(I_{off}); (**d**) log(I_{on}) for TFET at $V_{DS} = 0.3$ V considering WFV and fin LER.

For TFETs, the I_{off} distribution with WFV is boarder (worse) than that with fin LER since WFV leads to fluctuation in the energy bands and alters the critical tunneling path, and the effect decreases with increasing V_{GS}. The metal grains with various WF form the up and down energy bands that boost the band-to-band generation, resulting in large I_{off} distribution. Therefore, the variability of I_{off} is larger than I_{on}, and the correlation between I_{on} and I_{off} is weak. On the other hand, for fin LER, both I_{on} and I_{off} are degraded as fin width (W_{Fin}) increases due to the weaker electrostatic control of the channel from both gates, and the degradations of I_{on} and I_{off} track W_{Fin} with exponential-like behavior, especially for I_{off} which dramatically increases with decreasing W_{Fin}. Comparing with fin LER, the μ/σ (I_{on}) of WFV is better, and the μ/σ (I_{off}) of WFV is comparable to LER. In addition, WFV causes larger σ (I_{off}) than LER. Overall, comparing FinFET and TFET, the impacts due to WFV on I_{on} and I_{off} are quite different. The μ/σ (I_{off}) of TFET is worse while μ/σ (I_{on}) of TFET is better. In addition, the I_{off} distribution of TFET skews to high values, and not as symmetrical as the I_{off} distribution for FinFET, resulting in larger μ (I_{off}). On the other hand, the variation of TFET considering fin LER is slight better than FinFET.

3.2. C_g Variability

Figure 10 shows the impacts of WFV and fin LER on C_g-V_{GS} dispersions of TFET and FinFET devices at $V_{DS} = 0.3$ V. Figure 11 illustrates the probability distributions of $C_{g,ave}$ (the average capacitance across the gate-bias range from 0 to $V_{DD} = 0.3$ V) at $V_{DS} = V_{DD}$. For both TFET and FinFET, the C_g variation by WFV becomes more significant at larger V_{GS}. In contrast, the variation due to fin LER is more severe when V_{GS} is small. Note that $C_{g,ave}$ is extracted only for the range from $V_{GS} = 0$ V to 0.3 V. The μ/σ (WFV) are much better compared with μ/σ (LER). For TFET with WFV and FinFET with fin LER, the $C_{g,ave}$ skews to high values, resulting in larger μ than the nominal cases.

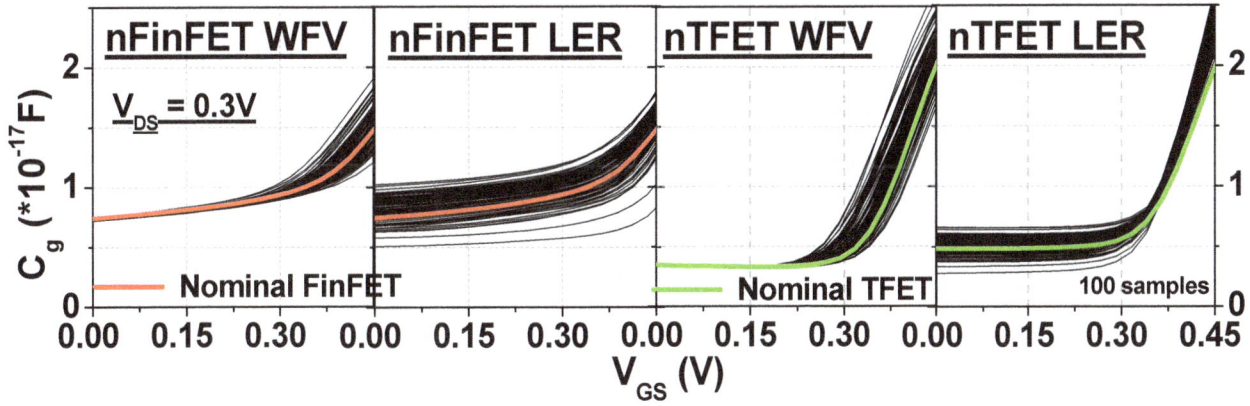

Figure 10. Simulated C_g-V_{GS} characteristics at V_{DS} = 0.3 V for TFET and FinFET with WFV and fin LER.

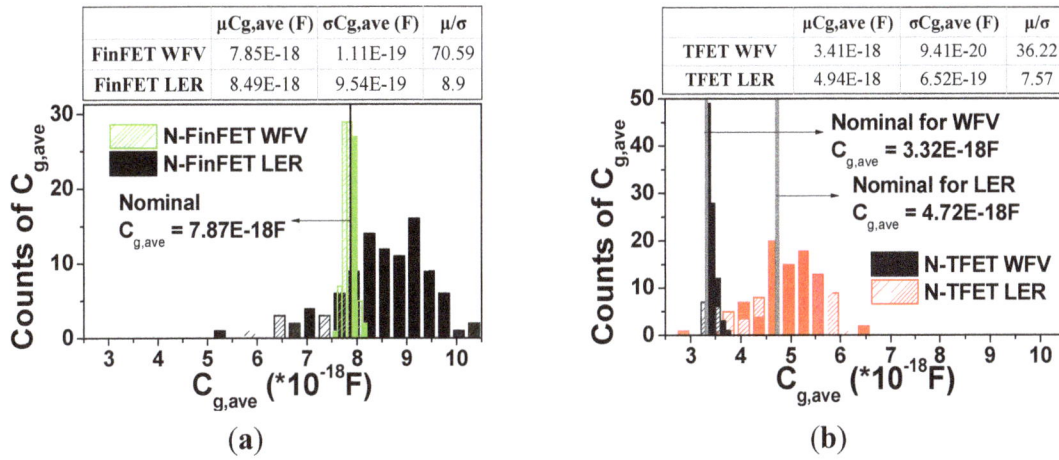

Figure 11. Probability distribution of $C_{g,ave}$ for (a) FinFET and (b) TFET at V_{DS} = 0.3 V considering WFV and fin LER.

4. Impacts of WFV and Fin LER on CLA Circuits

4.1. Delay Variability

The switching delay is commonly calculated as $\tau = (C_g V_{DD})/I_{on}$. Due to the strong bias dependence of gate capacitance (C_g), the average capacitance ($C_{g,ave}$) across the gate-bias range from 0 to V_{DD} (0.3 V in this case) at $V_{DS} = V_{DD}$ is determined for approximation: $\tau = (C_{g,ave} V_{DD})/I_{on}$.

The transient waveforms and the probability distributions of delays for 32-bit CLA of TFET and FinFET with WFV and fin LER are shown in Figures 12 and 13. As can be seen, the μ/σ (Delay) of TFET is better than FinFET in both cases (with WFV and fin LER). For both TFET and FinFET, the μ/σ (WFV) is better than μ/σ (LER). The variability of delay correlates with aforementioned I_{on} and $C_{g,ave}$ variations in Section 3. The smaller I_{on} of FinFET significantly degrades its μ/σ (Delay).

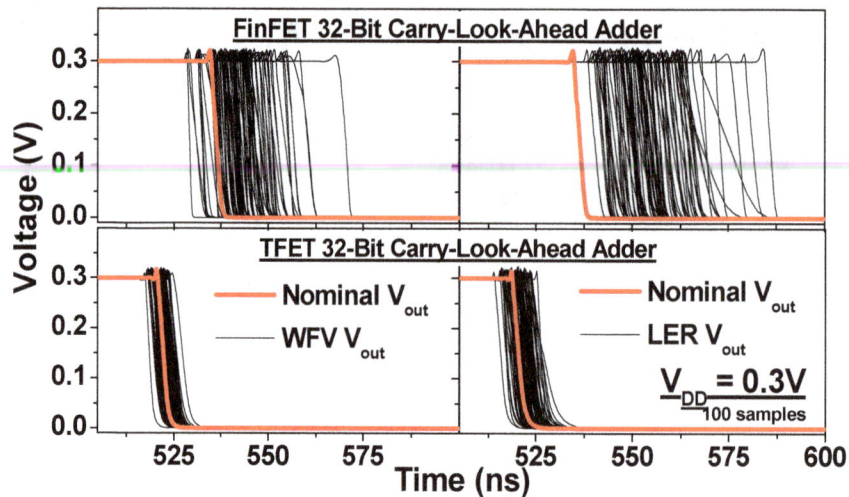

Figure 12. Transient waveforms of 32-bit CLA for TFET and FinFET at V_{DD} = 0.3 V considering WFV and fin LER.

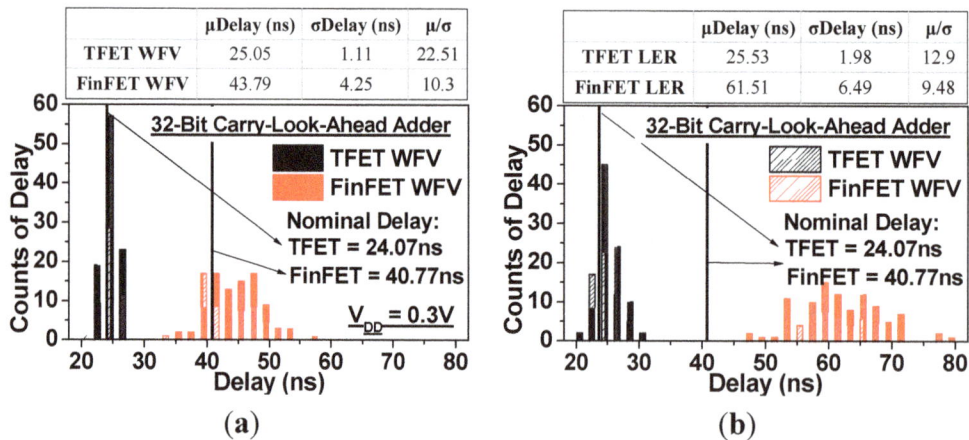

Figure 13. Probability distribution of delay for 32-bit CLA with (**a**) WFV; (**b**) fin LER for TFET and FinFET at V_{DD} = 0.3 V.

Figure 14 presents the delay for 32-bit CLA of TFET and Fin FET *versus* V_{DD} from 0.15 V to 0.35 V for the nominal cases and the cases considering WFV and fin LER (at 0.2 V and 0.3 V). The delay variability of all cases becomes worse with decreasing V_{DD} due to decreasing I_{DS}. The delay and its variability of TFET are significantly better than FinFET at low V_{DD} due to its larger I_{DS} and smaller $C_{g,ave}$ variation compared with FinFET.

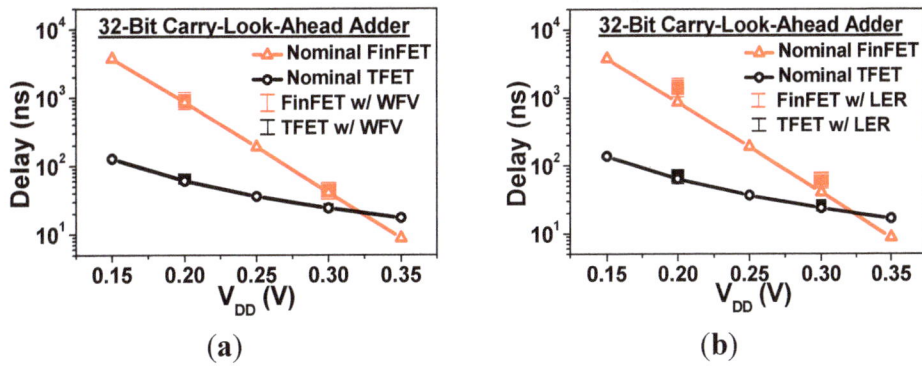

Figure 14. Delay for 32-bit CLA of TFET and FinFET *versus* V_{DD} from 0.15 V to 0.35 V for the nominal cases and the cases considering (**a**) WFV and (**b**) fin LER (0.2 V and 0.3 V).

4.2. PDP Variability

PDP is a figure of merit representing the power-performance trade-off. At a given operation frequency, PDP is calculated as PDP = $(C_g V^2_{DD} f) \times t_{delay} \approx C_{g,ave} V^2_{DD} f t_{delay}$. If the frequency is scaled up to the maximum operation frequency (*i.e.*, $f = 1/t_{delay}$), then PDP = $(C_g V^2_{DD})$ would represent the energy dissipated in a switching event.

The probability distributions of PDP 32-bit CLA of TFET and FinFET for the nominal cases and the cases with WFV and fin LER are shown in Figure 15. The μ/σ (WFV) is better than μ/σ (LER) for both TFET and FinFET, and the distributions of TFET with WFV and that of FinFET with fin LER skew to larger values.

	μPDP (J)	σPDP (J)	μ/σ
TFET WFV	3.01E-17	2.05E-18	14.68
FinFET WFV	3.12E-17	3.03E-18	10.31

	μPDP (J)	σPDP (J)	μ/σ
TFET LER	3.14E-17	2.68E-18	11.7
FinFET LER	4.68E-17	5.22E-18	8.97

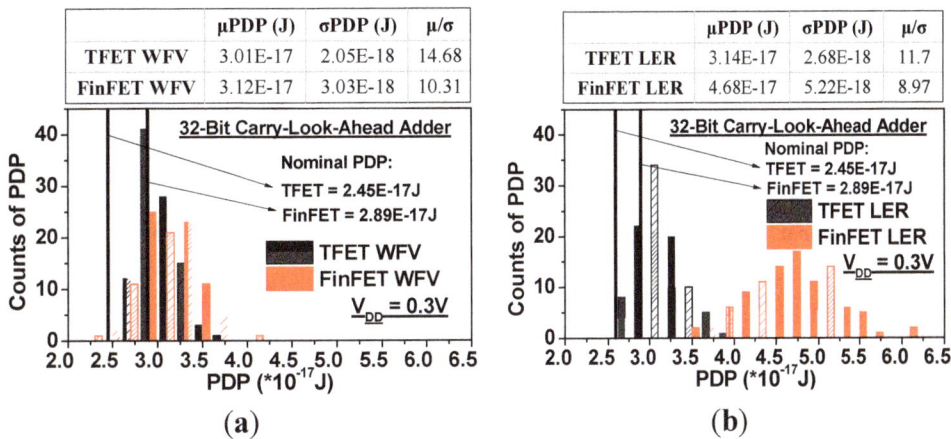

Figure 15. Probability distribution of PDP for 32-bit CLA with (**a**) WFV; (**b**) fin LER for TFET and FinFET at V_{DD} = 0.3 V.

Figure 16 shows the PDP for 32-bit CLA of TFET and FinFET *versus* V_{DD} from 0.15 V to 0.35 V for the nominal cases and the cases considering WFV and fin LER (at 0.2 V and 0.3 V). As can be seen, TFET PDP is much better than FinFET at low V_{DD} due to the fact that $C_{g,ave}$ variation of FinFET is larger and skewed to high values compared with TFET. Notice that the PDP of TFET is still better than FinFET considering random variations.

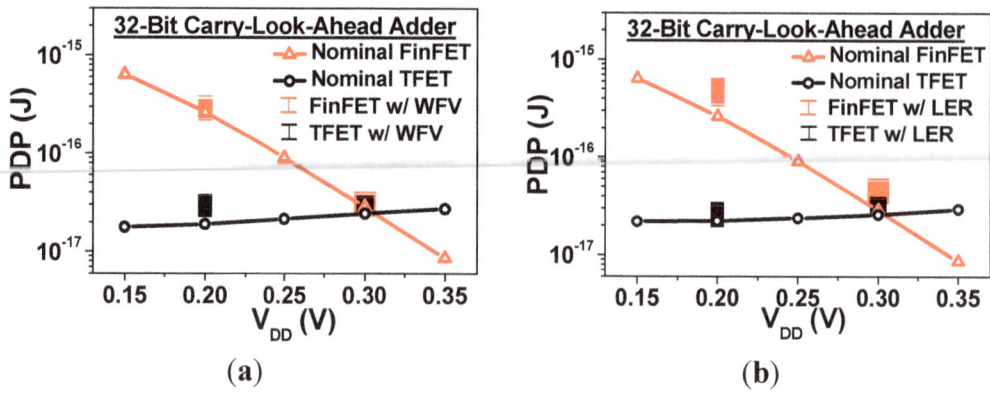

(a)   **(b)**

Figure 16. PDP for 32-bit CLA of TFET and FinFET *versus* V_{DD} from 0.15 V to 0.35 V for the nominal cases and the cases considering (**a**) WFV and (**b**) fin LER (0.2 V and 0.3 V).

4.3. Leakage Power Variability

The probability distributions of leakage power for 32-bit CLA of TFET and FinFET for the nominal cases and the cases with WFV and fin LER at $V_{DD} = 0.3$ V are shown in Figure 17. The leakage power variation of TFET with both variation sources are much worse than FinFET, and the distributions skew to larger values, especially under WFV. This correlates to aforementioned I_{off} variations in Section 3.

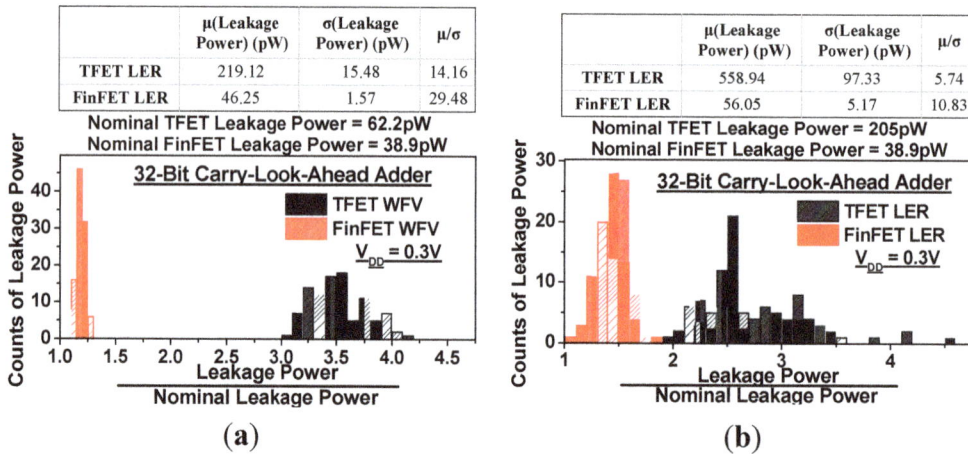

(a)   **(b)**

Figure 17. Probability distribution of leakage power for 32-bit CLA with (**a**) WFV; (**b**) fin LER for TFET and FinFET at $V_{DD} = 0.3$ V.

Figure 18 shows the leakage power for 32-bit CLA of TFET and FinFET *versus* V_{DD} from 0.15 V to 0.35 V for the nominal cases and the cases considering WFV and fin LER (at 0.2 V and 0.3 V). As the operating voltage is reduced, the leakage power decreases. Notice that the increase of leakage power by random variations is more significant than the influence by operating voltage for TFET.

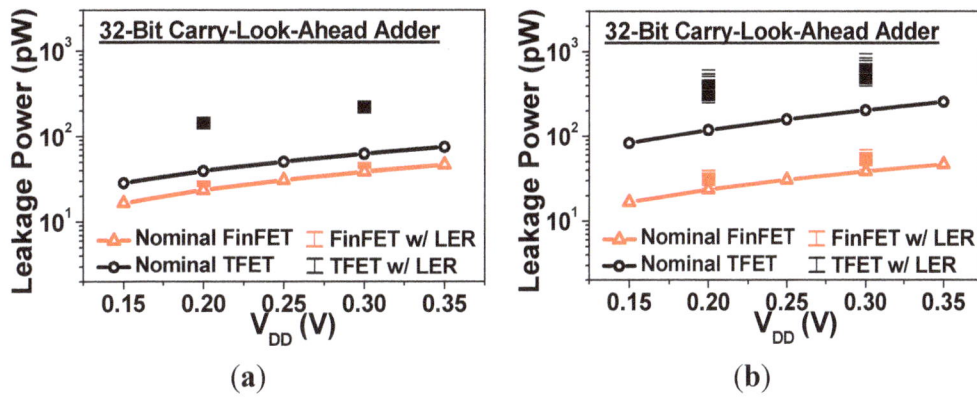

Figure 18. Leakage power for 32-bit CLA of TFET and FinFET *versus* V_{DD} from 0.15 V to 0.35 V for the nominal cases and the cases considering (**a**) work function variation (WFV) and (**b**) fin Line-Edge-Roughness (fin LER) (0.2 V and 0.3 V).

5. Conclusions

We investigate and compare the impacts of WFV and fin LER on TFET and FinFET I_{on}, I_{off} and $C_{g,ave}$ using atomistic 3D TCAD simulations with calibrated model and device parameters. Our studies indicate that considering WFV, FinFET has comparable I_{on} and I_{off} variability while TFET has smaller I_{on} variability and larger I_{off} variability. In addition, the band diagram dispersion caused by WFV increases the band-to-band generation for TFET in "OFF" state, leading to skewed I_{off} distribution to larger values. On the other hand, the impact of fin LER is similar for TFET and FinFET, resulting in comparable I_{on} and I_{off} variability. The $C_{g,ave}$ variability is worse with fin LER compared with WFV for both TFET and FinFET.

Using Verilog-A device models extracted from atomistic 3D TCAD simulations to capture the physical non-uniformities and variability, HSPICE circuit simulations are performed to assess the impacts of WFV and fin LER on TFET and FinFET 32-bit CLA. The results show that at low operating voltage (<0.3 V), the delay and PDP of TFET CLA are significantly better than the FinFET counterparts, even under the impacts of WFV and LER. However, the variability of leakage power for TFET CLA is worse than FinFET CLA, especially with WFV. The leakage power distribution of TFET CLA skews to larger values due to its worse I_{off} variability.

Acknowledgments

This work was supported in part by the Ministry of Science and Technology in Taiwan under Contract MOST 103-2221-E-009-196-MY2, and by the Ministry of Education in Taiwan under the ATU Program. The authors thank the National Center for High-Performance Computing in Taiwan for the software and facilities.

Author Contributions

Author Yin-Nien Chen contributed to the literature search and coordinated the the research, discussion and prepared the manuscript. Author Chien-Ju Chen contributed to the simulated works, discussion and the manuscript. Author Dr. Ming-Long Fan and Dr. Vita Pi-Ho Hu contributed to the technical suggestions and discussion. Author Prof. Pin Su guided this research work and contributed to technical discussions on device part about the impacts of intrinsic variations on TFET and FinFET devices. Author Prof. Ching-Te Chuang guided this research work and contributed to technical discussions on the circuit part and paper writing by reviewing all the results presented in this work and revising the technical writing and formatting of the manuscript.

Conflicts of Interest

The authors declare no conflict of interest.

References

1. Ionescu, A.M.; Riel, H. Tunnel field-effect transistors as enrgy-efficient electronics switches. *Nature* **2011**, *479*, 329–337.

2. Saripalli, V.; Datta, S.; Narayanan, V.; Kulkarni, J.P. Variation-tolerant ultra low-power heterojunction tunnel FET SRAM design. In Proceedings of the 2011 IEEE/ACM International Symposium on Nanoscale Architectures (NANOARCH), San Diego, CA, USA, 8–9 June 2011; pp. 45–52.

3. Cotter, M.; Liu, H.C.; Datta, S.; Narayanan, V. Evaluation of tunnel FET-based flip-flop designs for low power, high performance applications. In Proceedings of the 2013 14th International Symposium on Quality Electronic Design (ISQED), Santa Clara, CA, USA, 4–6 March 2013; pp. 430–437.

4. Datta, S.; Bijesh, R.; Liu, H.; Mohata, D.; Narayanan, V. Tunnel transistors for energy efficient computing. In Proceedings of the 2013 IEEE International Reliability Physics Symposium (IRPS), Anaheim, CA, USA, 14–18 April 2013; pp. 6A.3.1–6A.3.7.

5. Leung, G.; Chui, C.O. Stochastic Variability in Silicon Double-Gate Lateral Tunnel Field-Effect Transistors. *IEEE Trans. Electron Devices* **2012**, *60*, 84–91.

6. Fan, M.L.; Hu, V.P.H.; Chen, Y.N.; Su, P.; Chuang, C.T. Analysis of Single-Trap-Induced Random Telegraph Noise and its Interaction With Work Function Variation for Tunnel FET. *IEEE Trans. Electron Devices* **2013**, *60*, 2038–2044.

7. Choi, K.M.; Choi, W.Y. Work-function variation effects of tunneling field-effect transistors (TFETs). *IEEE Trans. Electron Device Lett.* **2013**, *34*, 942–944.

8. Damrongplasi, N.; Kim, N.S.; Shin, H.C.; Liu, T.J.K. Impact of Gate Line-Edge Roughness (LER) *vs.* Random Dopant Fluctuations (RDF) on Germanium-Source Tunnel FET Performance. *IEEE Trans. Nanotechnol.* **2013**, *12*, 1061–1067.

9. Avci, U.E.; Rios, R.; Kuhn, K.J.; Young, I.A. Comparison of performance, switching energy and process variations for the TFET and MOSFET in logic. In Proceedings of the 2011 Symposium on VLSI Technology (VLSIT), Honolulu, HI, USA, 14–16 June 2011; pp. 124–125.

10. Saripalli, V.; Mishra, A.; Datta, S.; Narayanan, V. An energy-efficient heterogeneous CMP based on hybrid TFET-CMOS cores. In Proceedings of the 48th ACM/EDAC/IEEE on Design Automation Conference (DAC), New York, NY, USA, 5–9 June 2011; pp. 729–734.

11. *Sentaurus TCAD Manual*; Sentaurus Device: Mountain View, CA, USA, 2011.

12. Liu, L.; Mohata, D.K.; Datta, S. Scaling Length Theory of Double-Gate Interband Tunnel Field-Effect Transistors. *IEEE Trans. Electron Devices* **2012**, *59*, 902–908.

13. Kotlyar, R.; Avci, U.E.; Cea, S.; Rios, R.; Linton, T.D.; Kuhn, K.J.; Young, I.A. Bandgap engineering of group IV materials for complementary n and p tunneling field effect transistors. *Appl. Phys. Lett.* **2013**, *102*, 106–113.

14. Pal, A.; Sachid, A.B.; Gossner, H.; Rao, V.R. Insights into design and optimization of TFET devices and circuits. *IEEE Trans. Electron Devices* **2011**, *58*, 1045–1053.

15. Chou, S.H.; Fan, M.L.; Su, P. Investigation and Comparison of Work Function Variation for FinFET and UTB SOI Devices Using a Voronoi Approach. *IEEE Trans. Electron Devices* **2013**, *60*, 1485–1489.

16. Asenov, A.; Kaya, S.; Brown, A.R. Intrinsic parameter fluctuations in decananometer MOSFETs introduced by gate line edge roughness. *IEEE Trans. Electron Devices* **2003**, *50*, 1254–1260.

13

Impact of Low-Variability SOTB Process on Ultra-Low-Voltage Operation of 1 Million Logic Gates [†]

Yasuhiro Ogasahara [1,*], Tadashi Nakagawa [1], Toshihiro Sekigawa [1], Toshiyuki Tsutsumi [2] and Hanpei Koike [1]

[1] Electroinformatics Group, Nanoelectronics Research Institute, National Institute of Advanced Industrial Science and Technology (AIST), 1-1-1 Umezono, Tsukuba 3058568, Japan;
E-Mails: nakagawa.tadashi@aist.go.jp (T.N.); t.sekigawa@aist.go.jp (T.S.); h.koike@aist.go.jp (H.K.)

[2] Computer Science Course, Fundamental Science and Technology, Graduate School of Science and Technology, Meiji University, 1-1-1 Higashi-Mita Tama, Kawasaki 2148571, Japan;
E-Mail: tsutsumi@cs.meiji.ac.jp

[†] This is an extended version of a paper that was presented at the IEEE S3S Conference 2014.

[*] Author to whom correspondence should be addressed; E-Mail: ys.ogasahara@aist.go.jp

Academic Editors: David Bol and Steven A. Vitale

Abstract: In this study, we demonstrate near-0.1 V minimum operating voltage of a low-variability Silicon on Thin Buried Oxide (SOTB) process for one million logic gates on silicon. Low process variability is required to obtain higher energy efficiency during ultra-low-voltage operation with steeper subthreshold slope transistors. In this study, we verify the decrease in operating voltage of logic circuits via a variability-suppressed SOTB process. In our measurement results with test chips fabricated in 65-nm SOTB and bulk processes, the operating voltage at which the first failure is observed was lowered from 0.2 to 0.125 V by introducing a low-variability SOTB process. Even at 0.115 V, over 40% yield can be expected as per our measurement results on SOTB test chips.

Keywords: SOTB; FD-SOI; ultra-low voltage; measurement on silicon

1. Introduction

Subthreshold designs yield high energy efficiency by lowering operating voltages of the given circuits. Such designs can be applied to low-power sensor nodes or other low-power applications [1]. On the other hand, several new-structure steep subthreshold slope (SS) FETs, such as FinFETs, Gate All Around (GAA) FETs, and tunnel FETs [2,3], have been discussed as further advancement of CMOS processes. These transistors have steeper SS values than conventional planar MOSFETs, and reduced leakage current at low voltages. Figure 1 [4] shows the energy/cycle of a 54-stage inverter chain under several SS conditions which was simulated following [5]. From the figure, the SS of the planar CMOS process is approximately 100 mV/dec, with the minimum energy/cycle obtained at about 0.22 V. On the other hand, $3\times$ and $8.5\times$ energy efficiency improvement was obtained at near 0.1 V under 65 mV/dec and 45 mV/dec conditions because of reduced leakage current. In general, when the SS values of transistors are improved, the minimum energy point is lowered.

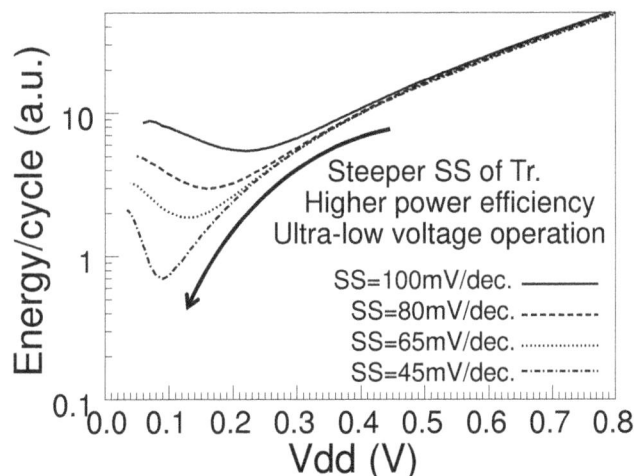

Figure 1. Energy efficiency simulation results of inverter chains for various SS values.

Lowering the operating voltage will be more effective for obtaining high energy efficiency in future processes: however, low-voltage operation is restricted by variability in transistor performance, and the impact of local random variation is serious in large-scale circuits [5]. In [6], Niiyama *et al.* reported the minimum operating voltage of ring oscillators (ROs) in a 90-nm bulk process. Although the minimum operating voltage of 11-stage ROs was 90 mV, that of 1 million-stage ROs reached 343 mV. Several studies have reported on the implementation of ultra-low-voltage operation circuits [7–10]; their operating ranges are from 0.175 V to 0.28 V depending on the scale of the circuit and process technology. SRAM has relatively high minimum operating voltage, and methods of lowering the minimum operating voltage via circuit structures are frequently discussed in the literature. A conventional 6T-SRAM has approximately 0.8 V minimum operating voltage [11], whereas 7T [12], 8T [13], 10T [14], and improved-6T [15] SRAMs operate at 0.44 V, 0.25 V, 0.16 V and 0.208 V, respectively.

In this study, we report notably decreased minimum operating voltages via variability suppressed Silicon on Thin Buried Oxide (SOTB) [16] technology with a million-gate logic circuit based on our previous research [17]. Figure 2 shows the structure of the SOTB transistor, which is formed on a thin silicon on insulator (SOI) layer and has an ultra-low dose channel. This ultra-low dose channel enables reductions in the random dopant variability and, as a result, decreased operating voltage is expected. Whereas a 0.37 V minimum operating voltage of 6T-SRAM in the SOTB process has been reported [11], we focus on logic circuits. We measure a 1 million-gate logic circuit consisting of 1011 ROs with 1001 stages on test chips fabricated in 65-nm SOTB and bulk processes, and demonstrate the notable impact of low-variability SOTB technology on ultra-low-voltage design. In addition, the operating voltage limit of logic circuits is the fundamental limit of the minimum operating voltage of SRAMs. Our measurement results on the operating voltage limit of logic circuits in bulk and SOTB processes suggest the goal of research on the reduction of the minimum operating voltage of SRAMs in the common bulk, SOTB, or other variability-suppressed processes.

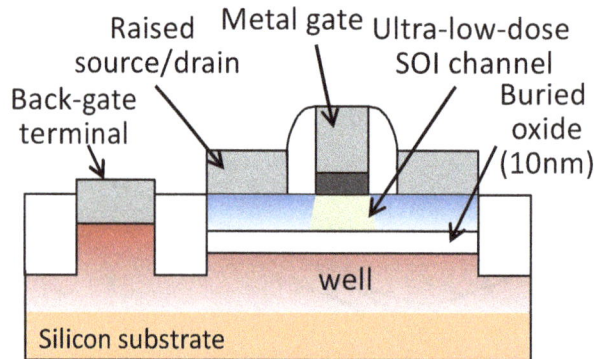

Figure 2. Schematic cross section of Silicon on Thin Buried Oxide (SOTB) transistor.

2. Implementation of Test Element Group (TEG)

We designed 337 RO patterns, and each RO pattern consists of 3 ROs. An RO pulse is counted by the counter on the chip. To measure ultra-low-voltage operation of ROs, we implemented three power supply and ground-line pairs for ROs, a counter and buffers. The buffers were located between the ROs and the counter.

Table 1 describes the details of the implemented 337 ROs. We selected INV, NAND, and NOR gates as measurement target gates. Our nominal design of INV, NAND and NOR cells had 1:1, 1:1 and 2:1 P:N (pMOS width:nMOS width) ratios, respectively. In addition to basic logic cells, complex gates are commonly included in standard cell libraries, and these complex gates frequently include serially stacked transistors, transmission gates, or an unbalanced P:N ratio. The impact of these designs in a low-voltage design is different from that of a MOS width (W) change. Although we observe ultra-low-voltage operation of 1 million logic gates with this RO Test Element Group (TEG), this TEG is also aimed to be designed for observation of layout dependence [18]. Gates in which shallow trench isolation (STI) stress effect and inverse narrow channel effect were escalated were included in the RO patterns. The stress caused by STI causes a mobility shift of pMOS and nMOS transistors depending on the distance between the transistor gate and STI edge. The inverse narrow channel effect is the reduction of V_{th} observed in narrow channel device. These effects can cause V_{th} or mobility variation, and are beneficial for estimating low-voltage behavior of logic circuits.

Table 1. Number of designed ring oscillator (RO) patterns; 3 ROs for each RO design were placed on a chip.

Included Gate(s)	Number of RO Designs Containing							
	Single Type			2 Types			3 Types	
	Basic	STI	Narrow Channel	Basic	STI	Narrow Channel	STI	Narrow Channel
INV	1	7	6	-	22	21	-	20
NAND	1	7	13	-	22	18	-	-
NOR	1	7	10	-	22	18	-	-
INV & NAND	-	-	-	1	29	15	-	-
INV & NOR	-	-	-	1	29	15	-	-
NAND & NOR	-	-	-	1	29	21	-	-

Changes in STI effect were implemented with active area design. Each standard cell consisted of an even number of pMOS or nMOS transistor, and their sources or drains, which are connected to ground or power, were connected to neighboring gates. When a source or drain area of a pMOS or nMOS transistor is connected to that of neighboring gates, STI is not inserted between the gates, and the impact of STI in the gate width direction can be weakened. The strength of the inverse narrow channel effect can be escalated by narrowing the transistor width. Table 2 shows implemented STI conditions in the gates. We designed four STI conditions, *i.e.*, (pMOS STI weak, nMOS STI weak), (pMOS STI weak, nMOS STI strong), (pMOS STI strong, nMOS STI weak) and (pMOS STI strong, nMOS STI strong), for INV, NAND and NOR cells. In addition to changing STI conditions, simple transistor width changes were also implemented such that the impact of STI could be compensated. The STI stress effect is known as nMOS mobility degradation and pMOS mobility enhancement, and (pMOS nominal W, nMOS W +10%) and (pMOS nominal W, nMOS W +20%) patterns were additionally designed for (pMOS STI weak, nMOS STI strong) and (pMOS STI strong, nMOS STI weak) designs.

We also designed the gates with narrow channel transistors as shown in Table 3. We set 1/n× width per finger and n× number of fingers for cells to observe narrow channel effects, and enhanced the narrow channel effect without changing the total transistor width in each cell. We used this approach because simply narrowing the width causes a reduction of driving current of transistors, and the narrow channel effect and reduction of width would otherwise be confused in our measurement results. Implemented (width per finger, number of fingers) patterns were (1/1.5× width per finger, 1.5× number of fingers), (1/2× width per finger, 2× number of fingers), and (1/3× width per finger, 3× number of fingers). The pMOS and nMOS conditions of narrow channel cells includes (pMOS nominal width, nMOS narrow width) and (pMOS narrow width, nMOS narrow width).

An RO consisting of only one type of cell may report too optimistic a minimum operating voltage and, in addition to ROs with one type of cell, ROs which include two or three types of cells were also designed. The details of the number of RO patterns are shown in Table 1. 53 patterns of ROs included a single type of cell, whereas 284 patterns were composed of two or three cells. Figure 3 shows the connection of cells when an RO included two or three types of cells. When two types of cells were included in an RO, the cells were organized in an alternating pattern. When an RO had three types of cells, these cells repetitively appear in the order shown in Figure 3. Six RO patterns were designed only with basic INV, NAND, and NOR cells. Moreover, 174 RO patterns included cells where STI conditions were modified. Among the 174 STI RO patterns, 21 consisted of a single type of cell and 153 ROs include two types of cells. Cells with narrow channel conditions were inserted in 157 RO patterns, among which were 29 ROs with a single type of cell, 108 ROs with two types of cells, and 20 ROs with three types of INV cells. Furthermore, 108 ROs with two types of narrow channel cells include 21 ROs with two types of INV cells, 18 ROs with two types of NAND cells, 18 ROs with two types of NOR cells, 15 ROs with INV and NAND cells, 15 ROs with INV and NOR cells, and 21 ROs with NAND and NOR cells.

Table 2. Implementation of escalation of shallow trench isolation (STI) stress effect in the gates.

Gate Design	pMOS		nMOS	
	Active Area	Width	Active Area	Width
nominal design	connected to neighboring gate	nominal	connected to neighboring gate	nominal
STI stress #1	NOT connected to neighboring gate	nominal	connected to neighboring gate	nominal
STI stress #2	NOT connected to neighboring gate	nominal	connected to neighboring gate	+10%
STI stress #3	NOT connected to neighboring gate	nominal	connected to neighboring gate	+20%
STI stress #4	connected to neighboring gate	nominal	NOT connected to neighboring gate	nominal
STI stress #5	connected to neighboring gate	nominal	NOT connected to neighboring gate	+10%
STI stress #6	connected to neighboring gate	nominal	NOT connected to neighboring gate	+20%
STI stress #7	NOT connected to neighboring gate	nominal	NOT connected to neighboring gate	nominal

Table 3. Implementation of escalation of inverse narrow channel effect in the gates.

Gate Design	pMOS		nMOS	
	Width	**Number of Fingers**	**Width**	**Number of Fingers**
nominal	nominal	nominal	nominal	nominal
narrow channel #1	nominal	nominal	$1/1.5\times$	$1.5\times$
narrow channel #2	$1/1.5\times$	$1.5\times$	$1/1.5\times$	$1.5\times$
narrow channel #3	nominal	nominal	$1/2\times$	$2\times$
narrow channel #4	$1/2\times$	$2\times$	$1/2\times$	$2\times$
narrow channel #5	nominal	nominal	$1/3\times$	$3\times$
narrow channel #6	$1/3\times$	$3\times$	$1/3\times$	$3\times$

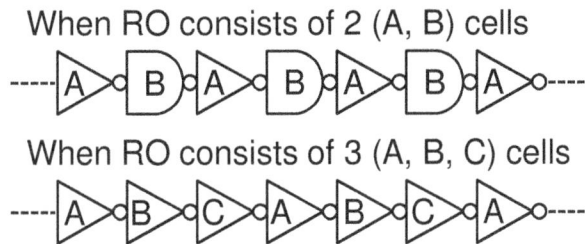

Figure 3. The gate connection in RO when RO includes two or three cell patterns.

Each RO consisted of 1000-stage measurement target gates and 1 NAND gate for controlling oscillation. There were trade-offs between implementation costs of glue logic and granularity of yield evaluation. The 1-million gate measurement structure can be implemented with a 1000-to-1 selector and 1000 output wires from ROs by adopting 1001-stage ROs. The implemented 1011-to-1 selector consists of 1010 multiplexer cells, and does not include a notably large number of cells in comparison with 1001-stage RO.

Although the routing of 1011 wires was a bottleneck of the layout of the selector circuit and required 0.964 mm^2 area on the test chip with the automated place and route software [19], the area cost of routing was acceptable because we had 12.5 mm^2 area on the chip. Had we adopted 101-stage ROs, we would have required a 10000-to-1 selector, which consists of 9999 multiplexer cells, and 10,000 wires. Compared to the chosen implementation of the 1011-to-1 selector and 1011 wires, the area cost for the implementation of 9999 cells and 10,000 wires would have been unacceptable for our test chip. Consequently, we avoid the huge area cost of a selector circuit and wiring cost of RO control and output signals by adopting 1001-stage ROs. Though the yield of a small-size circuit becomes difficult to be identified when too large RO size is adopted, yields of 1 million gates, 100,000 gates and 10,000 gates can easily be calculated from the measurement results of the 1001-stage ROs. These circuit scales are suitable for practical use of Application Specific Integrated Circuits (ASIC) or System on Chip (SoC) applications.

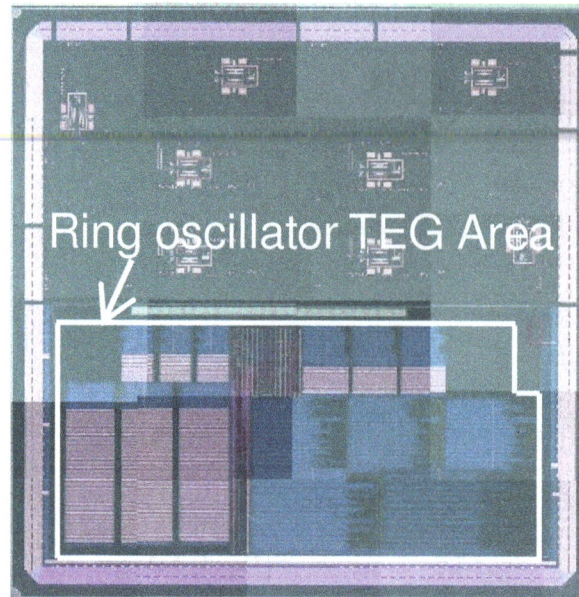

Figure 4. A micrograph of the test chip.

The test chips were fabricated in 65-nm SOTB and bulk processes. Figure 4 shows the micrograph of the fabricated test chip. The size of the test chips is 5.8 mm × 5.8 mm, with an area of approximately 5.0 mm × 2.5 mm allocated for the RO TEG.

3. Measurement Setup

We measured 4 bulk chips and 4 SOTB chips. Supply voltages and back-gate bias voltages were supplied by source measure units. Measurement was controlled by an external signal input generated via a pattern generator, and the values of the counter, which counted RO pulses, was read by an external logic analyzer. All of the 1011 ROs on each chip were measured. RO pulses were counted over a 125 ms time window, and each RO was measured 5 times. The average results excluding the fastest and slowest were recorded as measured values. Supply voltages were varied from 0.1 V to 0.55 V, and the RO period and tendency of operation failure were observed. Back-gate bias voltages were set to V_{dd} and V_{ss} for pMOS and nMOS transistors, respectively, in the RO period measurement, which is described in Section 4. Adequate reverse bias voltages were set for the observation of operation failure, which is detailed in Section 5.

4. Variability in SOTB and Bulk Processes

In this section, we evaluate the process variability of bulk and SOTB processes based on measured RO periods. RO periods on SOTB and bulk chips were measured at 0.4 V and 0.55 V V_{dd} respectively. As V_{th} of SOTB and bulk transistors were different, we compared RO periods of SOTB and bulk chips at supply voltages where the same RO period was obtained. The periods of basic ROs of SOTB at 0.4 V and bulk at 0.55 V were approximately 500 ns.

Figure 5 shows a histogram of measured RO periods. RO periods were normalized by averaging RO periods of each pattern. Each pattern had 3 ROs on each chip, and 12 ROs of each pattern on 4 chips were used for calculating the average value, and periods of 12 ROs were normalized by the calculated average of each pattern. The probabilities of SOTB were notably higher than that of bulk in the RO periods ranging from 0.998 to 1.002, and clearly lower in the range from 0.992 to 0.996 and the range from 1.004 to 1.018. The distribution of RO periods was more concentrated on the average period in the SOTB process than that of the bulk process and, thus, we confirmed smaller local random variability in the SOTB process. Although notable differences were not observed in ranges below 0.990 and above 1.020 between SOTB and bulk results, distributions in these ranges can be considered to be caused by the impact of global variability, which is not improved by SOTB technology.

We also calculated the standard deviations (σ) of RO periods for each of the 337 RO patterns on each chip, and Figure 6 shows the histogram of the calculated σ values. For each RO pattern, 3 ROs were implemented on each chip, and standard deviations of each pattern on each chip were calculated to eliminate the impact of global variability from standard deviations. Each standard deviation value was normalized by the average period of 3 ROs of each pattern on each chip. The peaks of σ distribution of SOTB and bulk were 0.2% and 0.5% respectively. The σ distribution of SOTB was larger than that of bulk in the range from 0.1% to 0.3%, and smaller in the range from 0.4% to 1%. The σ values of SOTB concentrate in a smaller range, and small random variability of SOTB was also confirmed, as in the results of average RO period. Both Figures 5 and 6 indicate low variability of the SOTB process in comparison with the bulk process.

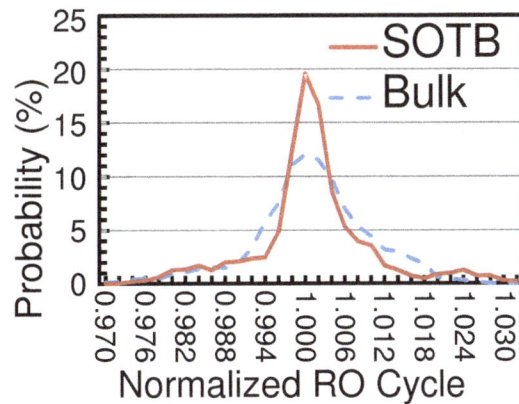

Figure 5. Measured RO periods on SOTB and bulk chips at 0.4 V and 0.55 V respectively.

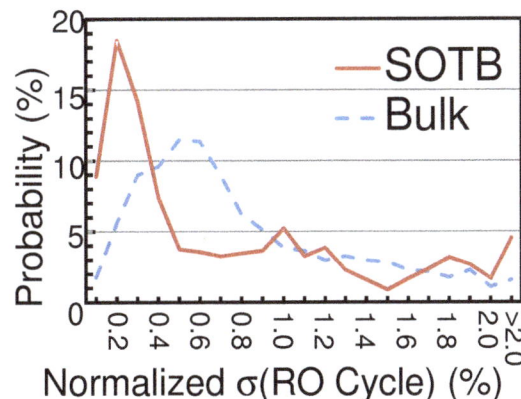

Figure 6. Standard deviations of period of each RO pattern on each chip.

5. Low-Voltage Operation

1011 ROs with 1001-stages on test chips fabricated both in bulk and SOTB processes were measured. The V_{th} mismatch between pMOS and nMOS transistors can be one of the factors that deteriorates circuit operating voltages, and considerable efforts are therefore devoted to minimize this V_{th} mismatch in process development; however, at near $V_{dd} = 0.1$ V, the margin of V_{th} mismatch is very small, and even the slight V_{th} mismatch has a serious impact. We tested several back-gate bias conditions, identifying back-gate bias conditions that minimized operation failure for bulk and SOTB chips. Our results for the applied back-gate bias conditions include:

- Bulk-a: $VBP = V_{dd}$, $VBN = 0\,V$
- Bulk-b: $VBP = V_{dd} + 1.5\,V$, $VBN = 0\,V$
- SOTB-a: $VBP = V_{dd} + 0.8\,V$, $VBN = -0.8\,V$
- SOTB-b: $VBP = V_{dd} + 1.0\,V$, $VBN = -0.6\,V$

VBP and VBN are pMOS and nMOS back-gate bias voltages. Furthermore, in conditions b, back-gate bias voltages were applied to compensate for V_{th} mismatch, whereas in conditions a, the back-gate bias voltages were applied equally to pMOS and nMOS transistors, and the back-gate bias voltages were not applied to compensate for V_{th} mismatch. When reverse body bias voltages were not applied, the RO period on bulk and SOTB chips were 307.5 µs and 4.36 µs, respectively, at $V_{dd} = 0.2$ V. The RO period in Bulk-b, SOTB-a and SOTB-b conditions increased to 902.5 µs, 11.1 µs and 14.0 µs at $V_{dd} = 0.2$ V, respectively.

Figures 7 and 8 show the measurement result of the RO operation. In these figures, the x-axis is the supply voltage and the y-axis is the ratio of the number of ROs that failed to operate. The yield of 10 k-gate and 100 k-gate circuits can be calculated as $(1-Y)^{10}$ and $(1-Y)^{100}$, respectively where Y is the failure ratio. When the back-gate bias voltages were equally applied for pMOS and nMOS transistors, the first failures were observed at 0.25 V and 0.15 V for bulk and SOTB, respectively. The failure rates at this voltage enabled a practical yield of a 100 k-gate circuit. A 0.072 failure ratio of SOTB at 0.125 V corresponds to approximately 47.3% yield of a 10 k-gate circuit, whereas 0.225 V is required for the bulk process to obtain this yield. Adjusting the back-gate bias lowered the first failure voltages to 0.2 V and 0.125 V for bulk and SOTB, respectively. Over 40% yield for a 10 k-gate circuit is achieved at 0.115 V in the SOTB process whereas 0.175 V was required for the bulk process.

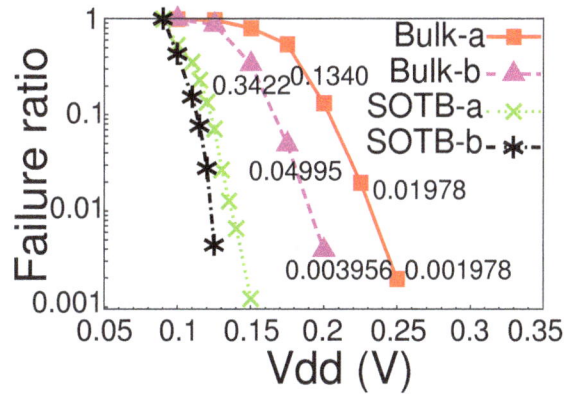

Figure 7. Measurement results of the RO operation. Y-axis is the ratio of the number of ROs which failed to operate.

These measurement results confirmed that the low variability of SOTB notably lowers the minimum operating voltage of a large scale logic circuit, and the low-variability of SOTB technology enables near-0.1 V operation of logic circuits. At near 0.1 V, the margin against V_{th} mismatch is seriously small in comparison with 0.2 V- to 0.4 V-class low-voltage operation. Even V_{th} control by the foundries for low-voltage operation will not sufficiently eliminate this V_{th} mismatch. Measurement results also indicate that back-gate biasing is effective for obtaining operation margin in extremely low-voltage and small-V_{th}-margin conditions.

Here, the limit of the operating voltage on the bulk process was 0.175 V–0.25 V according to our measurement results. The limit of operating voltage of several improved SRAMs reported in [13–15], 0.16 V–0.25 V, has already reached our measured limit of the operating voltage of the logic circuits. The operating voltage of SRAM cannot be fundamentally reduced beyond that of logic circuits, and further lowering of the operating voltage of SRAM is estimated to be difficult.

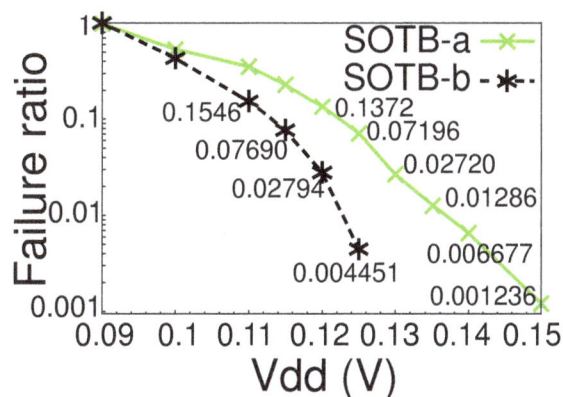

Figure 8. Measurement results of SOTB process at $V_{dd} = 0.09$ V–0.15 V.

We also focused on the tendency of 1011 ROs to operate successfully. Here we numbered 1011 ROs serially, and 3 ROs on each pattern on each chip were identified by the location. We measured 4 bulk chips and 4 SOTB chips, and the number of chips where specific ROs operate is from 0 to 4, depending on supply voltages. For example, in the case of bulk 0.15 V operation, the total failure ratio was 0.455 with 6.8% of ROs operating on all 4 chips, 13.6% of ROs operating on 3 chips, 44.8% of ROs operating on 2 chips, 24.1% of ROs operating on 1 chip, and 10.6% of ROs failing on all 4 chips. The number of chips with a specific RO operating successfully strongly depends on the total failure ratio; therefore, the total failure ratio was also considered in our evaluation. In Figures 9 and 10, the tendencies of bulk-b and SOTB-b conditions were demonstrated, with x-axis representing the total failure ratio, and the y-axis representing the ratio of the number of ROs operating on 0, 1, 2, 3, and 4 chips.

In Figures 9 and 10, the number of ROs operating on 1, 2, or 3 chips in the bulk test chip is larger than that in the SOTB test chip, and number of ROs operating on 0 or 4 chips in the bulk test chip is smaller than that in the SOTB test chip. This tendency indicates that RO operation failure tends to be random in bulk chips, and deterministic in SOTB chips. In particular, the cause of RO operation failure in bulk chips can be considered to be random variability. Conversely, in SOTB chips, the bottleneck of the operating voltage can be thought to be a gate design problem rather than random variability. These results indicate that random variability of the SOTB process is smaller than that estimated from the observed operating voltage and that the operating voltage of the SOTB process can be further decreased by optimizing the gate design.

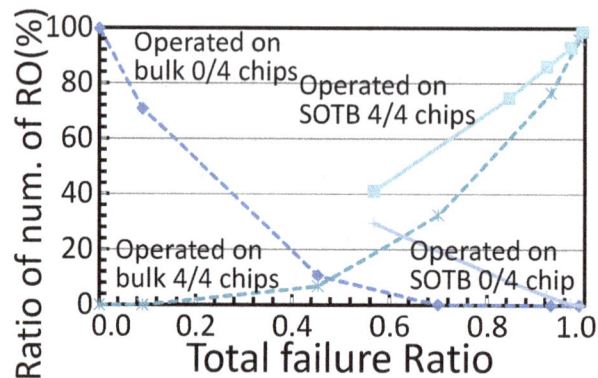

Figure 9. Distribution of the number of ROs operated on 0 or 4 chips; the x-axis is the total operation ratio illustrated in Figures 7 and 8.

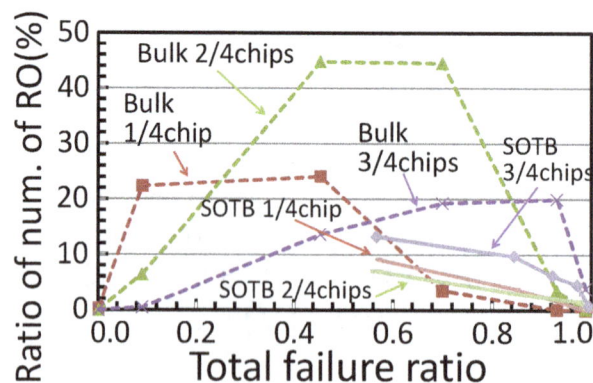

Figure 10. Distribution of the number of ROs operated on 1, 2, or 3 chips; the x-axis is the total operation ratio shown in Figures 7 and 8.

6. Conclusions

In this study, we implemented RO TEG chips in 65-nm SOTB and bulk processes, and measured the minimum operating voltages of logic gates. The SOTB process achieved fine yield with supply voltages as low as 0.11 V–0.15 V depending on the circuit size and back-gate bias conditions, whereas the bulk process required 0.175 V–0.25 V supply voltages. The low variability of the SOTB process significantly contributed to decreasing the minimum operating voltage. The minimum operating voltage achieved by the SOTB process was close to the voltage at which the minimum energy/cycle is expected with 65 mV/dec and 45 mV/dec SS transistors in simulation. Measurement results also indicated that applying a back-gate bias voltage was effective at reducing V_{th} mismatch beyond that of general process control, which is vital at ultra-low-voltages.

Acknowledgments

A part of this work was performed as the "Ultra-Low Voltage Device Project" funded and supported by the Ministry of Economy, Trade and Industry (METI) and the New Energy and Industrial Technology Development Organization (NEDO).

Author Contributions

Yasuhiro Oagsahara had a key role in this paper: generating ideas of measurement, designing and measuring TEG chips, and preparing the manuscript. Tadashi Nakagawa, Toshihiro Sekigawa, Toshiyuki Tsutsumi and Hanpei Koike discussed the results of this paper, and contributed to revising the technical writing and formatting of the manuscript.

Conflicts of Interest

The authors declare no conflict of interest.

References

1. Fuketa, H.; Yasufuku, T.; Iida, S.; Takamiya, M.; Nomura, M.; Shinohara, H.; Sakurai, T. Device-Circuit Interactions in Extremely Low Voltage CMOS Designs. In Proceedings of the Digest of Technical Papers of IEEE International Electron Devices Meeting, Washington, DC, USA, 5–7 December 2011; pp. 559–562.
2. Bohr, M. The New Era of Scaling in an SoC World. In Proceedings of the Digest of Technical Papers of IEEE International Solid-State Circuits Conference, San Francisco, CA, USA, 8–12 February 2009; pp. 23–28.
3. Bohr, M. The Evolution of Scaling from the Homogeneous Era to the Heterogeneous Era. In Proceedings of the IEEE International Electron Devices Meeting, Washington, DC, USA, 5–7 December 2011; pp. 1–6.
4. Ogasahara, Y.; Ma, C.; Hioki, M.; Nakagawa, T.; Sekigawa, T.; Tsutsumi, T.; Koike, H.; Tada, M.; Sakamoto, T. Utility of High on-Off Ratio, High Off Resistance Rewritable Device to EEPROM for Ultra-Low Voltage Operation of Steep Subthreshold Slope FETs. In Proceedings of the IEEE International Memory Workshop, Taipei, Taiwan, 18–21 May 2014; pp. 111–114.
5. Sakurai, T. Designing Ultra-Low Voltage Logic. In Proceedings of the IEEE/ACM International Symposium on Low Power Electronics and Design, Fukuoka, Japan, 1–3 August 2011; pp. 57–58.

6. Niiyama, T.; Piao, Z.; Ishida, K.; Murakata, M.; Takamiya, M.; Sakurai, T. Increasing Minimum Operating Voltage (VDDmin) with Number of CMOS Logic Gates and Experimental Verification with up to 1Mega-Stage Ring Oscillators. In Proceedings of the IEEE/ACM International Symposium on Low Power Electronics and Design, Bangalore, India, 11–13 August 2008; pp. 117–122.

7. Kao, J.T.; Miyazaki, M.; Chandrakasan, A.P. A 175-mV Multiply-Accumulate Unit Using an Adaptive Supply Voltage and Body Bias Architecture. *IEEE J. Solid-State Circuits* **2002**, *11*, 1545–1554.

8. Wang A.; Chandrakasan, A. A 180-mV Subthreshold FFT Processor Using a Minimum Energy Design Methodology. *IEEE J. Solid-State Circuits* **2005**, *1*, 310–317.

9. Zhai, B.; Pant, S.; Nazhandali, L.; Hanson, S.; Olson, J.; Reeves, A.; Minuth, M.; Helfand, R.; Austin, T.; Sylvester, D.; *et al.* Energy-Efficient Subthreshold Processor Design. *IEEE Trans. Very Large Scale Integr. Syst.* **2009**, *8*, 1127–1137.

10. Hsu, S.; Agarwal, A.; Anders, M.; Mathew, S.; Kaul, H.; Sheikh, F.; Krishnamurthy, R. A 280 mV-to-1.1 V 256 b Reconfigurable SIMD Vector Permutation Engine with 2-Dimensional Shuffle in 22 nm CMOS. In Proceedings of the Digest of Technical Papers of IEEE International Solid-State Circuits Conference, San Francisco, CA, USA, 19–23 February 2012; pp. 178–180.

11. Yamamoto, Y.; Makiyama, H.; Shinohara, H.; Iwamatsu, T.; Oda, H.; Kamohara, S.; Sugii, N.; Yamaguchi, Y.; Mizutani, T.; Hiramoto, T.; *et al.* Ultralow-Voltage Operation of Silicon-on-Thin-BOX (SOTB) 2Mbit SRAM Down to 0.37 V Utilizing Adaptive Back Bias. In Proceedings of the Digest of Technical Papers of Symposium on VLSI Technology, Kyoto, Japan, 11–14 June 2013; pp. 212–213.

12. Takeda, K.; Hagihara, Y.; Aimoto, Y.; Nomura, M.; Nakazawa, Y.; Ishii, T.; Kobatake, H. A Read-Static-Noise-Margin-Free SRAM Cell for Low-VDD and High-Speed Applications. *IEEE J. Solid-State Circuits* **2008** , *1*, 113–121.

13. Sinangil, M.E.; Verma, N.; Chandrakasan, A.P. A Reconfigurable 8T Ultra-Dynamic Voltage Scalable (U-DVS) SRAM in 65 nm CMOS. *IEEE J. Solid-State Circuits* **2009**, *11*, 3163–3173.

14. Chang, I.J.; Kim, J.-J.; Park, S.P.; Roy, K. A 32 kb 10 T Sub-Threshold SRAM Array With Bit-Interleaving and Differential Read Scheme in 90 nm CMOS. *IEEE J. Solid-State Circuits* **2009**, *2*, 650–658.

15. Zhai, B.; Hanson, S.; Blaauw, D.; Sylvester, D. A Variation-Tolerant Sub-200 mV 6-T Subthreshold SRAM. *IEEE J. Solid-State Circuits* **2008**, *10*, 2338–2348.

16. Sugii, N.; Tsuchiya, R.; Ishigaki, T.; Morita, Y.; Yoshimoto, H.; Kimura, S. Local V_{th} Variability and Scalability in Silicon-on-Thin-BOX (SOTB) CMOS with Small Random-Dopant Fluctuation. *IEEE Trans. Electron Devices 2010*, *4*, 835–845.

17. Ogasahara, Y.; Hioki, M.; Nakagawa, T.; Sekigawa, T.; Tsutsumi, T.; Koike, H. Near-0.1 V Ultra-Low Voltage Operation of SOTB 1 M Logic Gates. In Proceedings of the IEEE SOI-3D-Subthreshold Microelectronics Technology Unified Conference, Millbrae, CA, USA, 6–9 October 2014; pp. 1–3.

18. Ogasahara, Y.; Hioki, M.; Nakagawa, T.; Sekigawa, T.; Tsutsumi, T.; Koike, H. Measurement of V_{th} Variation due to STI Stress and Inverse Narrow Channel Effect at Ultra-Low Voltage in a Variability-Suppressed Process. In Proceedings of the IEEE International Conference on Microelectronic Test Structures, Phoenix, AZ, USA, 23–26 March 2015; pp. 126–130.

19. Cadence Design Systems, Inc. *EDI System User Guide Product Version 11.1*; Cadence Design Systems, Inc.: San Jose, CA, USA, 2012.

Permissions

List of Contributors

Naser Khosro Pour
Ecole Polytechnique Fédérale de Lausanne, CH-1015 Lausanne, Switzerland

François Krummenacher
Ecole Polytechnique Fédérale de Lausanne, CH-1015 Lausanne, Switzerland

Maher Kayal
Ecole Polytechnique Fédérale de Lausanne, CH-1015 Lausanne, Switzerland

Pascal Meinerzhagen
Institute of Electrical Engineering, Ecole Polytechnique F´ed´erale de Lausanne, Station 11, Lausanne, VD 1015, Switzerland

Adam Teman
VLSI Systems Center, Ben-Gurion University of the Negev, POB 653, Be'er Sheva 84105, Israel

Robert Giterman
VLSI Systems Center, Ben-Gurion University of the Negev, POB 653, Be'er Sheva 84105, Israel

Andreas Burg
Institute of Electrical Engineering, Ecole Polytechnique F´ed´erale de Lausanne, Station 11, Lausanne, VD 1015, Switzerland

Alexander Fish
Faculty of Engineering, Bar-Ilan University, Ramat-Gan 52900, Israel

Aatmesh Shrivastava
The Charles L. Brown Department of Electrical and Computer Engineering, University of Virginia, Charlottesville, VA 22904, USA

Benton H. Calhoun
The Charles L. Brown Department of Electrical and Computer Engineering, University of Virginia, Charlottesville, VA 22904, USA

Samer Houri
CEA-LETI, Minatec Campus, 17 Rue des Martyrs, Grenoble 38054, France

Christophe Poulain
CEA-LETI, Minatec Campus, 17 Rue des Martyrs, Grenoble 38054, France

Alexandre Valentian
CEA-LETI, Minatec Campus, 17 Rue des Martyrs, Grenoble 38054, France

Hervé Fanet
CEA-LETI, Minatec Campus, 17 Rue des Martyrs, Grenoble 38054, France

Yasuhisa Omura
Organization for Research and Development of Innovative Science and Technology (ORDIST), Kansai University, Yamate-cho, Suita 564-8680, Japan
Graduate School of Science and Engineering, Kansai University, Yamate-cho, Suita 564-8680, Japan

Daiki Sato
Graduate School of Science and Engineering, Kansai University, Yamate-cho, Suita 564-8680, Japan

Nobuyuki Sugii
Low-Power Electronics Association & Project, Tsukuba, Ibaraki 305-8569, Japan

Yoshiki Yamamoto
Low-Power Electronics Association & Project, Tsukuba, Ibaraki 305-8569, Japan

Hideki Makiyama
Low-Power Electronics Association & Project, Tsukuba, Ibaraki 305-8569, Japan

Tomohiro Yamashita
Low-Power Electronics Association & Project, Tsukuba, Ibaraki 305-8569, Japan

Hidekazu Oda
Low-Power Electronics Association & Project, Tsukuba, Ibaraki 305-8569, Japan

Shiro Kamohara
Low-Power Electronics Association & Project, Tsukuba, Ibaraki 305-8569, Japan

Yasuo Yamaguchi
Low-Power Electronics Association & Project, Tsukuba, Ibaraki 305-8569, Japan

Koichiro Ishibashi
Department of Engineering Science, Graduate School of Informatics and Engineering Departments, The University of Electro-Communications, Chofu, Tokyo 182-8585, Japan

Tomoko Mizutani
Institute of Industrial Science, The University of Tokyo, Meguro, Tokyo 153-8505, Japan

Toshiro Hiramoto
Institute of Industrial Science, The University of Tokyo, Meguro, Tokyo 153-8505, Japan

Arijit Banerjee
The Charles L. Brown Department of Electrical and Computer Engineering, University of Virginia, Charlottesville, VA 22904, USA

Benton H. Calhoun
The Charles L. Brown Department of Electrical and Computer Engineering, University of Virginia, Charlottesville, VA 22904, USA

Guerric de Streel
ICTEAM institute, Université catholique de Louvain, Place du Levant 3, 1348 Louvain-la-Neuve, Belgium

David Bol
ICTEAM institute, Université catholique de Louvain, Place du Levant 3, 1348 Louvain-la-Neuve, Belgium

Panagiotis Bertsias
Electronics Laboratory, Physics Department, University of Patras, Rio Patras GR-26504, Greece

Costas Psychalinos
Electronics Laboratory, Physics Department, University of Patras, Rio Patras GR-26504, Greece

Markus Hiienkari
Technology Research Center, University of Turku, Joukahaisenkatu 1C, 20520 Turku, Finland

Jukka Teittinen
Technology Research Center, University of Turku, Joukahaisenkatu 1C, 20520 Turku, Finland

Lauri Koskinen
Technology Research Center, University of Turku, Joukahaisenkatu 1C, 20520 Turku, Finland

Matthew Turnquist
Department of Micro and Nanosciences, Aalto University, Otakaari 5A, 02150 Espoo, Finland

Jani Mäkipää
VTT Technical Research Centre of Finland, Tietotie 3, 02150 Espoo, Finland

Arto Rantala
VTT Technical Research Centre of Finland, Tietotie 3, 02150 Espoo, Finland

Matti Sopanen
VTT Technical Research Centre of Finland, Tietotie 3, 02150 Espoo, Finland

Mikko Kaltiokallio
TDK, Keilaranta 8, 02601 Espoo, Finland

Liang Zhou
Advanced Micro Devices, Inc., Sunnyvale, CA 94089, USA

Ravi Parameswaran
Department of Electrical Engineering, University of Arkansas, Fayetteville, AR 72701, USA

Farhad A. Parsan
Department of Electrical Engineering, University of Arkansas, Fayetteville, AR 72701, USA

Scott C. Smith
Department of Electrical & Computer Engineering, North Dakota State University, Fargo, ND 58108, USA

Jia Di
Department of Computer Science & Computer Engineering, University of Arkansas, Fayetteville, AR 72701, USA

Yin-Nien Chen
Department of Electronics Engineering and Institute of Electronics, National Chiao-Tung University, 1001 University Road, Hsinchu 300, Taiwan

Chien-Ju Chen
Department of Electronics Engineering and Institute of Electronics, National Chiao-Tung University, 1001 University Road, Hsinchu 300, Taiwan

Ming-Long Fan
Department of Electronics Engineering and Institute of Electronics, National Chiao-Tung University, 1001 University Road, Hsinchu 300, Taiwan

Vita Pi-Ho Hu
Department of Electronics Engineering and Institute of Electronics, National Chiao-Tung University, 1001 University Road, Hsinchu 300, Taiwan

Pin Su
Department of Electronics Engineering and Institute of Electronics, National Chiao-Tung University, 1001 University Road, Hsinchu 300, Taiwan

Ching-Te Chuang
Department of Electronics Engineering and Institute of Electronics, National Chiao-Tung University, 1001 University Road, Hsinchu 300, Taiwan

Yasuhiro Ogasahara
Electroinformatics Group, Nanoelectronics Research Institute, National Institute of Advanced
Industrial Science and Technology (AIST), 1-1-1 Umezono, Tsukuba 3058568, Japan

Tadashi Nakagawa
Electroinformatics Group, Nanoelectronics Research Institute, National Institute of Advanced
Industrial Science and Technology (AIST), 1-1-1 Umezono, Tsukuba 3058568, Japan

Toshihiro Sekigawa
Electroinformatics Group, Nanoelectronics Research Institute, National Institute of Advanced
Industrial Science and Technology (AIST), 1-1-1 Umezono, Tsukuba 3058568, Japan

Toshiyuki Tsutsumi
Computer Science Course, Fundamental Science and Technology, Graduate School of Science and Technology, Meiji University, 1-1-1 Higashi-Mita Tama, Kawasaki 2148571, Japan

Hanpei Koike
Electroinformatics Group, Nanoelectronics Research Institute, National Institute of Advanced
Industrial Science and Technology (AIST), 1-1-1 Umezono, Tsukuba 3058568, Japan

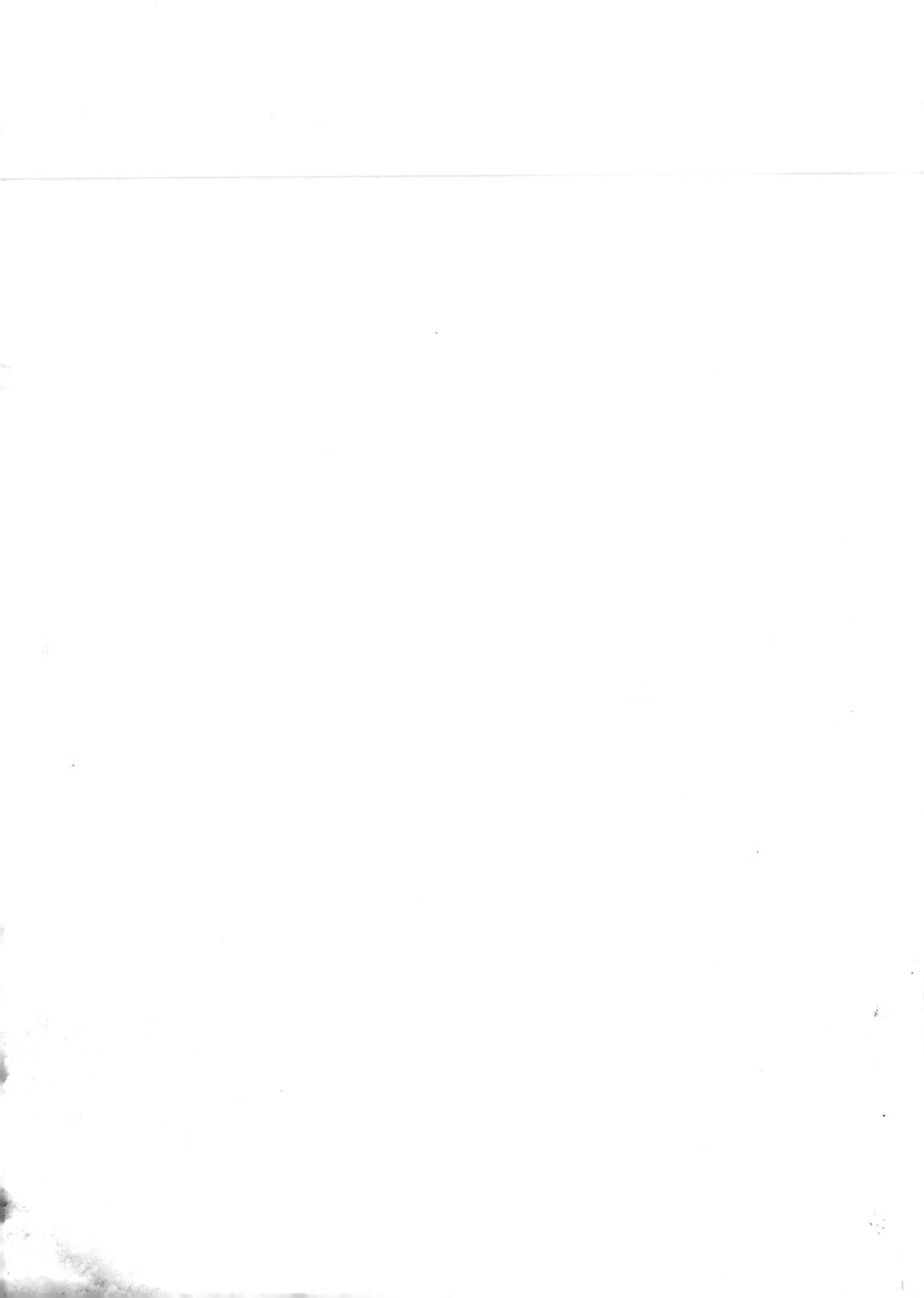